ENVIRONMENTAL BIOCHEMISTRY

Edited by
Levitah Castil Mapatac

ArclerPress

Environmental Biochemistry

Edited by Levitah Castil Mapatac

Edition 2018

ArclerPress

Published by
Arcler Press LLC
708 3rd Avenue, 6th Floor
New York
NY 10017
United States of America
www.arclerpress.com

ISBN: 978-1-68094-524-9 (Hardback)

Printed in United States of America

Library of Congress Control Number: 2017933192

About the Editor

Levitah Castil Mapatac

Levitah C. Mapatac is an Associate Professor, researcher, writer and a PhD holder from the University in Science and Technology of Southern Philippines. Her interest varies from different chemistry concepts and applications and currently doing research in traditional medicines, water waste pollution and environmental chemistry in the aspect of contamination pathways, profilling and rehabilatation of mining waste.

Table of Contents

2 FOOD CONTAMINANT AND FOODBORNE ILLNESS

3 DRINKING WATER AND WATER BORNE DISEASES

4 OXYGENATION AND AIR BORNE DISEASES

5 NITROGEN AND NITROGENOUS COMPOUNDS IN THE ENVIRONMENT

6 CARBOHYDRATES BIOCHEMICAL ASPECTS

7 AMINO ACID BIOCHEMICAL DESCRIPTION OF AMINO ACIDS

8 ENZYME BIOCHEMICAL DESCRIPTION OF ENZYMES

List of Abbreviations

ATP	Adenosine Triphosphate
ALT	Altenuene
AOH	Alternariol
AGWP	anthropogenic Global Warming Potential
ARMAN	Archaeal Richmond Mine Acidophilic Nanoorganisms
ARMAN	Archaeal Richmond Mine Acidophilic Nanoorganisms
CFU	Colony-Forming Units
DON	Deoxynivalenol
DAS	Diacetoxyscirpenol
DO	Dissolved Oxygen
EGT	Endosymbiotic Gene Transfer
EHEC	Enterohemorrhagic
EPEC	Enteropathogenic
ES	Enzyme-Substrate Complex
GABA	Gamma-Amino-Butyric Acid
GI	Glycemic Index
LEM	Leukoencephalomalacia
LOR	Limit of Reporting
OTA	Ochratoxin A
OH	One Hydroxyl
PCR	Polymerase Chain Reaction
RAD	Radicinin

rRNA	Ribosomal RNA
Sec	Selenocysteine
SET	Serial Endosymbiotic Theory
TeA	Tenuazonic Acid
TVC	Total Viable Count

Preface

Biodegradation and microbial transformation reactions are the most important processes determining environmental fate and ultimate elimination of environmental pollutants. The environmental biochemistry is concerned in studying microbial metabolism of pollutants with emphasis on elucidation of metabolites and critical metabolic reactions. On a molecular level, enzymes catalyzing key steps in the metabolism of pollutants with regard to genetic, kinetic, and structural criteria are analyzed. It will help environmental chemists to develop strategies, tools, and methods for improvement and advancement of drinking water and waste water purification technologies, bioremediation processes, construction and selection of bacteria with specific physiological characteristics, and environmentally friendly industrial production processes for chemicals. Environmental toxicology studies on dose-response relationships of heavy metals requires experiments on laboratory animals exposed to "low doses" of trace elements which should reflect "present or actual environmental levels" characteristic of polluted environments. Today daily intakes of trace elements by man are of the order of few micrograms or nanograms thus requiring the use of extremely sensitive analytical techniques to determine the very low amounts of heavy metals in tissues and cellular components. Environmental chemists draw on a range of concepts from chemistry and various environmental sciences to assist in their study of what is happening to a chemical species in the environment. Important general concepts from chemistry include understanding chemical reactions and equations, solutions, units, sampling, and analytical techniques.

Content Coverage

Chapter One presents the introduction of microorganisms that are very diverse and include all bacteria, archaea and most protozoa. This group also contains some species of fungi, algae, and certain microscopic animals, such as rotifers. Microorganisms live in every part of the biosphere, including soil, hot springs, "seven miles deep" in the ocean, "40 miles high" in the atmosphere and inside rocks far down within the Earth's crust.

Chapter Two focuses on foodborne illness that results from consuming food contaminated with pathogenic microorganisms, or toxic compounds released by microorganisms. The disease-causing microorganisms are most often bacteria, but can also include fungi, viruses and protozoa.

Chapter Three outlines the waterborne diseases caused by drinking contaminated or dirty water. Contaminated water can cause many types of diarrheal diseases, including Cholera, and other serious illnesses such as Guinea worm disease, Typhoid, and Dysentery.

Chapter Four highlights about environmental oxygenation that can be important to the sustainability of a particular ecosystem. Deoxygenation increases the relative population of anaerobic organisms such as plants and some bacteria, resulting in fish kills and other adverse events.

Chapter Five focuses on nitrogen efficiency, which forms as about 78% of Earth's atmosphere, making it the most abundant uncombined element. Nitrogen occurs in all organisms, primarily in amino acids (and thus proteins), in the nucleic acids (DNA and RNA) and in the energy transfer molecule adenosine triphosphate. The human body contains about 3% nitrogen by mass, the fourth most abundant element in the body after oxygen, carbon, and hydrogen.

Chapter Six emphasizes on carbohydrates, which provide energy for humans and animals, and cellulose which make up many plant structures.

Chapter Seven focuses on many important proteinogenic and non-proteinogenic amino acids that play critical non-protein roles within the body. Nine proteinogenic amino acids are called essential for humans because they cannot be created from other compounds by the human body and so must be taken in as food.

Chapter Eight presents a devotion on enzymes which enable chemical reactions in the body to occur many times more quickly than they otherwise would. In some cases, a reaction will not take place at all unless a specific enzyme is present.

CHAPTER 1

Microorganisms: General Characteristics and Human Health

Chapter Contents

- **Introduction**
- **Prokaryote**
- **Bacteria**
- **Archaea**
- **Eukaryote**
- **Protist**
- **Fungus**
- **Microorganisms Used in Food and Beverage Preparation**

Living organism (such as a bacteria, fungi, viruses) too small to be seen with naked eye but visible under a microscope. Also called microbe.

INTRODUCTION

A microorganism or microbe is a microscopic living organism, which may be single-celled or multicellular. The study of microorganisms is called microbiology, a subject that began with the discovery of microorganisms in 1674 by Antonie van Leeuwenhoek, using a microscope of his own design.

Microorganisms are very diverse and include all bacteria, archaea and most protozoa. This group also contains some species of fungi, algae, and certain microscopic animals, such as rotifers. Many macroscopic animals and plants have microscopic juvenile stages. Some microbiologists also classify viruses (and viroids) as microorganisms, but others consider these as nonliving. In July 2016, scientists reported identifying a set of 355 genes from the last universal common ancestor of all life, including microorganisms, living on Earth.

Microorganisms live in every part of the biosphere, including soil, hot springs, "seven miles deep" in the ocean, "40 miles high" in the atmosphere and inside rocks far down within the Earth's crust. Microorganisms, under certain test conditions, have been observed to thrive in the vacuum of outer space. According to some estimates, microorganisms outweigh "all other living things combined thousands of times over". The mass of prokaryote microorganisms — which includes bacteria and archaea, but not the nucleated eukaryote microorganisms — may be as much as 0.8 trillion tons of carbon (of the total biosphere mass, estimated at between 1 and 4 trillion tons). On 17 March 2013, researchers reported data that suggested microbial life forms thrive in the Mariana Trench. the deepest spot in the Earth's oceans. Other researchers reported related studies that microorganisms thrive inside rocks up to 580 m (1,900 ft; 0.36 mi) below the sea floor under 2,590 m (8,500 ft; 1.61 mi) of ocean off the coast of the northwestern United States, as well as 2,400 m (7,900 ft; 1.5 mi) beneath the seabed off Japan. On 20 August 2014, scientists confirmed the existence of microorganisms living 800 m (2,600 ft; 0.50 mi) below the ice of Antarctica. According to one researcher, "You can find microbes everywhere — they're extremely adaptable to conditions, and survive wherever they are."

Microorganisms are crucial to nutrient recycling in ecosystems as they act as decomposers. As some microorganisms can fix nitrogen, they are a vital part of the nitrogen cycle, and recent studies indicate that airborne microorganisms may play a role in precipitation and weather. Microorganisms are also exploited in biotechnology, both in traditional food and beverage preparation, and in modern technologies based on genetic engineering. A small proportion of microorganisms are pathogenic, causing disease and even death in plants and animals.

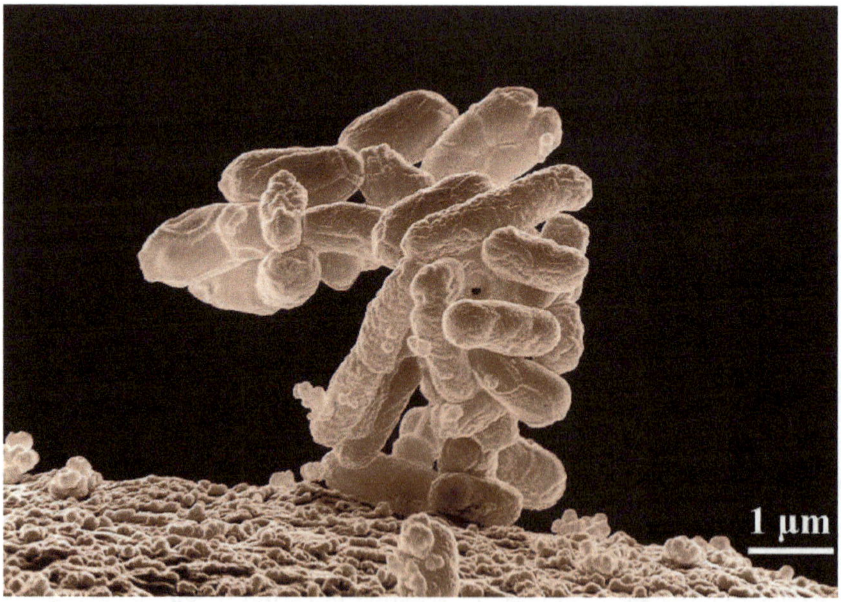

A cluster of Escherichia coli bacteria magnified 10,000 times

Evolution

Single-celled microorganisms were the first forms of life to develop on Earth, approximately 3–4 billion years ago. Further evolution was slow, and for about 3 billion years in the Precambrian eon, all organisms were microscopic. So, for most of the history of life on Earth, the only forms of life were microorganisms.Bacteria, algae and fungi have been identified in amber that is 220 million years old, which shows that the morphology of microorganisms has changed little since the Triassic period. The newly discovered biological role played by nickel, however — especially that engendered by volcanic eruptions from the Siberian Traps (site of the modern city of Norilsk) — is thought to have accelerated the evolution of methanogens towards the end of the Permian–Triassic extinction event.

Microorganisms tend to have a relatively fast rate of evolution. Most microorganisms can reproduce rapidly, and bacteria are also able to freely exchange genes through conjugation, transformation and transduction, even between widely divergent species. This horizontal gene transfer, coupled with a high mutation rate and many other means of genetic variation, allows microorganisms to swiftly evolve (via natural selection) to survive in new environments and respond to environmental stresses. This rapid evolution is important in medicine, as it has led to the recent development of "super-bugs", pathogenic bacteria that are resistant to modern antibiotics.

Pre-Microbiology

The possibility that microorganisms exist was discussed for many centuries before their discovery in the 17th century. The existence of unseen microbiological life was postulated by Jainism, which is based on Mahavira's teachings as early as 6th century BCE. Paul Dundas notes that Mahavira asserted the existence of unseen microbiological creatures living in earth, water, air and fire. The Jain scriptures also describe nigodas, which are sub-microscopic creatures living in large clusters and having a very short life, which are said to pervade every part of the universe, even the tissues of plants and animals. The earliest known idea to indicate the possibility of diseases spreading by yet unseen organisms was that of the Roman scholar Marcus Terentius Varro in a 1st-century BC book titled *On Agriculture* in which he warns against locating a homestead near swamps:

... and because there are bred certain minute creatures that cannot be seen by the eyes, which float in the air and enter the body through the mouth and nose and they cause serious diseases.

In *The Canon of Medicine* (1020), Abū Alī ibn Sīnā (Avicenna) hypothesized that tuberculosis and other diseases might be contagious.

In 1546, Girolamo Fracastoro proposed that epidemic diseases were caused by transferable seedlike entities that could transmit infection by direct or indirect contact, or even without contact over long distances.

All these early claims about the existence of microorganisms were speculative and while grounded on indirect observations, they were not based on direct observation of microorganisms or systematized empirical investigation, e.g. experimentation.

Microorganisms were neither proven, observed, nor accurately described until the 17th century. The reason for this was that all these early studies lacked the microscope.

History of Discovery

Antonie van Leeuwenhoek, the first person to observe microorganisms using a microscope

Lazzaro Spallanzani showed that boiling a broth stopped it from decaying.

Louis Pasteur showed that Spallanzani's findings held even if air could enter through a filter that kept particles out.

Robert Koch showed that microorganisms caused disease.

Antonie Van Leeuwenhoek (1632–1723) was one of the first people to observe microorganisms, using microscopes of his own design. Robert Hooke, a contemporary of Leeuwenhoek, also used microscopes to observe microbial life; his 1665 book *Micrographia* describes these observations and coined the term *cell.*

Before Leeuwenhoek's discovery of microorganisms in 1675, it had been a mystery why grapes could be turned into wine, milk into cheese, or why food would spoil. Leeuwenhoek did not make the connection between these processes and microorganisms, but using a microscope, he did establish that there were signs of life that were not visible to the naked eye. Leeuwenhoek's discovery, along with subsequent observations by Spallanzani and Pasteur, ended the long-held belief that life spontaneously appeared from non-living substances during the process of spoilage.

Lazzaro Spallanzani (1729–1799) found that boiling broth would sterilise it, killing any microorganisms in it. He also found that new microorganisms could only settle in a broth if the broth was exposed to air.

Louis Pasteur (1822–1895) expanded upon Spallanzani's findings by exposing boiled broths to the air, in vessels that contained a filter to prevent all particles from passing through to the growth medium, and also in vessels with no filter at all, with air being admitted via a curved tube that would not allow dust particles to come in contact with the broth. By boiling the broth beforehand, Pasteur ensured that no microorganisms survived within the broths at the beginning of his experiment. Nothing grew in the broths in the course of Pasteur's experiment. This meant that the living organisms that grew in such broths came from outside, as spores on dust, rather than spontaneously generated within the broth. Thus, Pasteur dealt the death blow to the theory of spontaneous generation and supported germ theory.

In 1876, Robert Koch (1843–1910) established that microorganisms can cause disease. He found that the blood of cattle which were infected with anthrax always had large numbers of *Bacillus anthracis*. Koch found that he could transmit anthrax from one animal to another by taking a small sample of blood from the infected animal and injecting it into a healthy one, and this caused the healthy animal to become sick. He also found that he could grow the bacteria in a nutrient broth, then inject it into a healthy animal, and cause illness. Based on these experiments, he devised criteria for establishing a causal link between a microorganism and a disease and these are now known as Koch's postulates. Although these postulates cannot be applied in all cases, they do retain historical importance to the development of scientific thought and are still being used today.

On 8 November 2013, scientists reported the discovery of what may be the earliest signs of life on Earth—the oldest complete fossils of a microbial mat (associated with sandstone in Western Australia) estimated to be 3.48 billion years old.

Classification and Structure

Microorganisms can be found almost anywhere in the taxonomic organization of life on the planet. Bacteria and archaea are almost always microscopic, while a number of eukaryotes are also microscopic, including most protists, some fungi, as well as some animals and plants. Viruses are generally regarded as not living and therefore not considered as microorganisms, although the field of microbiology also encompasses the study of viruses.

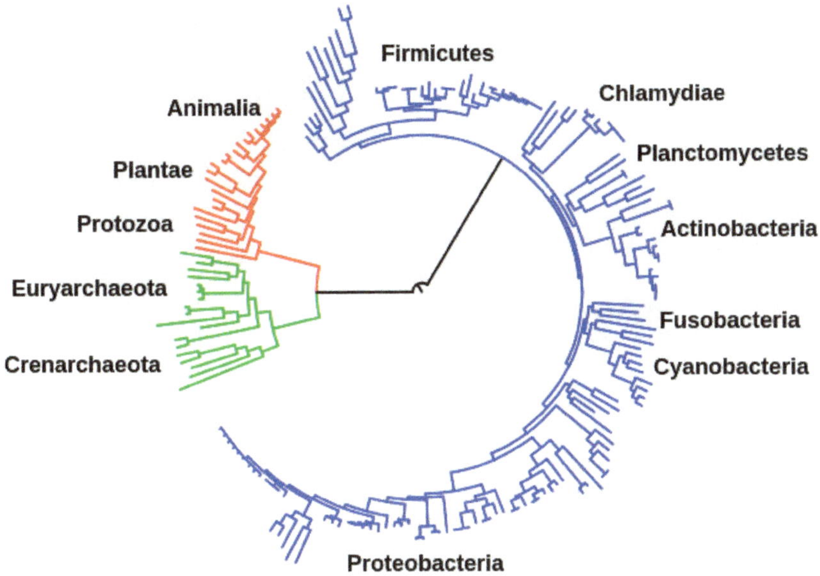

Evolutionary tree showing the common ancestry of all three domains of life. Bacteria are colored blue, eukaryotes red, and archaea green. Relative positions of some phyla are shown around the tree.

PROKARYOTE

A prokaryote is a unicellular organism that lacks a membrane-bound nucleus (karyon), mitochondria, or any other membrane-bound organelle. The word prokaryote comes from the Greek πρό (pro) "before" and καρυόν (karyon) "nut or kernel". Prokaryotes can be divided into two domains, Archaea and bacteria. In contrast, species with nuclei and organelles are placed in the domain Eukaryota.

In the prokaryotes, all the intracellular water-soluble components (proteins, DNA and metabolites) are located together in the cytoplasm enclosed by the cell membrane, rather than in separate cellular compartments. Bacteria, however, do possess protein-based bacterial microcompartments, which are thought to act as primitive organelles enclosed in protein shells. Some prokaryotes, such as cyanobacteria may form large colonies. Others, such as myxobacteria, have multicellular stages in their life cycles.

Molecular studies have provided insight into the evolution and interrelationships of the three domains of biological species. Eukaryotes are organisms, including humans, whose cells have a well defined membrane-bound nucleus (containing chromosomal DNA) and organelles. The division between prokaryotes and eukaryotes reflects the existence of two very different levels of cellular organization. Distinctive types of prokaryotes include extremophiles and methanogens; these are common in some extreme environments.

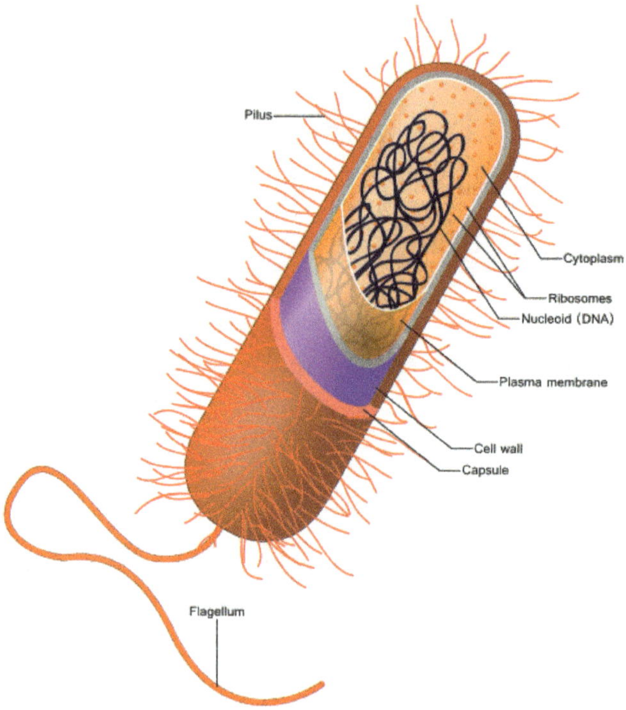

A diagram of a typical prokaryotic bacteria cell.

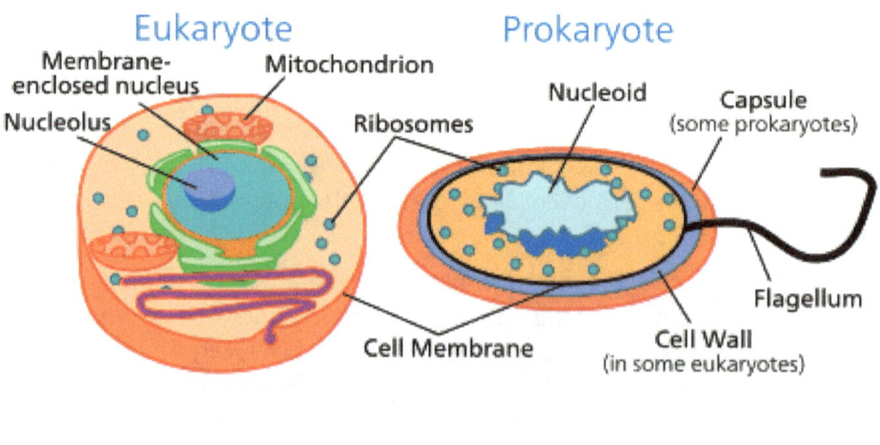

Comparison of eukaryotes vs. prokaryotes

Structure

Prokaryotes have a prokaryotic cytoskeleton, albeit more primitive than that of the eukaryotes. Besides homologues of actin and tubulin (MreB and FtsZ), the helically arranged building-block of the flagellum, flagellin, is one of the most significant cytoskeletal proteins of bacteria, as it provides structural backgrounds of chemotaxis, the basic cell physiological response of bacteria. At least some prokaryotes also contain intracellular structures that can be seen as primitive organelles. Membranous organelles (or intracellular membranes) are known in some groups of prokaryotes, such as vacuoles or membrane systems devoted to special metabolic properties, such as photosynthesis or chemolithotrophy. In addition, some species also contain carbohydrate-enclosed microcompartments, which have distinct physiological roles (e.g. carboxysomes or gas vacuoles).

Most prokaryotes are between 1 μm and 10 μm, but they can vary in size from 0.2 μm (*Mycoplasma genitalium*) to 750 μm (*Thiomargarita namibiensis*).

Prokaryotic cell structure
Flagellum (only in some types of prokaryotes) Long, whip-like protrusion that aids cellular locomotion.
Cell membrane Surrounds the cell's cytoplasm and regulates the flow of substances in and out of the cell.
Cell wall (except genera Mycoplasma and Thermoplasma) Outer covering of most cells that protects the bacterial cell and gives it shape.
Cytoplasm A gel-like substance composed mainly of water that also contains enzymes, salts, cell components, and various organic molecules.
Ribosome Cell structures responsible for protein production.
Nucleoid Area of the cytoplasm that contains the single bacterial DNA molecule.
Glycocalyx (only in some types of prokaryotes) A glycoprotein-polysaccharide covering that surrounds the cell membranes.
Inclusions

Morphology

Prokaryotic cells have various shapes; the four basic shapes of bacteria are:

- Cocci – spherical
- Bacilli – rod-shaped
- Spirochaete – spiral-shaped
- Vibrio – comma-shaped

The archaeon Haloquadratum has flat square-shaped cells.

Reproduction

Bacteria and archaea reproduce through asexual reproduction, usually by binary fission. Genetic exchange and recombination still occur, but this is a form of horizontal gene transfer and is not a replicative process, simply involving the transference of DNA between two cells, as in bacterial conjugation.

DNA Transfer

DNA transfer between prokaryotic cells occurs in bacteria and archaea, although it has been mainly studied in bacteria. In bacteria, gene transfer occurs by three processes. These are (1) bacterial virus (bacteriophage)-mediated transduction, (2) plasmid-mediated conjugation, and (3) natural transformation. Transduction of bacterial genes by bacteriophage appears to reflect an occasional error during intracellular assembly of virus particles, rather than an adaptation of the host bacteria. The transfer of bacterial DNA is under the control of the bacteriophage's genes rather than bacterial genes. Conjugation in the well-studied *E. coli* system is controlled by plasmid genes, and is an adaptation for distributing copies of a plasmid from one bacterial host to another. Infrequently during this process, a plasmid may integrate into the host bacterial chromosome, and subsequently transfer part of the host bacterial DNA to another bacterium. Plasmid mediated transfer of host bacterial DNA (conjugation) also appears to be an accidental process rather than a bacterial adaptation.

Natural bacterial transformation involves the transfer of DNA from one bacterium to another through the intervening medium. Unlike transduction and conjugation, transformation is clearly a bacterial adaptation for DNA transfer, because it depends on numerous bacterial gene products that specifically interact to perform this complex process. For a bacterium to bind, take up and recombine donor DNA into its own chromosome, it must first enter a special physiological state called competence. About 40 genes are required in *Bacillus subtilis* for the development of competence. The length of DNA transferred during *B. subtilis* transformation can be as much as a third to the whole chromosome. Transformation is a common mode of DNA transfer, and 67 prokaryotic species are thus far known to be naturally competent for transformation.

Among archaea, *Halobacterium volcanii* forms cytoplasmic bridges between cells that appear to be used for transfer of DNA from one cell to another. Another archaeon, *Sulfolobus solfataricus*, transfers DNA between cells by direct contact. Frols et al. found that exposure of *S. solfataricus* to DNA damaging agents induces cellular aggregation, and suggested that cellular aggregation may enhance DNA transfer among cells to provide increased repair of damaged DNA via homologous recombination.

Sociality

While prokaryotes are considered strictly unicellular, most can form stable aggregate communities. When such communities are encased in a stabilizing polymer matrix ("slime"), they may be called "biofilms". Cells in biofilms often show distinct patterns of gene expression (phenotypic differentiation) in time and space. Also, as with multicellular eukaryotes, these changes in expression often appear to result from cell-to-cell signaling, a phenomenon known as quorum sensing.

Biofilms may be highly heterogeneous and structurally complex and may attach to solid surfaces, or exist at liquid-air interfaces, or potentially even liquid-liquid interfaces. Bacterial biofilms are often made up of microcolonies (approximately dome-shaped masses of bacteria and matrix) separated by "voids" through which the medium (e.g., water) may flow easily. The microcolonies may join together above the substratum to form a continuous layer, closing the network of channels separating microcolonies. This structural complexity—combined with observations that oxygen limitation (a ubiquitous challenge for anything growing in size beyond the scale of diffusion) is at least partially eased by movement of medium throughout the biofilm—has led some to speculate that this may constitute a circulatory system and many researchers have started calling prokaryotic communities multicellular (for example). Differential cell expression, collective behavior, signaling, programmed cell death, and (in some cases) discrete biological dispersal events all seem to point in this direction. However, these colonies are seldom if ever founded by a single founder (in the way that animals and plants are founded by single cells), which presents a number of theoretical issues. Most explanations of co-operation and the evolution of multicellularity have focused on high relatedness between members of a group (or colony, or whole organism). If a copy of a gene is present in all members of a group, behaviors that promote cooperation between members may permit those members to have (on average) greater fitness than a similar group of selfish individuals (see inclusive fitness and Hamilton's rule).

Should these instances of prokaryotic sociality prove to be the rule rather than the exception, it would have serious implications for the way we view prokaryotes in general, and the way we deal with them in medicine. Bacterial biofilms may be 100 times more resistant to antibiotics than free-living unicells and may be nearly impossible to remove from surfaces once they have colonized them. Other aspects of bacterial cooperation—such as bacterial conjugation and quorum-sensing-mediated pathogenicity, present additional challenges to researchers and medical professionals seeking to treat the associated diseases.

Environment

Prokaryotes have diversified greatly throughout their long existence. The metabolism of prokaryotes is far more varied than that of eukaryotes, leading to many highly distinct prokaryotic types. For example, in addition to using photosynthesis or organic compounds for energy, as eukaryotes do, prokaryotes may obtain energy from inorganic compounds such as hydrogen sulfide. This enables prokaryotes to thrive in harsh environments as cold as the snow surface of Antarctica, studied in cryobiology or as hot as undersea hydrothermal vents and land-based hot springs.

Prokaryotes live in nearly all environments on Earth. Some archaea and bacteria thrive in harsh conditions, such as high temperatures (thermophiles) or high salinity (halophiles). Organisms such as these are referred to as extremophiles. Many archaea grow as plankton in the oceans. Symbiotic prokaryotes live in or on the bodies of other organisms, including humans.

Classification

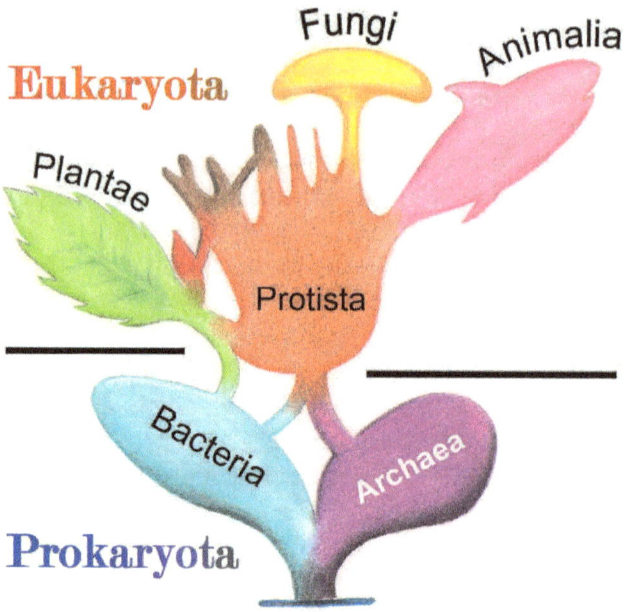

Phylogenetic and symbiogenetic tree of living organisms, showing the origins of eukaryotes and prokaryotes

In 1977, Carl Woese proposed dividing prokaryotes into the Bacteria and Archaea (originally Eubacteria and Archaebacteria) because of the major differences in the structure and genetics between the two groups of organisms. Archaea were originally thought to be extremophiles, living only in inhospitable conditions such as extremes of temperature, pH, and radiation but have since been found in all types of habitats. The resulting arrangement of Eukaryota (also called "Eukarya"), Bacteria, and Archaea is called the three-domain system, replacing the traditional two-empire system.

One criticism of the dichotomous prokaryote-eukaryote distinction points out that the word "prokaryote" is based on what these organisms are not (they are not eukaryotic), rather than what they are (either archaea or bacteria).

Evolution

The current model of the evolution of the first living organisms is that these were some form of prokaryotes, which may have evolved out of protocells. In general, the eukaryotes are thought to have evolved later in the history of life.

However, some authors have questioned this conclusion, arguing that the current set of prokaryotic species may have evolved from more complex eukaryotic ancestors through a process of simplification.

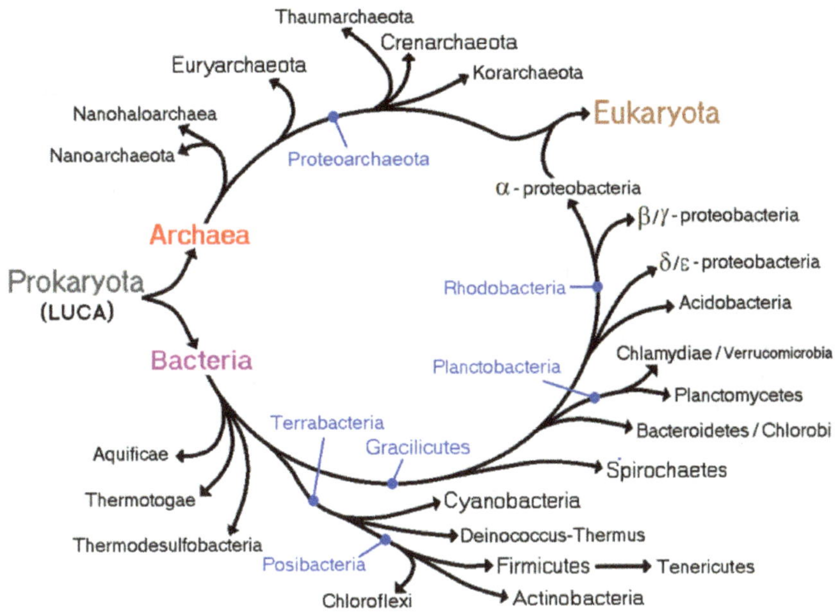

Phylogenetic ring showing the diversity of prokaryotes, and symbiogenetic origins of eukaryotes

Others have argued that the three domains of life arose simultaneously, from a set of varied cells that formed a single gene pool.

This controversy was summarized in 2005:

There is no consensus among biologists concerning the position of the eukaryotes in the overall scheme of cell evolution. Current opinions on the origin and position of eukaryotes span a broad spectrum including the views that eukaryotes arose first in evolution and that prokaryotes descend from them, that eukaryotes arose contemporaneously with eubacteria and archeabacteria and hence represent a primary line of descent of equal age and rank as the prokaryotes, that eukaryotes arose through a symbiotic event entailing an endosymbiotic origin of the nucleus, that eukaryotes arose without endosymbiosis, and that eukaryotes arose through a symbiotic event entailing a simultaneous endosymbiotic origin of the flagellum and the nucleus, in addition to many other models, which have been reviewed and summarized elsewhere.

The oldest known fossilized prokaryotes were laid down approximately 3.5 billion years ago, only about 1 billion years after the formation of the Earth's crust. Eukaryotes only appear in the fossil record later, and may have formed from endosymbiosis of multiple prokaryote ancestors. The oldest known fossil eukaryotes are about 1.7 billion years old. However, some genetic evidence suggests eukaryotes appeared as early as 3 billion years ago.

While Earth is the only place in the universe where life is known to exist, some have suggested that there is evidence on Mars of fossil or living prokaryotes.However, this possibility remains the subject of considerable debate and skepticism.

Relationship to Eukaryotes

The division between prokaryotes and eukaryotes is usually considered the most important distinction or difference among organisms. The distinction is that eukaryotic cells have a "true" nucleus containing their DNA, whereas prokaryotic cells do not have a nucleus. Both eukaryotes and prokaryotes contain large RNA/protein structures called ribosomes, which produce protein.

Another difference is that ribosomes in prokaryotes are smaller than in eukaryotes. However, two organelles found in many eukaryotic cells, mitochondria and chloroplasts, contain ribosomes similar in size and makeup to those found in prokaryotes. This is one of many pieces of evidence that mitochondria and chloroplasts are themselves descended from free-living bacteria. This theory holds that early eukaryotic cells took in primitive prokaryotic cells by phagocytosis and adapted themselves to incorporate their structures, leading to the mitochondria we see today.

The genome in a prokaryote is held within a DNA/protein complex in the cytosol called the nucleoid, which lacks a nuclear envelope. The complex contains a single, cyclic, double-stranded molecule of stable chromosomal DNA, in contrast to the multiple linear, compact, highly organized chromosomes found in eukaryotic cells. In addition, many important genes of prokaryotes are stored in separate circular DNA structures called plasmids.

Prokaryotes lack mitochondria and chloroplasts. Instead, processes such as oxidative phosphorylation and photosynthesis take place across the prokaryotic cell membrane. However, prokaryotes do possess some internal structures, such as prokaryotic cytoskeletons. It has been suggested that the bacterial order Planctomycetes have a membrane around their nucleoid and contain other membrane-bound cellular structures. However, further investigation revealed that Planctomycetes cells are not compartmentalized or nucleated and like the other bacterial membrane systems are all interconnected.

Prokaryotic cells are usually much smaller than eukaryotic cells. Therefore, prokaryotes have a larger surface-area-to-volume ratio, giving them a higher metabolic rate, a higher growth rate, and as a consequence, a shorter generation time than eukaryotes.

Prokaryotes are organisms that lack a cell nucleus and the other membrane bound organelles. They are almost always unicellular, although some species such as myxobacteria can aggregate into complex structures as part of their life cycle.

Consisting of two domains, bacteria and archaea, the prokaryotes are the most diverse and abundant group of organisms on Earth and inhabit practically all environments where the temperature is below +140 °C. They are found in water, soil, air, animals' gastrointestinal tracts, hot springs and even deep beneath the Earth's crust in rocks. Practically all surfaces that have not been specially sterilized are covered by prokaryotes. The number of prokaryotes on Earth is estimated to be around five million trillion trillion, or 5×10^{30}, accounting for at least half the biomass on Earth.

The biodiversity of the prokaryotes is unknown, but may be very large. A May 2016 estimate, based on laws of scaling from known numbers of species against the size of organism, gives an estimate of perhaps 1 trillion species on the planet, of which most would be microorganisms. Currently, only one-thousandth of one percent of that total have been described.

BACTERIA

Bacteria constitute a large domain of prokaryoticmicroorganisms. Typically a few micrometres in length, bacteria have a number of shapes, ranging from spheres to rods and spirals. Bacteria were among the first life forms to appear on Earth, and are present in most of its habitats. Bacteria inhabit soil, water, acidic hot springs, radioactive waste, and the deep portions of Earth›s crust. Bacteria also live in symbiotic and parasitic relationships with plants and animals.

There are typically 40 million bacterial cells in a gram of soil and a million bacterial cells in a millilitre of fresh water. There are approximately 5×10^{30} bacteria on Earth, forming a biomass which exceeds that of all plants and animals. Bacteria are vital in recycling nutrients, with many of the stages in nutrient cycles dependent on these organisms, such as the fixation of nitrogen from the atmosphere and putrefaction. In the biological communities surrounding hydrothermal vents and cold seeps, bacteria provide the nutrients needed to sustain life by converting dissolved compounds, such as hydrogen sulphide and methane, to energy. On 17 March 2013, researchers reported data that suggested bacterial life forms thrive in the Mariana Trench, which with a depth of up to 11 kilometres is the deepest part of the Earth's oceans. Other researchers reported related studies that microbes thrive inside rocks up to 580 metres below the sea floor under 2.6 kilometres of ocean off the coast of the northwestern United States. According to one of the researchers, "You can find microbes everywhere—they're extremely adaptable to conditions, and survive wherever they are."

Most bacteria have not been characterised, and only about half of the bacterial phyla have species that can be grown in the laboratory. The study of bacteria is known as bacteriology, a branch of microbiology.

There are approximately ten times as many bacterial cells in the human flora as there are human cells in the body, with the largest number of the human flora being in the gut flora, and a large number on the skin. The vast majority of the bacteria in the body are rendered harmless by the protective effects of the immune system, and some are beneficial. However, several species of bacteria are pathogenic and cause infectious diseases, including cholera, syphilis, anthrax, leprosy, and bubonic plague. The most common fatal bacterial diseases are respiratory infections, with tuberculosis alone killing about 2 million people per year, mostly in sub-Saharan Africa. In developed countries, antibiotics are used to treat bacterial infections and are also used in farming, making antibiotic resistance a growing problem. In industry, bacteria are important in sewage treatment and the breakdown of oil spills, the production of cheese and yogurt through fermentation, and the recovery of gold, palladium, copper and other metals in the mining sector, as well as in biotechnology, and the manufacture of antibiotics and other chemicals.

Once regarded as plants constituting the class *Schizomycetes*, bacteria are now classified as prokaryotes. Unlike cells of animals and other eukaryotes, bacterial cells do not contain a nucleus and rarely harbour membrane-bound organelles. Although the term *bacteria* traditionally included all prokaryotes, the scientific classification changed after the discovery in the 1990s that prokaryotes consist of two very different groups of organisms that evolved from an ancient common ancestor. These evolutionary domains are called *Bacteria* and *Archaea*.

Etymology

The word *bacteria* is the plural of the New Latin *bacterium*, which is the latinisation of the Greek βακτήριον (*bakterion*), the diminutive of βακτηρία (*bakteria*), meaning "staff, cane", because the first ones to be discovered were rod-shaped.

Origin and early Evolution

The ancestors of modern bacteria were unicellular microorganisms that were the first forms of life to appear on Earth, about 4 billion years ago. For about 3 billion years, most organisms were microscopic, and bacteria and archaea were the dominant forms of life. In 2008, fossils of macroorganisms were discovered and named as the Francevillian biota. Although bacterial fossils exist, such as stromatolites, their lack of distinctive morphology prevents them from being used to examine the history of bacterial evolution, or to date the time of origin of a particular bacterial species. However, gene sequences can be used to reconstruct the bacterial phylogeny, and these studies indicate that bacteria diverged first from the archaeal/eukaryotic lineage.Bacteria were also involved in the second great evolutionary divergence, that of the archaea and eukaryotes. Here, eukaryotes resulted from the entering of ancient bacteria into endosymbiotic associations with the ancestors of eukaryotic cells, which were themselves possibly related to the Archaea. This involved the engulfment by proto-eukaryotic cells of alphaproteobacterial symbionts to form either mitochondria or hydrogenosomes, which are still found in all known Eukarya (sometimes in highly reduced form, e.g. in ancient "amitochondrial" protozoa). Later on, some eukaryotes that already contained mitochondria also engulfed cyanobacterial-like organisms. This led to the formation of chloroplasts in algae and plants. There are also some algae that originated from even later endosymbiotic events. Here, eukaryotes engulfed a eukaryotic algae that developed into a "second-generation" plastid. This is known as secondary endosymbiosis.

Morphology

Bacteria display a wide diversity of shapes and sizes, called *morphologies*. Bacterial cells are about one-tenth the size of eukaryotic cells and are typically 0.5–5.0 micrometres in length. However, a few species are visible to the unaided eye—for example, *Thiomargarita namibiensis* is up to half a millimetre long and *Epulopiscium fishelsoni* reaches 0.7 mm. Among the smallest bacteria are members of the genus *Mycoplasma*, which measure only 0.3 micrometres, as small as the largest viruses. Some bacteria may be even smaller, but these ultramicrobacteria are not well-studied.

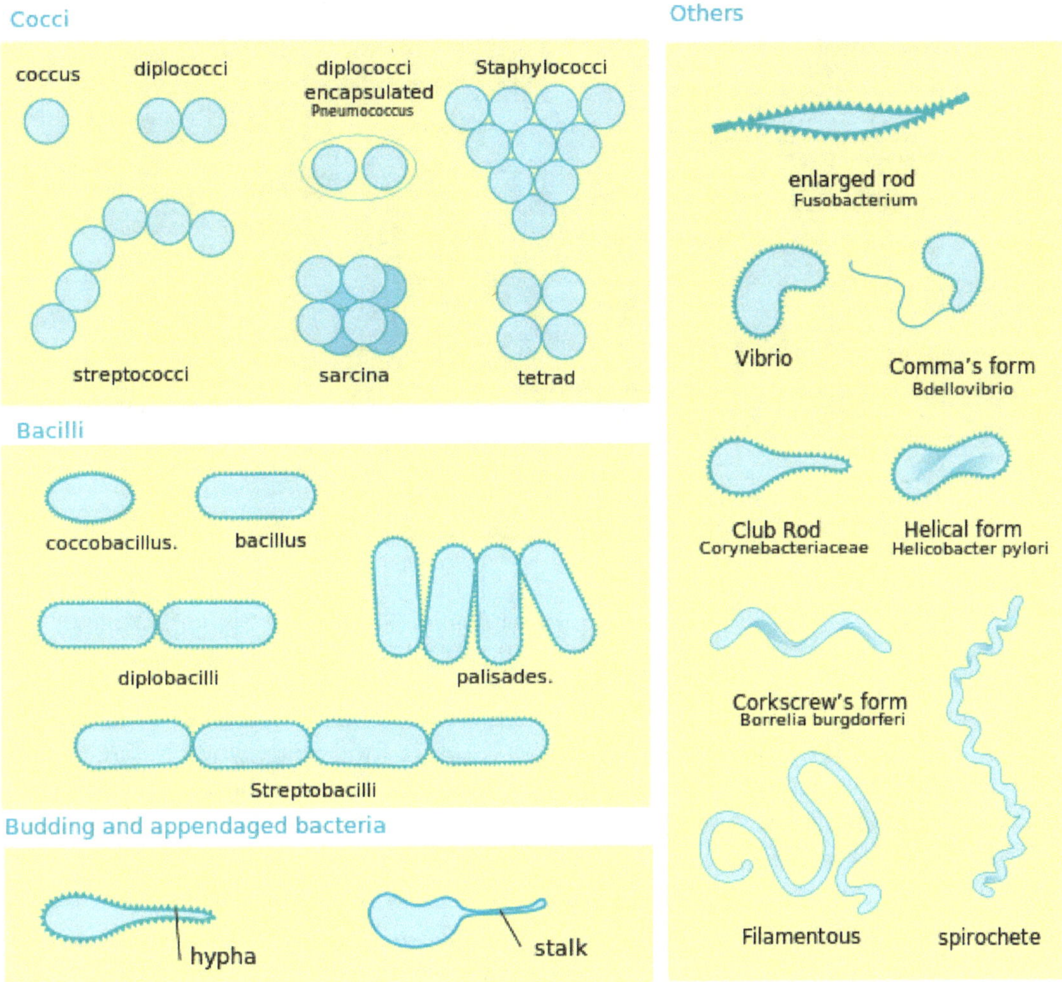

Bacteria display many cell morphologies and arrangements.

Most bacterial species are either spherical, called *cocci* (*sing*. coccus, from Greek *kókkos*, grain, seed), or rod-shaped, called *bacilli* (*sing*. bacillus, from Latin *baculus*, stick). Elongation is associated with swimming. Some bacteria, called *vibrio*, are shaped like slightly curved rods or comma-shaped; others can be spiral-shaped, called *spirilla*, or tightly coiled, called *spirochaetes*. A small number of species even have tetrahedral or cuboidal shapes. More recently, some bacteria were discovered deep under Earth's crust that grow as branching filamentous types with a star-shaped cross-section. The large surface area to volume ratio of this morphology may give these bacteria an advantage in nutrient-poor environments. This wide variety of shapes is determined by the bacterial cell wall and cytoskeleton, and is important because it can influence the ability of bacteria to acquire nutrients, attach to surfaces, swim through liquids and escape predators.

A biofilm of thermophilic bacteria in the outflow of Mickey Hot Springs, Oregon, approximately 20 mm thick.

Many bacterial species exist simply as single cells, others associate in characteristic patterns: *Neisseria* form diploids (pairs), *Streptococcus* form chains, and *Staphylococcus* group together in "bunch of grapes" clusters. Bacteria can also be elongated to form filaments, for example the Actinobacteria. Filamentous bacteria are often surrounded by a sheath that contains many individual cells. Certain types, such as species of the genus *Nocardia*, even form complex, branched filaments, similar in appearance to fungal mycelia.

Bacteria often attach to surfaces and form dense aggregations called *biofilms* or bacterial mats. These films can range from a few micrometres in thickness to up to half a metre in depth, and may contain multiple species of bacteria, protists and archaea. Bacteria living in biofilms display a complex arrangement of cells and extracellular components, forming secondary structures, such as microcolonies, through which there are networks of channels to enable better diffusion of nutrients.In natural environments, such as soil or the surfaces of plants, the majority of bacteria are bound to surfaces in biofilms.Biofilms are also important in medicine, as these structures are often present during chronic bacterial infections or in infections of implanted medical devices, and bacteria protected within biofilms are much harder to kill than individual isolated bacteria.

Even more complex morphological changes are sometimes possible. For example, when starved of amino acids, Myxobacteria detect surrounding cells in a process known as quorum sensing, migrate towards each other, and aggregate to form fruiting bodies up to 500 micrometres long and containing approximately 100,000 bacterial cells. In these fruiting bodies, the bacteria perform separate tasks; this type of cooperation is a simple type of multicellular organisation. For example, about one in 10 cells migrate to the top of these fruiting bodies and differentiate into a specialised dormant state called myxospores, which are more resistant to drying and other adverse environmental conditions than are ordinary cells.

Cellular Structure

Intracellular Structures

The bacterial cell is surrounded by a cell membrane (also known as a lipid, cytoplasmic or plasma membrane). This membrane encloses the contents of the cell and acts as a barrier to hold nutrients, proteins and other essential components of the *cytoplasm* within the cell. As they are prokaryotes, bacteria do not usually have membrane-bound organelles in their cytoplasm, and thus contain few large intracellular structures. They lack a true nucleus, mitochondria, chloroplasts and the other organelles present in eukaryotic cells. Bacteria were once seen as simple bags of cytoplasm, but structures such as the *prokaryotic cytoskeleton* and the localisation of proteins to specific locations within the cytoplasm that give bacteria some complexity have been discovered. These subcellular levels of organisation have been called "bacterial hyperstructures".

Bacterial microcompartments, such as carboxysomes, provide a further level of organisation; they are compartments within bacteria that are surrounded by polyhedral protein shells, rather than by lipid membranes. These "polyhedral organelles" localise and compartmentalise bacterial metabolism, a function performed by the membrane-bound organelles in eukaryotes.

Many important biochemical reactions, such as energy generation, use concentration gradients across membranes. The general lack of internal membranes in bacteria means reactions such as electron transport occur across the cell membrane between the cytoplasm and the *periplasmic space*. However, in many photosynthetic bacteria the plasma membrane is highly folded and fills most of the cell with layers of light-gathering membrane. These light-gathering complexes may even form lipid-enclosed structures called chlorosomes in green sulfur bacteria.Other proteins import nutrients across the cell membrane, or expel undesired molecules from the cytoplasm.

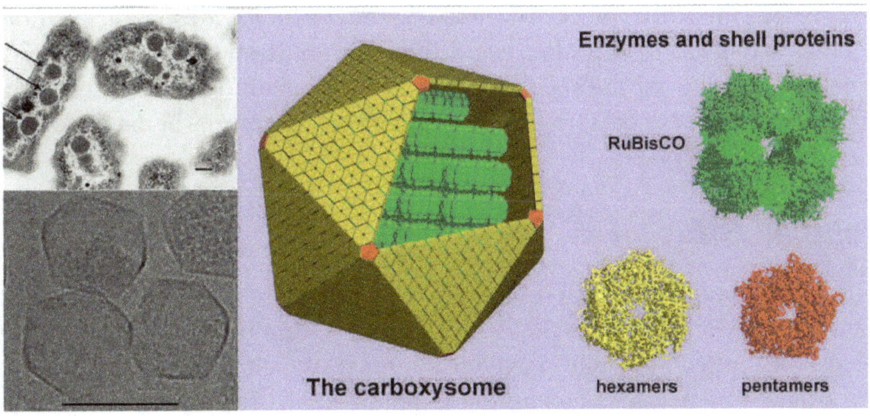

Carboxysomes are protein-enclosed bacterial organelles. Top left is an electron microscope image of carboxysomes in *Halothiobacillus neapolitanus*, below is an image of purified carboxysomes. On the right is a model of their structure. Scale bars are 100 nm.

Bacteria do not have a membrane-bound nucleus, and their genetic material is typically a single circular DNA chromosome located in the cytoplasm in an irregularly shaped body called the *nucleoid*. The nucleoid contains the chromosome with its associated proteins and RNA. The phylum Planctomycetes and candidate phylum Poribacteria may be exceptions to the general absence of internal membranes in bacteria, because they appear to have a double membrane around their nucleoids and contain other membrane-bound cellular structures. Like all living organisms, bacteria contain *ribosomes*, often grouped in chains called polyribosomes, for the production of proteins, but the structure of the bacterial ribosome is different from that of eukaryotes and Archaea. Bacterial ribosomes have a sedimentation rate of 70S (measured in Svedberg units): their subunits have rates of 30S and 50S. Some antibiotics bind specifically to 70S ribosomes and inhibit bacterial protein synthesis. Those antibiotics kill bacteria without affecting the larger 80S ribosomes of eukaryotic cells and without harming the host.

Some bacteria produce intracellular nutrient storage granules for later use, such as glycogen, polyphosphate, sulfur or polyhydroxyalkanoates. Certain bacterial species, such as the photosynthetic Cyanobacteria, produce internal gas vesicles, which they use to regulate their buoyancy—allowing them to move up or down into water layers with different light intensities and nutrient levels.*Intracellular membranes* called *chromatophores* are also found in membranes of phototrophic bacteria. Used primarily for photosynthesis, they contain bacteriochlorophyll pigments and carotenoids. An early idea was that bacteria might contain membrane folds termed mesosomes, but these were later shown to be artefacts produced by the chemicals used to prepare the cells for electron microscopy. *Inclusions* are considered to be nonliving components of the cell that do not possess metabolic activity and are not bounded by membranes. The most common inclusions are glycogen, lipid droplets, crystals, and pigments. *Volutin granules* are cytoplasmic inclusions of complexed inorganic polyphosphate. These granules are called *metachromatic granules* due to their displaying the metachromatic effect; they appear red or blue when stained with the blue dyes methylene blue or toluidine blue. *Gas vacuoles*, which are freely permeable to gas, are membrane-bound vesicles present in some species of *Cyanobacteria*. They allow the bacteria to control their buoyancy. *Microcompartments* are widespread, membrane-bound organelles that are made of a protein shell that surrounds and encloses various enzymes. *Carboxysomes* are bacterial microcompartments that contain enzymes involved in carbon fixation. *Magnetosomes* are bacterial microcompartments, present in magnetotactic bacteria, that contain magnetic crystals.

Extracellular Structures

In most bacteria, a *cell wall* is present on the outside of the cell membrane. The cell membrane and cell wall comprise the *cell envelope*. A common bacterial cell wall material is *peptidoglycan* (called "murein" in older sources), which is made from polysaccharide chains cross-linked by peptides containing D-amino acids.Bacterial cell walls are different from the cell walls of plants and fungi, which are made of cellulose and chitin, respectively. The cell wall of bacteria is also distinct from that of Archaea, which do not contain peptidoglycan. The cell wall is essential to the survival of many bacteria, and the antibiotic penicillin is able to kill bacteria by inhibiting a step in the synthesis of peptidoglycan.

There are broadly speaking two different types of cell wall in bacteria, a thick one in the gram-positives and a thinner one in the gram-negatives. The names originate from the reaction of cells to the Gram stain, a long-standing test for the classification of bacterial species.

Gram-positive bacteria possess a thick cell wall containing many layers of peptidoglycan and *teichoic acids*. In contrast, *gram-negative bacteria* have a relatively thin cell wall consisting of a few layers of peptidoglycan surrounded by a second lipid membrane containing *lipopolysaccharides* and lipoproteins. Lipopolysaccharides, also called *endotoxins*, are composed of polysaccharides and *lipid A* that is responsible for much of the toxicity of gram-negative bacteria. Most bacteria have the gram-negative cell wall, and only the Firmicutes and Actinobacteria have the alternative gram-positive arrangement. These two groups were previously known as the low G+C and high G+C gram-positive bacteria, respectively. These differences in structure can produce differences in antibiotic susceptibility; for instance, vancomycin can kill only gram-positive bacteria and is ineffective against gram-negative pathogens, such as *Haemophilus influenzae* or *Pseudomonas aeruginosa*. If the bacterial cell wall is entirely removed, it is called a *protoplast*, whereas if it is partially removed, it is called a *spheroplast*. *β*-Lactam antibiotics, such as penicillin, inhibit the formation of peptidoglycan cross-links in the bacterial cell wall. The enzyme lysozyme, found in human tears, also digests the cell wall of bacteria and is the body›s main defence against eye infections.

Acid-fast bacteria, such as *Mycobacteria*, are resistant to decolorisation by acids during staining procedures. The high mycolic acid content of *Mycobacteria*, is responsible for the staining pattern of poor absorption followed by high retention. The most common staining technique used to identify acid-fast bacteria is the Ziehl-Neelsen stain or acid-fast stain, in which the acid-fast bacilli are stained bright-red and stand out clearly against a blue background. *L-form bacteria* are strains of bacteria that lack cell walls. The main pathogenic bacteria in this class is *Mycoplasma* (not to be confused with *Mycobacteria*).

In many bacteria, an *S-layer* of rigidly arrayed protein molecules covers the outside of the cell. This layer provides chemical and physical protection for the cell surface and can act as a macromolecular diffusion barrier. S-layers have diverse but mostly poorly understood functions, but are known to act as virulence factors in *Campylobacter* and contain surface enzymes in *Bacillus stearothermophilus*.

Flagella are rigid protein structures, about 20 nanometres in diameter and up to 20 micrometres in length, that are used for motility. Flagella are driven by the energy released by the transfer of ions down an electrochemical gradient across the cell membrane.

Fimbriae (sometimes called "attachment pili") are fine filaments of protein, usually 2–10 nanometres in diameter and up to several micrometres in length. They are distributed over the surface of the cell, and resemble fine hairs when seen under the electron microscope. Fimbriae are believed to be involved in attachment to solid surfaces or to other cells, and are essential for the virulence of some bacterial pathogens. *Pili* (*sing.* pilus) are cellular appendages, slightly larger than fimbriae, that can transfer genetic material between bacterial cells in a process called conjugation where they are called *conjugation pili* or "sex pili" (see bacterial genetics, below). They can also generate movement where they are called *type IV pili* (see movement, below).

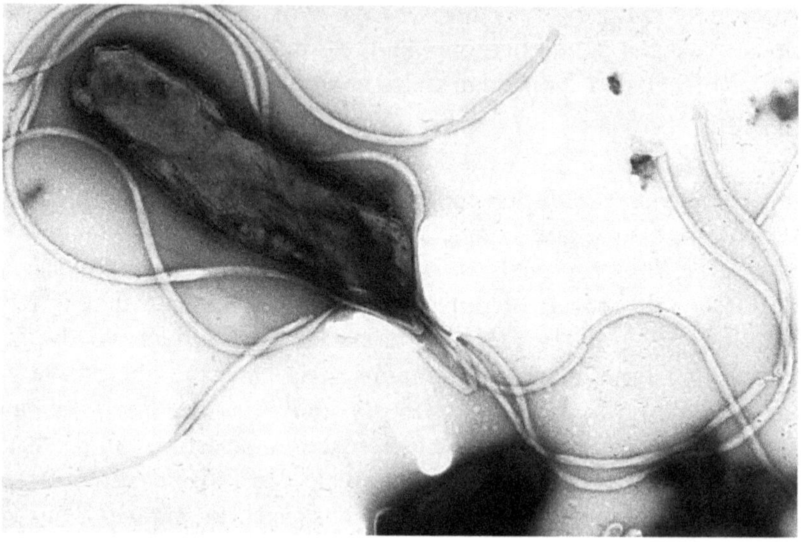

Helicobacter pylori electron micrograph, showing multiple flagella on the cell surface

Glycocalyx are produced by many bacteria to surround their cells, and vary in structural complexity: ranging from a disorganised *slime layer* of extra-cellular polymer to a highly structured *capsule*. These structures can protect cells from engulfment by eukaryotic cells such as macrophages (part of the human immune system). They can also act as antigens and be involved in cell recognition, as well as aiding attachment to surfaces and the formation of biofilms.

The assembly of these extracellular structures is dependent on bacterial secretion systems. These transfer proteins from the cytoplasm into the periplasm or into the environment around the cell. Many types of secretion systems are known and these structures are often essential for the virulence of pathogens, so are intensively studied.

Endospores

Certain genera of gram-positive bacteria, such as *Bacillus*, *Clostridium*, *Sporohalobacter*, *Anaerobacter*, and *Heliobacterium*, can form highly resistant, dormant structures called *endospores*. In almost all cases, one endospore is formed and this is not a reproductive process, although *Anaerobacter* can make up to seven endospores in a single cell. Endospores have a central core of cytoplasm containing DNA and ribosomes surrounded by a cortex layer and protected by an impermeable and rigid coat. Dipicolinic acid is a chemical compound that composes 5% to 15% of the dry weight of bacterial spores. It is implicated as responsible for the heat resistance of the endospore.

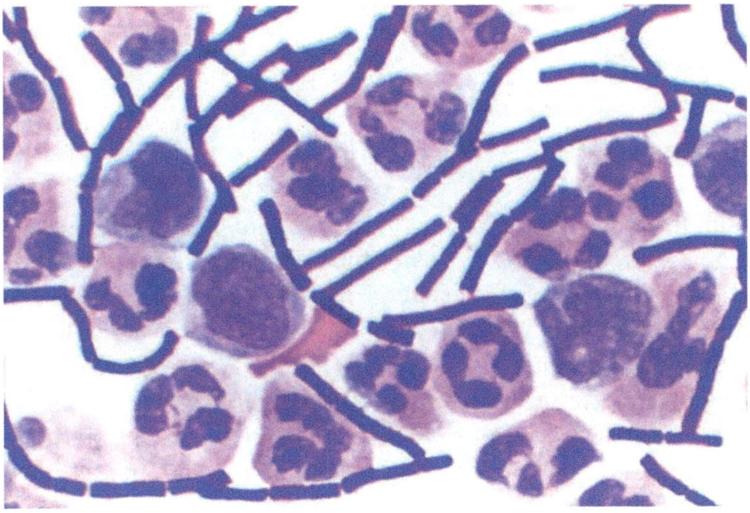

Bacillus anthracis (stained purple) growing in cerebrospinal fluid

Endospores show no detectable metabolism and can survive extreme physical and chemical stresses, such as high levels of UV light, gamma radiation, detergents, disinfectants, heat, freezing, pressure, and desiccation. In this dormant state, these organisms may remain viable for millions of years, and endospores even allow bacteria to survive exposure to the vacuum and radiation in space. According to scientist Dr. Steinn Sigurdsson, "There are viable bacterial spores that have been found that are 40 million years old on Earth—and we know they›re very hardened to radiation." Endospore-forming bacteria can also cause disease: for example, anthrax can be contracted by the inhalation of *Bacillus anthracis* endospores, and contamination of deep puncture wounds with *Clostridium tetani* endospores causes tetanus.

Metabolism

Bacteria exhibit an extremely wide variety of metabolic types. The distribution of metabolic traits within a group of bacteria has traditionally been used to define their taxonomy, but these traits often do not correspond with modern genetic classifications. Bacterial metabolism is classified into nutritional groups on the basis of three major criteria: the kind of energy used for growth, the source of carbon, and the electron donors used for growth. An additional criterion of respiratory microorganisms are the electron acceptors used for aerobic or anaerobic respiration.

Nutritional types in bacterial metabolism			
Nutritional type	**Source of energy**	**Source of carbon**	**Examples**
Phototrophs	Sunlight	Organic compounds (photoheterotrophs) or carbon fixation (photoautotrophs)	Cyanobacteria, Green sulfur bacteria, Chloroflexi, or Purple bacteria

Lithotrophs	Inorganic compounds	Organic compounds (lithoheterotrophs) or carbon fixation (lithoautotrophs)	Thermodesulfobacteria, *Hydrogenophilaceae*, or Nitrospirae
Organotrophs	Organic compounds	Organic compounds (chemoheterotrophs) or carbon fixation (chemoautotrophs)	*Bacillus*, *Clostridium* or *Enterobacteriaceae*

Carbon metabolism in bacteria is either *heterotrophic*, where organic carbon compounds are used as carbon sources, or *autotrophic*, meaning that cellular carbon is obtained by fixing carbon dioxide. Heterotrophic bacteria include parasitic types. Typical autotrophic bacteria are phototrophic cyanobacteria, green sulfur-bacteria and some purple bacteria, but also many chemolithotrophic species, such as nitrifying or sulfur-oxidising bacteria. Energy metabolism of bacteria is either based on *phototrophy*, the use of light through photosynthesis, or based on *chemotrophy*, the use of chemical substances for energy, which are mostly oxidised at the expense of oxygen or alternative electron acceptors (aerobic/anaerobic respiration).

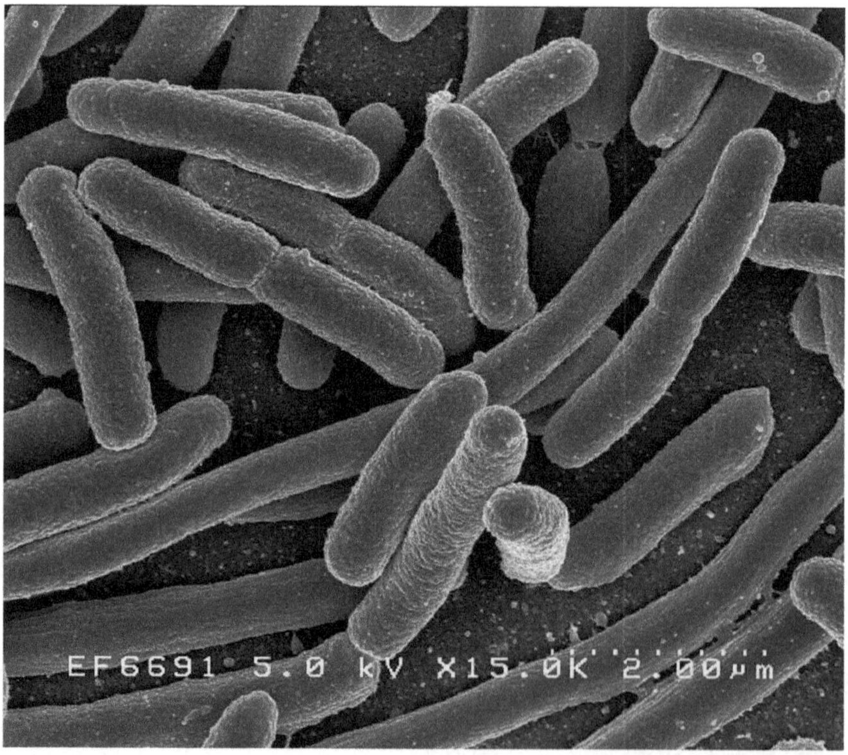

Filaments of photosyntheticcyanobacteria,

Bacteria are further divided into *lithotrophs* that use inorganic electron donors and *organotrophs* that use organic compounds as electron donors. Chemotrophic organisms use the respective electron donors for energy conservation (by aerobic/anaerobic respiration or fermentation) and biosynthetic reactions (e.g., carbon dioxide fixation), whereas phototrophic organisms use them only for biosynthetic purposes. Respiratory organisms use chemical compounds as a source of energy by taking electrons from the reduced substrate and transferring them to a terminal electron acceptor in a redox reaction. This reaction releases energy that can be used to synthesise ATP and drive metabolism. In *aerobic organisms*, oxygen is used as the electron acceptor. In *anaerobic organisms* other inorganic compounds, such as nitrate, sulfate or carbon dioxide are used as electron acceptors. This leads to the ecologically important processes of denitrification, sulfate reduction, and acetogenesis, respectively.

Another way of life of chemotrophs in the absence of possible electron acceptors is fermentation, wherein the electrons taken from the reduced substrates are transferred to oxidised intermediates to generate reduced fermentation products (e.g., lactate, ethanol, hydrogen, butyric acid). Fermentation is possible, because the energy content of the substrates is higher than that of the products, which allows the organisms to synthesise ATP and drive their metabolism.

These processes are also important in biological responses to pollution; for example, sulfate-reducing bacteria are largely responsible for the production of the highly toxic forms of mercury (methyl- and dimethylmercury) in the environment. Non-respiratory anaerobes use fermentation to generate energy and reducing power, secreting metabolic by-products (such as ethanol in brewing) as waste. Facultative anaerobes can switch between fermentation and different terminal electron acceptors depending on the environmental conditions in which they find themselves.

Lithotrophic bacteria can use inorganic compounds as a source of energy. Common inorganic electron donors are hydrogen, carbon monoxide, ammonia (leading to nitrification), ferrous iron and other reduced metal ions, and several reduced sulfur compounds. In unusual circumstances, the gas methane can be used by methanotrophic bacteria as both a source of electrons and a substrate for carbon anabolism. In both aerobic phototrophy and chemolithotrophy, oxygen is used as a terminal electron acceptor, whereas under anaerobic conditions inorganic compounds are used instead. Most lithotrophic organisms are autotrophic, whereas organotrophic organisms are heterotrophic.

In addition to fixing carbon dioxide in photosynthesis, some bacteria also fix nitrogen gas (nitrogen fixation) using the enzyme nitrogenase. This environmentally important trait can be found in bacteria of nearly all the metabolic types listed above, but is not universal.

Regardless of the type of metabolic process they employ, the majority of bacteria are able to take in raw materials only in the form of relatively small molecules, which enter the cell by diffusion or through molecular channels in cell membranes. The Planctomycetes are the exception (as they are in possessing membranes around their nuclear material). It has recently been shown that *Gemmata obscuriglobus* is able to take in large molecules via a process that in some ways resembles endocytosis, the process used by eukaryotic cells to engulf external items.

Growth and Reproduction

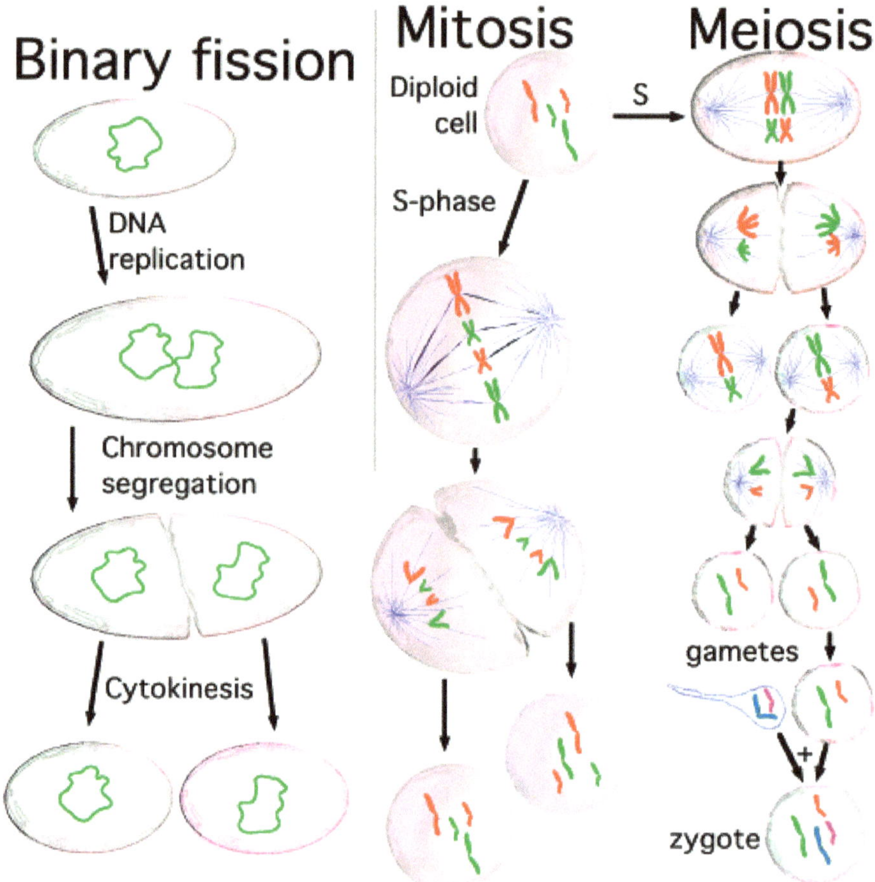

Many bacteria reproduce through binary fission, which is compared to mitosis and meiosis in this image.

Unlike in multicellular organisms, increases in cell size (cell growth) and reproduction by cell division are tightly linked in unicellular organisms. Bacteria grow to a fixed size and then reproduce through *binary fission*, a form of asexual reproduction. Under optimal conditions, bacteria can grow and divide extremely rapidly, and bacterial populations can double as quickly as every 9.8 minutes. In cell division, two identical clone daughter cells are produced. Some bacteria, while still reproducing asexually, form more complex reproductive structures that help disperse the newly formed daughter cells. Examples include fruiting body formation by *Myxobacteria* and aerial hyphae formation by *Streptomyces*, or budding. Budding involves a cell forming a protrusion that breaks away and produces a daughter cell.

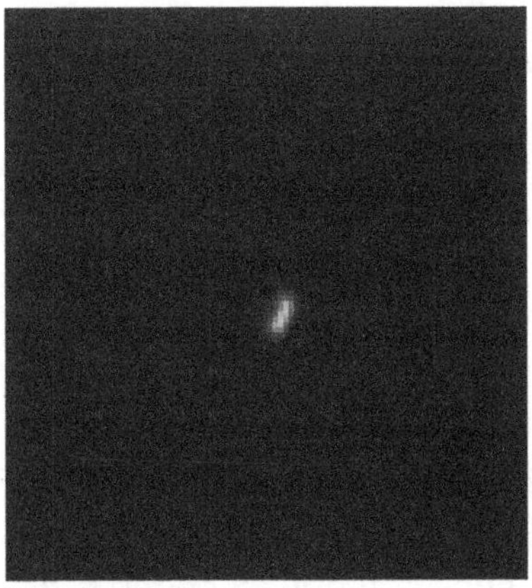

A colony of *Escherichia coli.*

In the laboratory, bacteria are usually grown using solid or liquid media. Solid *growth media*, such as agar plates, are used to isolate pure cultures of a bacterial strain. However, liquid growth media are used when measurement of growth or large volumes of cells are required. Growth in stirred liquid media occurs as an even cell suspension, making the cultures easy to divide and transfer, although isolating single bacteria from liquid media is difficult. The use of selective media (media with specific nutrients added or deficient, or with antibiotics added) can help identify specific organisms.

Most laboratory techniques for growing bacteria use high levels of nutrients to produce large amounts of cells cheaply and quickly. However, in natural environments, nutrients are limited, meaning that bacteria cannot continue to reproduce indefinitely. This nutrient limitation has led the evolution of different growth strategies (see r/K selection theory). Some organisms can grow extremely rapidly when nutrients become available, such as the formation of algal (and cyanobacterial) blooms that often occur in lakes during the summer. Other organisms have adaptations to harsh environments, such as the production of multiple antibiotics by *Streptomyces* that inhibit the growth of competing microorganisms. In nature, many organisms live in communities (e.g., biofilms) that may allow for increased supply of nutrients and protection from environmental stresses. These relationships can be essential for growth of a particular organism or group of organisms (syntrophy).

Bacterial growth follows four phases. When a population of bacteria first enter a high-nutrient environment that allows growth, the cells need to adapt to their new environment. The first phase of growth is the *lag phase*, a period of slow growth when the cells are adapting to the high-nutrient environment and preparing for fast growth. The lag phase has high biosynthesis rates, as proteins necessary for rapid growth are produced. The second phase of growth is the *log phase*, also known as the *logarithmic or exponential phase*. The log phase is marked by rapid exponential growth. The rate at which cells grow during this phase is known as the *growth rate* (k), and the time it takes the cells to double is known as the *generation time* (g). During log phase, nutrients are metabolised at maximum speed until one of the nutrients is depleted and starts limiting growth. The third phase of growth is the *stationary phase* and is caused by depleted nutrients. The cells reduce their metabolic activity and consume non-essential cellular proteins. The stationary phase is a transition from rapid growth to a stress response state and there is increased expression of genes involved in DNA repair, antioxidant metabolism and nutrient transport. The final phase is the *death phase* where the bacteria run out of nutrients and die.

Genomes

The genomes of thousands of bacterial species have been sequenced, with at least 9,000 sequences completed and more than 42,000 left as "permanent" drafts (as of Sep 2016).

Most bacteria have a single circular chromosome that can range in size from only 160,000 base pairs in the endosymbiotic bacteria *Candidatus Carsonella ruddii*, to 12,200,000 base pairs in the soil-dwelling bacteria *Sorangium cellulosum*. The genes in bacterial genomes are usually a single continuous stretch of DNA and although several different types of introns do exist in bacteria, these are much rarer than in eukaryotes. Some bacteria, including the Spirochaetes of the genus *Borrelia* are a notable exception to this arrangement. *Borrelia burgdorferi*, the cause of Lyme disease, contains a single linear chromosome and several linear and circular plasmids.

Plasmids are small extra-chromosomal DNAs that may contain genes for antibiotic resistance or virulence factors. Plasmids replicate independently of chromosomes, so it is possible that plasmids could be lost in bacterial cell division. Against this possibility is the fact that a single bacterium can contain hundreds of copies of a single plasmid.

Genetics

Bacteria, as sexual organisms, inherit identical copies of their parent's genes (i.e., they are clonal). However, all bacteria can evolve by selection on changes to their genetic material DNA caused by genetic recombination or mutations. Mutations come from errors made during the replication of DNA or from exposure to mutagens. Mutation rates vary widely among different species of bacteria and even among different clones of a single species of bacteria. Genetic changes in bacterial genomes come from either random mutation during replication or "stress-directed mutation", where genes involved in a particular growth-limiting process have an increased mutation rate.

DNA Transfer

Some bacteria also transfer genetic material between cells. This can occur in three main ways. First, bacteria can take up exogenous DNA from their environment, in a process called *transformation*. Genes can also be transferred by the process of *transduction*, when the integration of a bacteriophage introduces foreign DNA into the chromosome. The third method of gene transfer is *conjugation*, whereby DNA is transferred through direct cell contact.

Transduction of bacterial genes by bacteriophage appears to be a consequence of infrequent errors during intracellular assembly of virus particles, rather than a bacterial adaptation. Conjugation, in the much-studied E. coli system is determined by plasmid genes, and is an adaptation for transferring copies of the plasmid from one bacterial host to another. It is seldom that a conjugative plasmid integrates into the host bacterial chromosome, and subsequently transfers part of the host bacterial DNA to another bacterium. Plasmid-mediated transfer of host bacterial DNA also appears to be an accidental process rather than a bacterial adaptation. Transformation, unlike transduction or conjugation, depends on numerous bacterial gene products that specifically interact to perform this complex process, and thus transformation is clearly a bacterial adaptation for DNA transfer. In order for a bacterium to bind, take up and recombine donor DNA into its own chromosome, it must first enter a special physiological state termed competence (Natural competence). In *Bacillus subtilis*, about 40 genes are required for the development of competence. The length of DNA transferred during *B. subtilis* transformation can be between a third of a chromosome up to the whole chromosome.Transformation appears to be common among bacterial species, and thus far at least 60 species are known to have the natural ability to become competent for transformation. The development of competence in nature is usually associated with stressful environmental conditions, and seems to be an adaptation for facilitating repair of DNA damage in recipient cells.

In ordinary circumstances, transduction, conjugation, and transformation involve transfer of DNA between individual bacteria of the same species, but occasionally transfer may occur between individuals of different bacterial species and this may have significant consequences, such as the transfer of antibiotic resistance. In such cases, gene acquisition from other bacteria or the environment is called *horizontal gene transfer* and may be common under natural conditions. Gene transfer is particularly important in antibiotic resistance as it allows the rapid transfer of resistance genes between different pathogens.

Bacteriophages

Bacteriophages are viruses that infect bacteria. Many types of bacteriophage exist, some simply infect and lyse their host bacteria, while others insert into the bacterial chromosome. A bacteriophage can contain genes that contribute to its host's phenotype: for example, in the evolution of *Escherichia coli* O157:H7 and *Clostridium botulinum*, the toxin genes in an integrated phage converted a harmless ancestral bacterium into a lethal pathogen. Bacteria resist phage infection through restriction modification systems that degrade foreign DNA, and a system that uses CRISPR sequences to retain fragments of the genomes of phage that the bacteria have come into contact with in the past, which allows them to block virus replication through a form of RNA interference. This CRISPR system provides bacteria with acquired immunity to infection.

Behaviour

Secretion

Bacteria frequently secrete chemicals into their environment in order to modify it favourably. The secretions are often proteins and may act as enzymes that digest some form of food in the environment.

Bioluminescence

A few bacteria have chemical systems that generate light. This bioluminescence often occurs in bacteria that live in association with fish, and the light probably serves to attract fish or other large animals.

Multicellularity

Bacteria often function as multicellular aggregates known as biofilms, exchanging a variety of molecular signals for inter-cell communication, and engaging in coordinated multicellular behaviour.

The communal benefits of multicellular cooperation include a cellular division of labour, accessing resources that cannot effectively be used by single cells, collectively defending against antagonists, and optimising population survival by differentiating into distinct cell types. For example, bacteria in biofilms can have more than 500 times increased resistance to antibacterial agents than individual "planktonic" bacteria of the same species.

One type of inter-cellular communication by a molecular signal is called quorum sensing, which serves the purpose of determining whether there is a local population density that is sufficiently high that it is productive to invest in processes that are only successful if large numbers of similar organisms behave similarly, as in excreting digestive enzymes or emitting light.

Quorum sensing allows bacteria to coordinate gene expression, and enables them to produce, release and detect autoinducers or pheromones which accumulate with the growth in cell population.

Movement

Many bacteria can move using a variety of mechanisms: flagella are used for swimming through fluids; bacterial gliding and twitching motility move bacteria across surfaces; and changes of buoyancy allow vertical motion.

Swimming bacteria frequently move near 10 body lengths per second and a few as fast as 100. This makes them at least as fast as fish, on a relative scale.

In bacterial gliding and twitching motility, bacteria use their *type IV pili* as a grappling hook, repeatedly extending it, anchoring it and then retracting it with remarkable force (>80 pN).

Our observations redefine twitching motility as a rapid, highly organized mechanism of bacterial translocation by which Pseudomonas aeruginosa can disperse itself over large areas to colonize new territories. It is also now clear, both morphologically and genetically, that twitching motility and social gliding motility, such as occurs in Myxococcus xanthus, are essentially the same process.

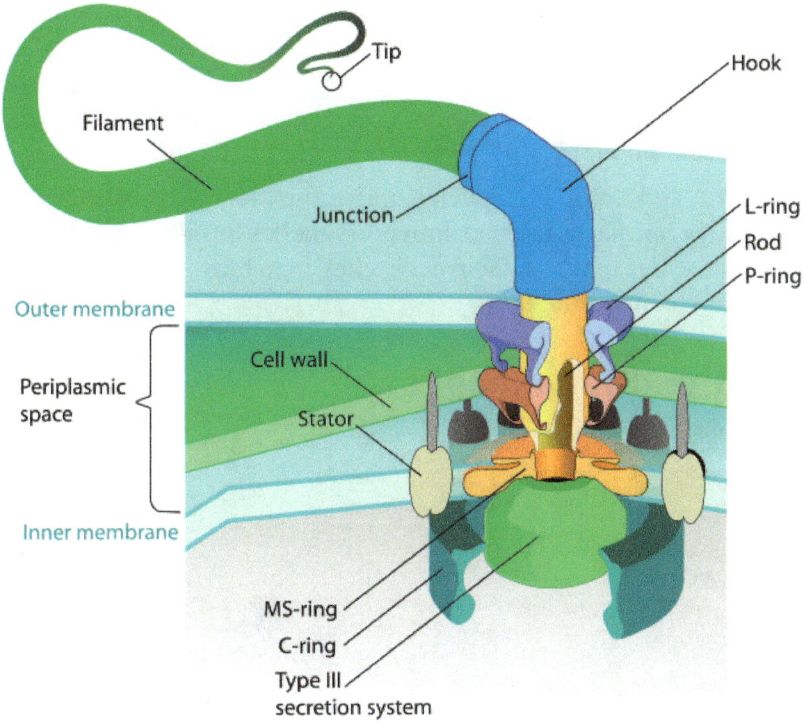

Flagellum of gram-negative bacteria. The base drives the rotation of the hook and filament.

—Semmler, Whitchurch & Mattick (1999)

Flagella are semi-rigid cylindrical structures that are rotated and function much like the propeller on a ship. Objects as small as bacteria operate a low Reynolds number and cylindrical forms are more efficient than the flat, paddle-like, forms appropriate at human-size scale.

Bacterial species differ in the number and arrangement of flagella on their surface; some have a single flagellum (*monotrichous*), a flagellum at each end (*amphitrichous*), clusters of flagella at the poles of the cell (*lophotrichous*), while others have flagella distributed over the entire surface of the cell (*peritrichous*). The bacterial flagella is the best-understood motility structure in any organism and is made of about 20 proteins, with approximately another 30 proteins required for its regulation and assembly. The flagellum is a rotating structure driven by a reversible motor at the base that uses the electrochemical gradient across the membrane for power. This motor drives the motion of the filament, which acts as a propeller.

Many bacteria (such as *E. coli*) have two distinct modes of movement: forward movement (swimming) and tumbling. The tumbling allows them to reorient and makes their movement a three-dimensional random walk. (See external links below for link to videos.) The flagella of a unique group of bacteria, the spirochaetes, are found between two membranes in the periplasmic space. They have a distinctive helical body that twists about as it moves.

Motile bacteria are attracted or repelled by certain stimuli in behaviours called **taxes**: these include chemotaxis, phototaxis, energy taxis, and magnetotaxis. In one peculiar group, the myxobacteria, individual bacteria move together to form waves of cells that then differentiate to form fruiting bodies containing spores. The myxobacteria move only when on solid surfaces, unlike *E. coli*, which is motile in liquid or solid media.

Several *Listeria* and *Shigella* species move inside host cells by usurping the cytoskeleton, which is normally used to move organelles inside the cell. By promoting actin polymerisation at one pole of their cells, they can form a kind of tail that pushes them through the host cell's cytoplasm.

Classification and Identification

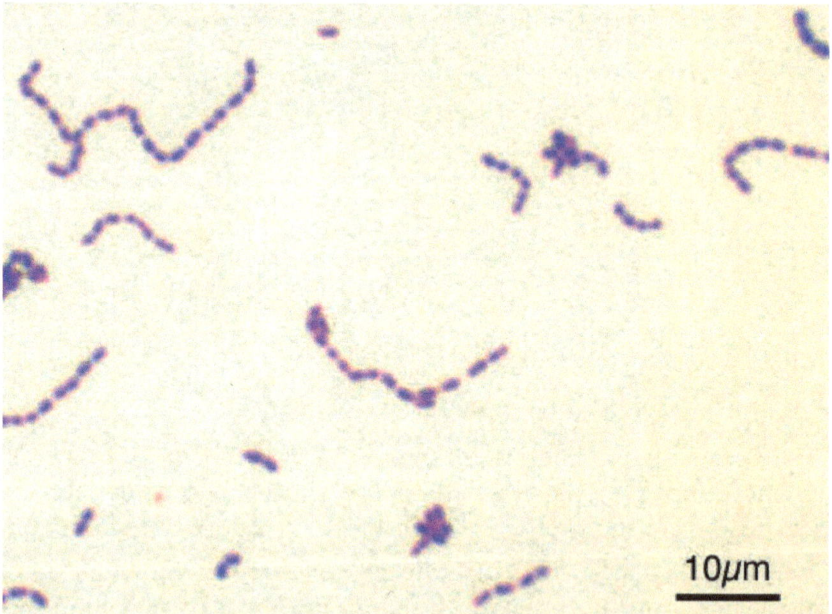

Streptococcus mutans visualised with a Gram stain

Classification seeks to describe the diversity of bacterial species by naming and grouping organisms based on similarities. Bacteria can be classified on the basis of cell structure, cellular metabolism or on differences in cell components, such as DNA, fatty acids, pigments, antigens and quinones. While these schemes allowed the identification and classification of bacterial strains, it was unclear whether these differences represented variation between distinct species or between strains of the same species. This uncertainty was due to the lack of distinctive structures in most bacteria, as well as lateral gene transfer between unrelated species. Due to lateral gene transfer, some closely related bacteria can have very different morphologies and metabolisms. To overcome this uncertainty, modern bacterial classification emphasises molecular systematics, using genetic techniques such as guanine cytosine ratio determination, genome-genome hybridisation, as well as sequencing genes that have not undergone extensive lateral gene transfer, such as the rRNA gene. Classification of bacteria is determined by publication in the International Journal of Systematic Bacteriology,and Bergey›s Manual of Systematic Bacteriology. The International Committee on Systematic Bacteriology (ICSB) maintains international rules for the naming of bacteria and taxonomic categories and for the ranking of them in the International Code of Nomenclature of Bacteria.

The term "bacteria" was traditionally applied to all microscopic, single-cell prokaryotes. However, molecular systematics showed prokaryotic life to consist of two separate domains, originally called *Eubacteria* and *Archaebacteria*, but now called *Bacteria* and *Archaea* that evolved independently from an ancient common ancestor. The archaea and eukaryotes are more closely related to each other than either is to the bacteria. These two domains, along with Eukarya, are the basis of the three-domain system, which is currently the most widely used classification system in microbiolology. However, due to the relatively recent introduction of molecular systematics and a rapid increase in the number of genome sequences that are available, bacterial classification remains a changing and expanding field. For example, a few biologists argue that the Archaea and Eukaryotes evolved from gram-positive bacteria.

The identification of bacteria in the laboratory is particularly relevant in medicine, where the correct treatment is determined by the bacterial species causing an infection. Consequently, the need to identify human pathogens was a major impetus for the development of techniques to identify bacteria.

The *Gram stain*, developed in 1884 by Hans Christian Gram, characterises bacteria based on the structural characteristics of their cell walls. The thick layers of peptidoglycan in the "gram-positive" cell wall stain purple, while the thin "gram-negative" cell wall appears pink. By combining morphology and Gram-staining, most bacteria can be classified as belonging to one of four groups (gram-positive cocci, gram-positive bacilli, gram-negative cocci and gram-negative bacilli). Some organisms are best identified by stains other than the Gram stain, particularly mycobacteria or *Nocardia*, which show acid-fastness on Ziehl–Neelsen or similar stains. Other organisms may need to be identified by their growth in special media, or by other techniques, such as serology.

Culture techniques are designed to promote the growth and identify particular bacteria, while restricting the growth of the other bacteria in the sample. Often these techniques are designed for specific specimens; for example, a sputum sample will be treated to identify organisms that cause pneumonia, while stool specimens are cultured on selective media to identify organisms that cause diarrhoea, while preventing growth of non-pathogenic bacteria. Specimens that are normally sterile, such as blood, urine or spinal fluid, are cultured under conditions designed to grow all possible organisms. Once a pathogenic organism has been isolated, it can be further characterised by its morphology, growth patterns (such as aerobic or anaerobic growth), patterns of hemolysis, and staining.

As with bacterial classification, identification of bacteria is increasingly using molecular methods. Diagnostics using DNA-based tools, such as polymerase chain reaction, are increasingly popular due to their specificity and speed, compared to culture-based methods. These methods also allow the detection and identification of "viable but nonculturable" cells that are metabolically active but non-dividing. However, even using these improved methods, the total number of bacterial species is not known and cannot even be estimated with any certainty. Following present classification, there are a little less than 9,300 known species of prokaryotes, which includes bacteria and archaea; but attempts to estimate the true number of bacterial diversity have ranged from 10^7 to 10^9 total species—and even these diverse estimates may be off by many orders of magnitude.

Interactions with other Organisms

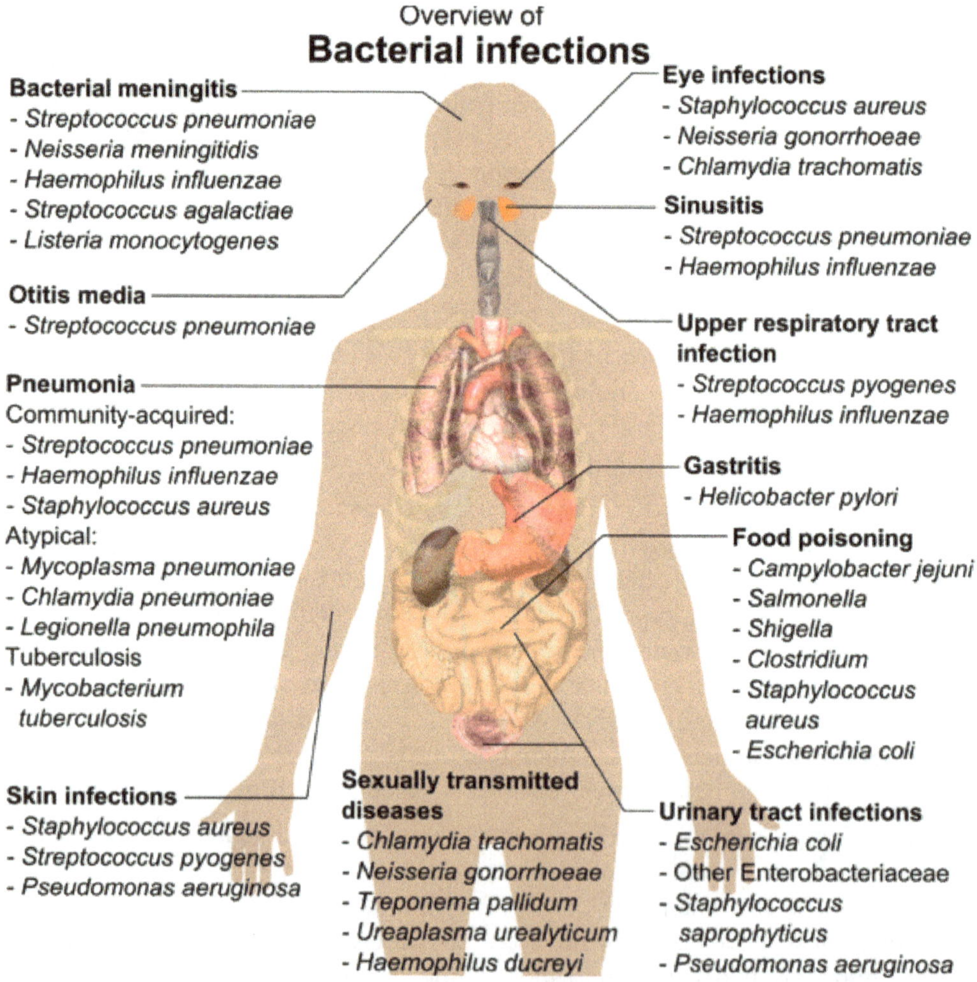

Overview of bacterial infections and main species involved.

Despite their apparent simplicity, bacteria can form complex associations with other organisms. These symbiotic associations can be divided into parasitism, mutualism and commensalism. Due to their small size, commensal bacteria are ubiquitous and grow on animals and plants exactly as they will grow on any other surface. However, their growth can be increased by warmth and sweat, and large populations of these organisms in humans are the cause of body odour.

Predators

Some species of bacteria kill and then consume other microorganisms, these species are called *predatory bacteria*. These include organisms such as *Myxococcus xanthus*, which forms swarms of cells that kill and digest any bacteria they encounter. Other bacterial predators either attach to their prey in order to digest them and absorb nutrients, such as *Vampirovibrio chlorellavorus*, or invade another cell and multiply inside the cytosol, such as *Daptobacter*. These predatory bacteria are thought to have evolved from saprophages that consumed dead microorganisms, through adaptations that allowed them to entrap and kill other organisms.

Mutualists

Certain bacteria form close spatial associations that are essential for their survival. One such mutualistic association, called interspecies hydrogen transfer, occurs between clusters of anaerobic bacteria that consume organic acids, such as butyric acid or propionic acid, and produce hydrogen, and methanogenic Archaea that consume hydrogen. The bacteria in this association are unable to consume the organic acids as this reaction produces hydrogen that accumulates in their surroundings. Only the intimate association with the hydrogen-consuming Archaea keeps the hydrogen concentration low enough to allow the bacteria to grow.

In soil, microorganisms that reside in the rhizosphere (a zone that includes the root surface and the soil that adheres to the root after gentle shaking) carry out nitrogen fixation, converting nitrogen gas to nitrogenous compounds. This serves to provide an easily absorbable form of nitrogen for many plants, which cannot fix nitrogen themselves. Many other bacteria are found as symbionts in humans and other organisms. For example, the presence of over 1,000 bacterial species in the normal human gut flora of the intestines can contribute to gut immunity, synthesise vitamins, such as folic acid, vitamin K and biotin, convert sugars to lactic acid (see *Lactobacillus*), as well as fermenting complex undigestible carbohydrates. The presence of this gut flora also inhibits the growth of potentially pathogenic bacteria (usually through competitive exclusion) and these beneficial bacteria are consequently sold as probiotic dietary supplements.

Pathogens

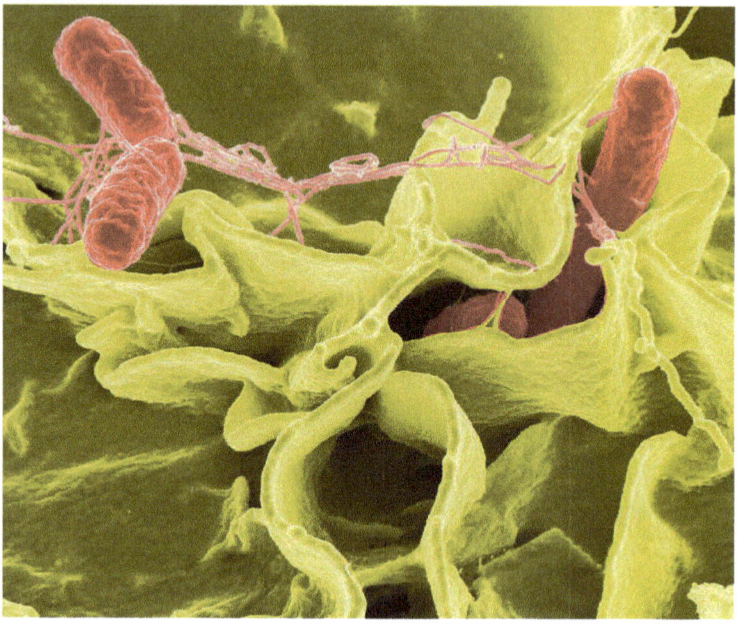

Colour-enhanced scanning electron micrograph showing *Salmonella typhimurium* (red) invading cultured human cells

If bacteria form a parasitic association with other organisms, they are classed as pathogens. Pathogenic bacteria are a major cause of human death and disease and cause infections such as tetanus, typhoid fever, diphtheria, syphilis, cholera, foodborne illness, leprosy and tuberculosis. A pathogenic cause for a known medical disease may only be discovered many years after, as was the case with *Helicobacter pylori* and peptic ulcer disease. Bacterial diseases are also important in agriculture, with bacteria causing leaf spot, fire blight and wilts in plants, as well as Johne's disease, mastitis, salmonella and anthrax in farm animals.

Each species of pathogen has a characteristic spectrum of interactions with its human hosts. Some organisms, such as *Staphylococcus* or *Streptococcus*, can cause skin infections, pneumonia, meningitis and even overwhelming sepsis, a systemic inflammatory response producing shock, massive vasodilation and death. Yet these organisms are also part of the normal human flora and usually exist on the skin or in the nose without causing any disease at all. Other organisms invariably cause disease in humans, such as the Rickettsia, which are obligate intracellular parasites able to grow and reproduce only within the cells of other organisms. One species of Rickettsia causes typhus, while another causes Rocky Mountain spotted fever. *Chlamydia*, another phylum of obligate intracellular parasites, contains species that can cause pneumonia, or urinary tract infection and may be involved in coronary heart disease. Finally, some species, such as *Pseudomonas aeruginosa*, *Burkholderia cenocepacia*, and *Mycobacterium avium*, are opportunistic pathogens and cause disease mainly in people suffering from immunosuppression or cystic fibrosis.

Bacterial infections may be treated with antibiotics, which are classified as bacteriocidal if they kill bacteria, or bacteriostatic if they just prevent bacterial growth. There are many types of antibiotics and each class inhibits a process that is different in the pathogen from that found in the host. An example of how antibiotics produce selective toxicity are chloramphenicol and puromycin, which inhibit the bacterial ribosome, but not the structurally different eukaryotic ribosome.Antibiotics are used both in treating human disease and in intensive farming to promote animal growth, where they may be contributing to the rapid development of antibiotic resistance in bacterial populations. Infections can be prevented by antiseptic measures such as sterilising the skin prior to piercing it with the needle of a syringe, and by proper care of indwelling catheters. Surgical and dental instruments are also sterilised to prevent contamination by bacteria. Disinfectants such as bleach are used to kill bacteria or other pathogens on surfaces to prevent contamination and further reduce the risk of infection.

Significance in Technology and Industry

Bacteria, often lactic acid bacteria, such as *Lactobacillus* and *Lactococcus*, in combination with yeasts and moulds, have been used for thousands of years in the preparation of fermented foods, such as cheese, pickles, soy sauce, sauerkraut, vinegar, wine and yogurt.

The ability of bacteria to degrade a variety of organic compounds is remarkable and has been used in waste processing and bioremediation. Bacteria capable of digesting the hydrocarbons in petroleum are often used to clean up oil spills. Fertiliser was added to some of the beaches in Prince William Sound in an attempt to promote the growth of these naturally occurring bacteria after the 1989 *Exxon Valdez* oil spill. These efforts were effective on beaches that were not too thickly covered in oil. Bacteria are also used for the bioremediation of industrial toxic wastes. In the chemical industry, bacteria are most important in the production of enantiomerically pure chemicals for use as pharmaceuticals or agrichemicals.

Bacteria can also be used in the place of pesticides in the biological pest control. This commonly involves *Bacillus thuringiensis* (also called BT), a gram-positive, soil dwelling bacterium. Subspecies of this bacteria are used as a Lepidopteran-specific insecticides under trade names such as Dipel and Thuricide. Because of their specificity, these pesticides are regarded as environmentally friendly, with little or no effect on humans, wildlife, pollinators and most other beneficial insects.

Because of their ability to quickly grow and the relative ease with which they can be manipulated, bacteria are the workhorses for the fields of molecular biology, genetics and biochemistry. By making mutations in bacterial DNA and examining the resulting phenotypes, scientists can determine the function of genes, enzymes and metabolic pathways in bacteria, then apply this knowledge to more complex organisms. This aim of understanding the biochemistry of a cell reaches its most complex expression in the synthesis of huge amounts of enzyme kinetic and gene expression data into mathematical models of entire organisms. This is achievable in some well-studied bacteria, with models of *Escherichia coli* metabolism now being produced and tested. This understanding of bacterial metabolism and genetics allows the use of biotechnology to bioengineer bacteria for the production of therapeutic proteins, such as insulin, growth factors, or antibodies.

Because of their importance for research in general, samples of bacterial strains are isolated and preserved in Biological Resource Centers. This ensures the availability of the strain to scientists worldwide.

History of Bacteriology

For the history of microbiology, see Microbiology. For the history of bacterial classification, see Bacterial taxonomy. For the natural history of Bacteria, see Last universal common ancestor.

Bacteria were first observed by the Dutch microscopist Antonie van Leeuwenhoek in 1676, using a single-lens microscope of his own design. He then published his observations in a series of letters to the Royal Society of London.Bacteria were Leeuwenhoek's most remarkable microscopic discovery. They were just at the limit of what his simple lenses could make out and, in one of the most striking hiatuses in the history of science, no one else would see them again for over a century. Only then were his by-then-largely-forgotten observations of bacteria—as opposed to his famous "animalcules" (spermatozoa)—taken seriously.

Christian Gottfried Ehrenberg introduced the word "bacterium" in 1828. In fact, his *Bacterium* was a genus that contained non-spore-forming rod-shaped bacteria, as opposed to *Bacillus*, a genus of spore-forming rod-shaped bacteria defined by Ehrenberg in 1835.

Louis Pasteur demonstrated in 1859 that the growth of microorganisms causes the fermentation process, and that this growth is not due to spontaneous generation. (Yeasts and moulds, commonly associated with fermentation, are not bacteria, but rather fungi.) Along with his contemporary Robert Koch, Pasteur was an early advocate of the germ theory of disease.

Robert Koch, a pioneer in medical microbiology, worked on cholera, anthrax and tuberculosis. In his research into tuberculosis Koch finally proved the germ theory, for which he received a Nobel Prize in 1905. In *Koch's postulates*, he set out criteria to test if an organism is the cause of a disease, and these postulates are still used today.

Though it was known in the nineteenth century that bacteria are the cause of many diseases, no effective antibacterial treatments were available. In 1910, Paul Ehrlich developed the first antibiotic, by changing dyes that selectively stained *Treponema pallidum*—the spirochaete that causes syphilis—into compounds that selectively killed the pathogen. Ehrlich had been awarded a 1908 Nobel Prize for his work on immunology, and pioneered the use of stains to detect and identify bacteria, with his work being the basis of the Gram stain and the Ziehl–Neelsen stain.

A major step forward in the study of bacteria came in 1977 when Carl Woese recognised that archaea have a separate line of evolutionary descent from bacteria.This new phylogenetic taxonomy depended on the sequencing of 16S ribosomal RNA, and divided prokaryotes into two evolutionary domains, as part of the three-domain system.

Almost all bacteria are invisible to the naked eye, with a few extremely rare exceptions, such as *Thiomargarita namibiensis*. They lack a nucleus and other membrane-bound organelles, and can function and reproduce as individual cells, but often aggregate in multicellular colonies. Their genome is usually a single loop of DNA, although they can also harbor small pieces of DNA called plasmids. These plasmids can be transferred between cells through bacterial conjugation. Bacteria are surrounded by a cell wall, which provides strength and rigidity to their cells. They reproduce by binary fission or sometimes by budding, but do not undergo meiotic sexual reproduction. However, many bacterial species can transfer DNA between individual cells by a process referred to as natural transformation. Some species form extraordinarily resilient spores, but for bacteria this is a mechanism for survival, not reproduction. Under optimal conditions bacteria can grow extremely rapidly and can double as quickly as every 20 minutes.

ARCHAEA

The **Archaea** constitute a domain and kingdom of single-celledmicroorganisms. These microbes (archaea; singular archaeon) are prokaryotes, meaning that they have no cell nucleus or any other membrane-bound organelles in their cells.

Archaea were initially classified as bacteria, receiving the name archaebacteria (in the Archaebacteria kingdom), but this classification is outdated. Archaeal cells have unique properties separating them from the other two domains of life, Bacteria and Eukaryota. The Archaea are further divided into multiple recognized phyla. Classification is difficult because the majority have not been isolated in the laboratory and have only been detected by analysis of their nucleic acids in samples from their environment.

Archaea and bacteria are generally similar in size and shape, although a few archaea have very strange shapes, such as the flat and square-shaped cells of *Haloquadratum walsbyi*. Despite this morphological similarity to bacteria, archaea possess genes and several metabolic pathways that are more closely related to those of eukaryotes, notably the enzymes involved in transcription and translation. Other aspects of archaeal biochemistry are unique, such as their reliance on ether lipids in their cell membranes, including archaeols. Archaea use more energy sources than eukaryotes: these range from organic compounds, such as sugars, to ammonia, metal ions or even hydrogen gas. Salt-tolerant archaea (the Haloarchaea) use sunlight as an energy source, and other species of archaea fix carbon; however, unlike plants and cyanobacteria, no known species of archaea does both. Archaea reproduce asexually by binary fission, fragmentation, or budding; unlike bacteria and eukaryotes, no known species forms spores.

Archaea were initially viewed as extremophiles living in harsh environments, such as hot springs and salt lakes, but they have since been found in a broad range of habitats, including soils, oceans, marshlands and the human colon, oral cavity, and skin. Archaea are particularly numerous in the oceans, and the archaea in plankton may be one of the most abundant groups of organisms on the planet. Archaea are a major part of Earth's life and may play roles in both the carbon cycle and the nitrogen cycle. No clear examples of archaeal pathogens or parasites are known, but they are often mutualists or commensals. One example is the methanogens that inhabit human and ruminant guts, where their vast numbers aid digestion. Methanogens are also used in biogas production and sewage treatment, and enzymes from extremophile archaea that can endure high temperatures and organic solvents are exploited in biotechnology.

Classification

New Domain

Archaea were first found in extreme environments, such as volcanic hot springs. Pictured here is Grand Prismatic Spring of Yellowstone National Park.

For much of the 20th century, prokaryotes were regarded as a single group of organisms and classified based on their biochemistry, morphology and metabolism. For example, microbiologists tried to classify microorganisms based on the structures of their cell walls, their shapes, and the substances they consume. In 1965, Emile Zuckerkandl and Linus Pauling proposed instead using the sequences of the genes in different prokaryotes to work out how they are related to each other. This approach, known as phylogenetics, is the main method used today.

Archaea were first classified as a separate group of prokaryotes in 1977 by Carl Woese and George E. Fox in phylogenetic trees based on the sequences of ribosomal RNA (rRNA) genes. These two groups were originally named the Archaebacteria and Eubacteria and treated as kingdoms or subkingdoms, which Woese and Fox termed *Urkingdoms*. Woese argued that this group of prokaryotes is a fundamentally different sort of life. To emphasize this difference, Woese later proposed a new natural system of organisms with three separate Domains: the Eukarya, the Bacteria and the Archaea, in what is now known as "The Woesian Revolution".

The word *archaea* comes from the Ancient Greek ρχαα, meaning "ancient things", as the first representatives of the domain Archaea were methanogens and it was assumed that their metabolism reflected Earth's primitive atmosphere and the organism's antiquity. For a long time, archaea were seen as extremophiles that only exist in extreme habitats such as hot springs and salt lakes. However, as new habitats were studied, more organisms were discovered. Extreme halophilic and hyperthermophilic microbes were also included in the Archaea. By the end of the 20th century, archaea had been identified in non-extreme environments as well. Today, they are known to be a large and diverse group of organisms that are widely distributed in nature and are common in all habitats. This new appreciation of the importance and ubiquity of archaea came from using polymerase chain reaction (PCR) to detect prokaryotes from environmental samples (such as water or soil) by multiplying their ribosomal genes. This allows the detection and identification of organisms that have not been cultured in the laboratory.

Current Classification

The ARMAN are a new group of archaea recently discovered in acid mine drainage.

The classification of archaea, and of prokaryotes in general, is a rapidly moving and contentious field. Current classification systems aim to organize archaea into groups of organisms that share structural features and common ancestors. These classifications rely heavily on the use of the sequence of ribosomal RNA genes to reveal relationships between organisms (molecular phylogenetics). Most of the culturable and well-investigated species of archaea are members of two main phyla, the Euryarchaeota and Crenarchaeota. Other groups have been tentatively created. For example, the peculiar species *Nanoarchaeum equitans*, which was discovered in 2003, has been given its own phylum, the Nanoarchaeota. A new phylum Korarchaeota has also been proposed. It contains a small group of unusual thermophilic species that shares features of both of the main phyla, but is most closely related to the Crenarchaeota. Other recently detected species of archaea are only distantly related to any of these groups, such as the Archaeal Richmond Mine acidophilic nanoorganisms (ARMAN), which were discovered in 2006 and are some of the smallest organisms known.

A superphylum - TACK - has been proposed that includes the Thaumarchaeota, Aigarchaeota, Crenarchaeota, and Korarchaeota. This superphylum may be related to the origin of eukaryotes.

Concept of Species

The classification of archaea into species is also controversial. Biology defines a species as a group of related organisms. The familiar exclusive breeding criterion (organisms that can breed with each other but not with others) is of no help here because archaea reproduce asexually.

Archaea show high levels of horizontal gene transfer between lineages. Some researchers suggest that individuals can be grouped into species-like populations given highly similar genomes and infrequent gene transfer to/from cells with less-related genomes, as in the genus *Ferroplasma*. On the other hand, studies in *Halorubrum* found significant genetic transfer to/from less-related populations, limiting the criterion's applicability. A second concern is to what extent such species designations have practical meaning.

Current knowledge on genetic diversity is fragmentary and the total number of archaeal species cannot be estimated with any accuracy. Estimates of the number of phyla range from 18 to 23, of which only 8 have representatives that have been cultured and studied directly. Many of these hypothesized groups are known from a single rRNA sequence, indicating that the diversity among these organisms remains obscure. The Bacteria also contain many uncultured microbes with similar implications for characterization.

Origin and Evolution

The age of the Earth is about 4.54 billion years. Scientific evidence suggests that life began on Earth at least 3.5 billion years ago. The earliest evidence for life on Earth is graphite found to be biogenic in 3.7 billion-year-old metasedimentary rocks discovered in Western Greenland and microbial matfossils found in 3.48 billion-year-old sandstone discovered in Western Australia. More recently, in 2015, "remains of biotic life" were found in 4.1 billion-year-old rocks in Western Australia. An unrelated researcher, S. Blair Hedges, wrote in an email about the study saying, "If life arose relatively quickly on Earth ... then it could be common in the universe."

Although probable prokaryotic cell fossils date to almost 3.5 billion years ago, most prokaryotes do not have distinctive morphologies and fossil shapes cannot be used to identify them as archaea. Instead, chemical fossils of unique lipids are more informative because such compounds do not occur in other organisms.Some publications suggest that archaeal or eukaryotic lipid remains are present in shales dating from 2.7 billion years ago; such data have since been questioned. Such lipids have also been detected in even older rocks from west Greenland. The oldest such traces come from the Isua district, which include Earth's oldest known sediments, formed 3.8 billion years ago. The archaeal lineage may be the most ancient that exists on Earth.

Woese argued that the bacteria, archaea, and eukaryotes represent separate lines of descent that diverged early on from an ancestral colony of organisms.One possibility is that this occurred before the evolution of cells, when the lack of a typical cell membrane allowed unrestricted lateral gene transfer, and that the common ancestors of the three domains arose by fixation of specific subsets of genes. It is possible that the last common ancestor of the bacteria and archaea was a thermophile, which raises the possibility that lower temperatures are "extreme environments" in archaeal terms, and organisms that live in cooler environments appeared only later. Since the Archaea and Bacteria are no more related to each other than they are to eukaryotes, the term *prokaryote*'s only surviving meaning is "not a eukaryote", limiting its value.

Comparison to other Domains

The following table compares some major characteristics of the three domains, to illustrate their similarities and differences. Many of these characteristics are also discussed below.

Property	Archaea	Bacteria	Eukarya
Cell Membrane	Ether-linked lipids, pseudopeptidoglycan	Ester-linked lipids, peptidoglycan	Ester-linked lipids, various structures
Gene Structure	Circular chromosomes, similar translation and transcription to Eukarya	Circular chromosomes, unique translation and transcription	Multiple, linear chromosomes, similar translation and transcription to Archaea
Internal Cell Structure	No membrane-bound organelles or nucleus	No membrane-bound organelles or nucleus	Membrane-bound organelles and nucleus
Metabolism	Various, with methanogenesis unique to Archaea	Various, including photosynthesis, aerobic and anaerobic respiration, fermentation, and autotrophy	Photosynthesis and cellular respiration
Reproduction	Asexual reproduction, horizontal gene transfer	Asexual reproduction, horizontal gene transfer	Sexual and asexual reproduction

Archaea were split off as a third domain because of the large differences in their ribosomal RNA structure. The particular RNA molecule sequenced, known as 16s rRNA, is present in all organisms and always has the same vital function, the production of proteins. Because this function is so central to life, organisms with mutations of its 16s rRNA are unlikely to survive, leading to great stability in the structure of this nucleotide over many generations. 16s rRNA is also large enough to retain organism-specific information, but small enough to be sequenced in a manageable amount of time. In 1977, Carl Woese, a microbiologist studying the genetic sequencing of organisms, developed a new sequencing method that involved splitting the RNA into fragments that could be sorted and compared to other fragments from other organisms. The more similar the patterns between species were, the more closely related the organisms.

Woese used his new rRNA comparison method to categorize and contrast different organisms. He sequenced a variety of different species and happened upon a group of methanogens that had vastly different patterns than any known prokaryotes or eukaryotes. These methanogens were much more similar to each other than they were to other organisms sequenced, leading Woese to propose the new domain of Archaea. One of the interesting results of his experiments was that the Archaea were more similar to eukaryotes than prokaryotes, even though they were more similar to prokaryotes in structure. This led to the conclusion that Archaea and Eukarya shared a more recent common ancestor than Eukarya and Bacteria in general. The development of the nucleus occurred after the split between Bacteria and this common ancestor. Although Archaea are prokaryotic, they are more closely related to Eukarya and thus cannot be placed within either the Bacteria or Eukarya domains.

One property unique to Archaea is the abundant use of ether-linked lipids in their cell membranes. Ether linkages are more chemically stable than the ester linkages found in Bacteria and Eukarya, which may be a contributing factor to the ability of many Archaea to survive in extreme environments that place heavy stress on cell membranes, such as extreme heat and salinity. Another unique feature of Archaea is that no other known organisms are capable of methanogenesis (the metabolic production of methane). Methanogenic Archaea play a pivotal role in ecosystems with organisms that derive energy from oxidation of methane, many of which are Bacteria, as they are often a major source of methane in such environments and can play a role as primary producers. Methanogens also play a critical role in the carbon cycle, breaking down organic carbon into methane, which is also a major greenhouse gas.

Relationship to Bacteria

The relationship between the three domains is of central importance for understanding the origin of life. Most of the metabolic pathways, which comprise the majority of an organism's genes, are common between Archaea and Bacteria, while most genes involved in genome expression are common between Archaea and Eukarya. Within prokaryotes, archaeal cell structure is most similar to that of gram-positive bacteria, largely because both have a single lipid bilayer and usually contain a thick sacculus (exoskeleton) of varying chemical composition. In some phylogenetic trees based upon different gene/protein sequences of prokaryotic homologs, the archaeal homologs are more closely related to those of gram-positive bacteria. Archaea and gram-positive bacteria also share conserved indels in a number of important proteins, such as Hsp70 and glutamine synthetase I; however, the phylogeny of these genes was interpreted to reveal interdomain gene transfer, and might not reflect the organismal relationship(s).

It has been proposed that the archaea evolved from gram-positive bacteria in response to antibiotic selection pressure. This is suggested by the observation that archaea are resistant to a wide variety of antibiotics that are primarily produced by gram-positive bacteria, and that these antibiotics primarily act on the genes that distinguish archaea from bacteria. The proposal is that the selective pressure towards resistance generated by the gram-positive antibiotics was eventually sufficient to cause extensive changes in many of the antibiotics' target genes, and that these strains represented the common ancestors of present-day Archaea. The evolution of Archaea in response to antibiotic selection, or any other competitive selective pressure, could also explain their adaptation to extreme environments (such as high temperature or acidity) as the result of a search for unoccupied niches to escape from antibiotic-producing organisms; Cavalier-Smith has made a similar suggestion. This proposal is also supported by other work investigating protein structural relationships and studies that suggest that gram-positive bacteria may constitute the earliest branching lineages within the prokaryotes.

Relation to Eukaryotes

The evolutionary relationship between archaea and eukaryotes remains unclear. Aside from the similarities in cell structure and function that are discussed below, many genetic trees group the two.

Complicating factors include claims that the relationship between eukaryotes and the archaeal phylum Crenarchaeota is closer than the relationship between the Euryarchaeota and the phylum Crenarchaeota and the presence of archaea-like genes in certain bacteria, such as *Thermotoga maritima*, from horizontal gene transfer. The standard hypothesis states that the ancestor of the eukaryotes diverged early from the Archaea, and that eukaryotes arose through fusion of an archaean and eubacterium, which became the nucleus and cytoplasm; this explains various genetic similarities but runs into difficulties explaining cell structure. An alternative hypothesis, the eocyte hypothesis, posits that Eukaryota emerged relatively late from the Archaea.

A recently discovered lineage of archaea, *Lokiarchaeum*, named for a hydrothermal vent called Loki's Castle in the Arctic Ocean, has been found to be most closely related to eukaryotes. It has been called a transitional organism between prokaryotes and eukaryotes.

Morphology

Individual archaea range from 0.1 micrometers (μm) to over 15 μm in diameter, and occur in various shapes, commonly as spheres, rods, spirals or plates. Other morphologies in the Crenarchaeota include irregularly shaped lobed cells in *Sulfolobus*, needle-like filaments that are less than half a micrometer in diameter in *Thermofilum*, and almost perfectly rectangular rods in *Thermoproteus* and *Pyrobaculum*. Archaea in the genus *Haloquadratum* such as *Haloquadratum walsbyi* are flat, square archaea that live in hypersaline pools. These unusual shapes are probably maintained both by their cell walls and a prokaryotic cytoskeleton. Proteins related to the cytoskeleton components of other organisms exist in archaea, and filaments form within their cells, but in contrast to other organisms, these cellular structures are poorly understood. In *Thermoplasma* and *Ferroplasma* the lack of a cell wall means that the cells have irregular shapes, and can resemble amoebae.

Some species form aggregates or filaments of cells up to 200 μm long. These organisms can be prominent in biofilms. Notably, aggregates of *Thermococcus coalescens* cells fuse together in culture, forming single giant cells. Archaea in the genus *Pyrodictium* produce an elaborate multicell colony involving arrays of long, thin hollow tubes called *cannulae* that stick out from the cells' surfaces and connect them into a dense bush-like agglomeration. The function of these cannulae is not settled, but they may allow communication or nutrient exchange with neighbors. Multi-species colonies exist, such as the "string-of-pearls" community that was discovered in 2001 in a German swamp. Round whitish colonies of a novel Euryarchaeota species are spaced along thin filaments that can range up to 15 centimetres (5.9 in) long; these filaments are made of a particular bacteria species.

Structure, Composition Development, and Operation

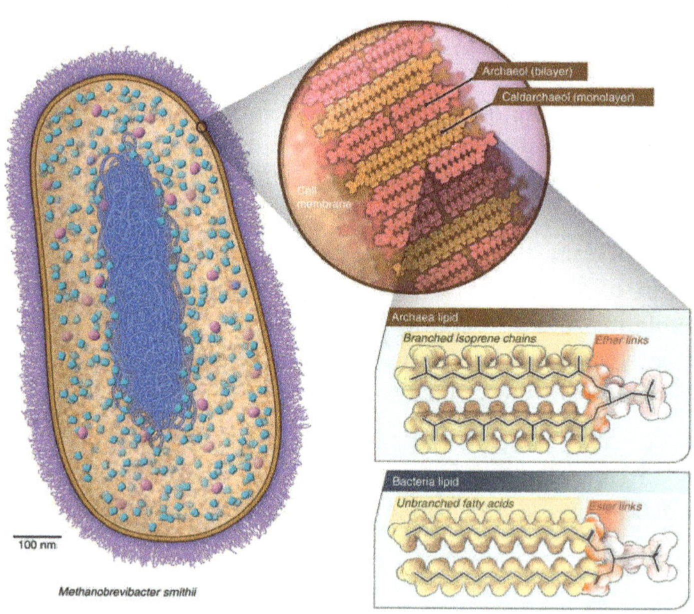

Diagrammatic view of *Methanobrevibacter smithii*, showing the cell membrane (ochre, with inset) and cell wall (purple).

Archaea and bacteria have generally similar cell structure, but cell composition and organization set the archaea apart. Like bacteria, archaea lack interior membranes and organelles. Like bacteria, archaea cell membranes are usually bounded by a cell wall and they swim using one or more flagella. Structurally, archaea are most similar to gram-positive bacteria. Most have a single plasma membrane and cell wall, and lack a periplasmic space; the exception to this general rule is *Ignicoccus*, which possess a particularly large periplasm that contains membrane-bound vesicles and is enclosed by an outer membrane.

Cell Wall and Flagella

Most archaea (but not *Thermoplasma* and *Ferroplasma*) possess a cell wall. In most archaea the wall is assembled from surface-layer proteins, which form an S-layer. An S-layer is a rigid array of protein molecules that cover the outside of the cell (like chain mail). This layer provides both chemical and physical protection, and can prevent macromolecules from contacting the cell membrane. Unlike bacteria, archaea lack peptidoglycan in their cell walls. Methanobacteriales do have cell walls containing pseudopeptidoglycan, which resembles eubacterial peptidoglycan in morphology, function, and physical structure, but pseudopeptidoglycan is distinct in chemical structure; it lacks D-amino acids and N-acetylmuramic acid.

Archaea flagella operate like bacterial flagella—their long stalks are driven by rotatory motors at the base. These motors are powered by the proton gradient across the membrane. However, archaeal flagella are notably different in composition and development. The two types of flagella evolved from different ancestors. The bacterial flagellum shares a common ancestor with the type III secretion system, while archaeal flagella appear to have evolved from bacterial type IV pili. In contrast to the bacterial flagellum, which is hollow and is assembled by subunits moving up the central pore to the tip of the flagella, archaeal flagella are synthesized by adding subunits at the base.

Membranes

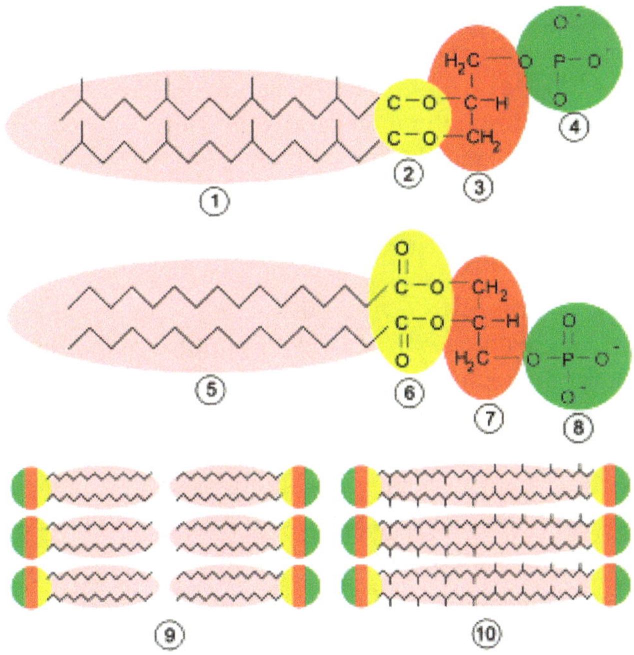

Membrane structures. **Top**, an archaeal phospholipid: **1**, isoprene chains; **2**, ether linkages; **3**, L-glycerol moiety; **4**, phosphate group. **Middle**, a bacterial or eukaryotic phospholipid: **5**, fatty acid chains; **6**, ester linkages; **7**, D-glycerol moiety; **8**, phosphate group. **Bottom**: **9**, lipid bilayer of bacteria and eukaryotes; **10**, lipid monolayer of some archaea.

Archaeal membranes are made of molecules that are distinctly different from those in all other life forms, showing that archaea are related only distantly to bacteria and eukaryotes. In all organisms, cell membranes are made of molecules known as phospholipids. These molecules possess both a polar part that dissolves in water (the phosphate "head"), and a "greasy" non-polar part that does not (the lipid tail). These dissimilar parts are connected by a glycerol moiety. In water, phospholipids cluster, with the heads facing the water and the tails facing away from it. The major structure in cell membranes is a double layer of these phospholipids, which is called a lipid bilayer.

The phospholipids of archaea are unusual in four ways:

- They have membranes composed of glycerol-ether lipids, whereas bacteria and eukaryotes have membranes composed mainly of glycerol-ester lipids. The difference is the type of bond that joins the lipids to the glycerol moiety; the two types are shown in yellow in the figure at the right. In ester lipids this is an ester bond, whereas in ether lipids this is an ether bond. Ether bonds are chemically more resistant than ester bonds. This stability might help archaea to survive extreme temperatures and very acidic or alkaline environments. Bacteria and eukaryotes do contain some ether lipids, but in contrast to archaea these lipids are not a major part of their membranes.
- The stereochemistry of the archaeal glycerol moiety is the reverse of that found in other organisms. The glycerol moiety can occur in two forms that are mirror images of one another, called the right-handed and left-handed forms or *enantiomers*. Just as a right hand does not fit easily into a left-handed glove, a right-handed phospholipid generally cannot be used or made by enzymes adapted for the left-handed form. This suggests that archaea use entirely different enzymes for synthesizing phospholipids than do bacteria and eukaryotes. Such enzymes developed very early in life's history, indicating an early split from the other two domains.
- Archaeal lipid tails differ from those of other organisms in that they are based upon long isoprenoid chains with multiple side-branches, sometimes with cyclopropane or cyclohexane rings. By contrast, the fatty acids in the membranes of other organisms have straight chains without side branches or rings. Although isoprenoids play an important role in the biochemistry of many organisms, only the archaea use them to make phospholipids. These branched chains may help prevent archaeal membranes from leaking at high temperatures.
- In some archaea, the lipid bilayer is replaced by a monolayer. In effect, the archaea fuse the tails of two phospholipid molecules into a single molecule with two polar heads (a bolaamphiphile); this fusion may make their membranes more rigid and better able to resist harsh environments. For example, the lipids in *Ferroplasma* are of this type, which is thought to aid this organism's survival in its highly acidic habitat.

Metabolism

Archaea exhibit a great variety of chemical reactions in their metabolism and use many sources of energy. These reactions are classified into nutritional groups, depending on energy and carbon sources. Some archaea obtain energy from inorganic compounds such as sulfur or ammonia (they are lithotrophs). These include nitrifiers, methanogens and anaerobic methane oxidisers. In these reactions one compound passes electrons to another (in a redox reaction), releasing energy to fuel the cell's activities. One compound acts as an electron donor and one as an electron acceptor. The energy released generates adenosine triphosphate (ATP) through chemiosmosis, in the same basic process that happens in the mitochondrion of eukaryotic cells.

Other groups of archaea use sunlight as a source of energy (they are phototrophs). However, oxygen–generating photosynthesis does not occur in any of these organisms. Many basic metabolic pathways are shared between all forms of life; for example, archaea use a modified form of glycolysis (the Entner–Doudoroff pathway) and either a complete or partial citric acid cycle. These similarities to other organisms probably reflect both early origins in the history of life and their high level of efficiency.

Nutritional types in archaeal metabolism

Nutritional type	Source of energy	Source of carbon	Examples
Phototrophs	Sunlight	Organic compounds	*Halobacterium*
Lithotrophs	Inorganic compounds	Organic compounds or carbon fixation	*Ferroglobus*, *Methanobacteria* or *Pyrolobus*
Organotrophs	Organic compounds	Organic compounds or carbon fixation	*Pyrococcus*, *Sulfolobus* or *Methanosarcinales*

Some Euryarchaeota are methanogens living in anaerobic environments, such as swamps. This form of metabolism evolved early, and it is even possible that the first free-living organism was a methanogen. A common reaction involves the use of carbon dioxide as an electron acceptor to oxidize hydrogen. Methanogenesis involves a range of coenzymes that are unique to these archaea, such as coenzyme M and methanofuran. Other organic compounds such as alcohols, acetic acid or formic acid are used as alternative electron acceptors by methanogens. These reactions are common in gut-dwelling archaea. Acetic acid is also broken down into methane and carbon dioxide directly, by *acetotrophic* archaea. These acetotrophs are archaea in the order Methanosarcinales, and are a major part of the communities of microorganisms that produce biogas.

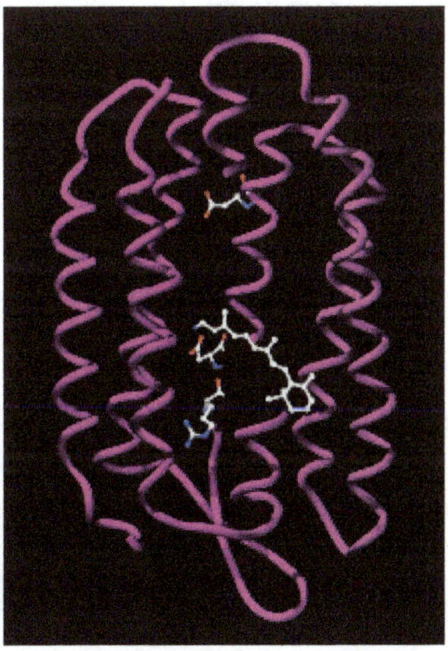

Bacteriorhodopsin from *Halobacterium salinarum*. The retinol cofactor and residues involved in proton transfer are shown as ball-and-stick models.

Other archaea use CO_2 in the atmosphere as a source of carbon, in a process called carbon fixation (they are autotrophs). This process involves either a highly modified form of the Calvin cycle or a recently discovered metabolic pathway called the 3-hydroxypropionate/4-hydroxybutyrate cycle. The Crenarchaeota also use the reverse Krebs cycle while the Euryarchaeota also use the reductive acetyl-CoA pathway. Carbon–fixation is powered by inorganic energy sources. No known archaea carry out photosynthesis. Archaeal energy sources are extremely diverse, and range from the oxidation of ammonia by the Nitrosopumilales to the oxidation of hydrogen sulfide or elemental sulfur by species of *Sulfolobus*, using either oxygen or metal ions as electron acceptors.

Phototrophic archaea use light to produce chemical energy in the form of ATP. In the Halobacteria, light-activated ion pumps like bacteriorhodopsin and halorhodopsin generate ion gradients by pumping ions out of the cell across the plasma membrane. The energy stored in these electrochemical gradients is then converted into ATP by ATP synthase. This process is a form of photophosphorylation. The ability of these light-driven pumps to move ions across membranes depends on light-driven changes in the structure of a retinol cofactor buried in the center of the protein.

Genetics

Archaea usually have a single circular chromosome, with as many as 5,751,492 base pairs in *Methanosarcina acetivorans*, the largest known archaeal genome. The tiny 490,885 base-pair genome of *Nanoarchaeum equitans* is one-tenth of this size and the smallest archaeal genome known; it is estimated to contain only 537 protein-encoding genes. Smaller independent pieces of DNA, called *plasmids*, are also found in archaea. Plasmids may be transferred between cells by physical contact, in a process that may be similar to bacterial conjugation.

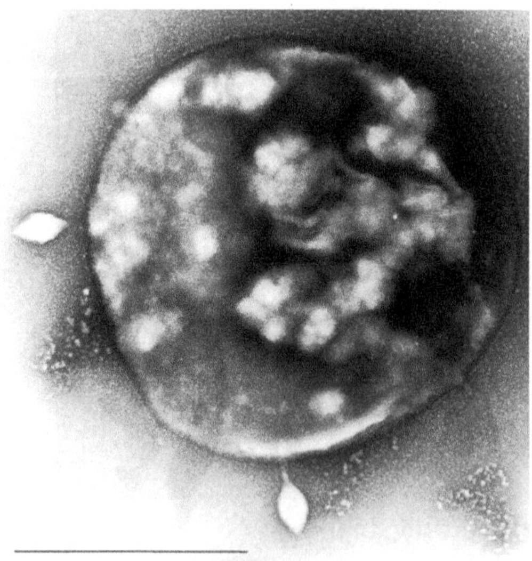

Sulfolobus infected with the DNA virus STSV1. Bar is 1 micrometer.

Archaea can be infected by double-stranded DNA viruses that are unrelated to any other form of virus and have a variety of unusual shapes, including bottles, hooked rods, or teardrops. These viruses have been studied in most detail in thermophilics, particularly the orders Sulfolobales and Thermoproteales. Two groups of single-stranded DNA viruses that infect archaea have been recently isolated. One group is exemplified by the *Halorubrum pleomorphic virus 1* ("Pleolipoviridae") infecting halophilic archaea and the other one by the Aeropyrum coil-shaped virus ("Spiraviridae") infecting a hyperthermophilic (optimal growth at 90–95 °C) host. Notably, the latter virus has the largest currently reported ssDNA genome. Defenses against these viruses may involve RNA interference from repetitive DNA sequences that are related to the genes of the viruses.

Archaea are genetically distinct from bacteria and eukaryotes, with up to 15% of the proteins encoded by any one archaeal genome being unique to the domain, although most of these unique genes have no known function. Of the remainder of the unique proteins that have an identified function, most belong to the Euryarchaea and are involved in methanogenesis. The proteins that archaea, bacteria and eukaryotes share form a common core of cell function, relating mostly to transcription, translation, and nucleotide metabolism. Other characteristic archaeal features are the organization of genes of related function—such as enzymes that catalyze steps in the same metabolic pathway into novel operons, and large differences in tRNA genes and their aminoacyl tRNA synthetases.

Transcription in archaea more closely resembles eukaryotic than bacterial transcription, with the archaeal RNA polymerase being very close to its equivalent in eukaryotes; while archaeal translation shows signs of both bacterial and eukaryal equivalents. Although archaea only have one type of RNA polymerase, its structure and function in transcription seems to be close to that of the eukaryotic RNA polymerase II, with similar protein assemblies (the general transcription factors) directing the binding of the RNA polymerase to a gene›s promoter. However, other archaeal transcription factors are closer to those found in bacteria. Post-transcriptional modification is simpler than in eukaryotes, since most archaeal genes lack introns, although there are many introns in their transfer RNA and ribosomal RNA genes, and introns may occur in a few protein-encoding genes.

Gene Transfer and Genetic Exchange

Halobacterium volcanii, an extreme halophilic archaeon, forms cytoplasmic bridges between cells that appear to be used for transfer of DNA from one cell to another in either direction.

When the hyperthermophilic archaea *Sulfolobus solfataricus* and *Sulfolobus acidocaldarius* are exposed to the DNA damaging agents UV irradiation, bleomycin or mitomycin C, species-specific cellular aggregation is induced. Aggregation in *S. solfataricus* could not be induced by other physical stressors, such as pH or temperature shift, suggesting that aggregation is induced specifically by DNA damage. Ajon et al. showed that UV-induced cellular aggregation mediates chromosomal marker exchange with high frequency in *S. acidocaldarius*. Recombination rates exceeded those of uninduced cultures by up to three orders of magnitude. Frols et al. and Ajon et al. hypothesized that cellular aggregation enhances species specific DNA transfer between *Sulfolobus* cells in order to provide increased repair of damaged DNA by means of homologous recombination. This response may be a primitive form of sexual interaction similar to the more well-studied bacterial transformation systems that are also associated with species specific DNA transfer between cells leading to homologous recombinational repair of DNA damage.

Reproduction

Archaea reproduce asexually by binary or multiple fission, fragmentation, or budding; meiosis does not occur, so if a species of archaea exists in more than one form, all have the same genetic material. Cell division is controlled in a cell cycle; after the cell's chromosome is replicated and the two daughter chromosomes separate, the cell divides. In the genus *Sulfolobus*, the cycle has characteristics that are similar to both bacterial and eukaryotic systems. The chromosomes replicate from multiple starting-points (origins of replication) using DNA polymerases that resemble the equivalent eukaryotic enzymes.

In euryarchaea the cell division protein FtsZ, which forms a contracting ring around the cell, and the components of the septum that is constructed across the center of the cell, are similar to their bacterial equivalents. In cren- and thaumarchaea, however, the cell division machinery Cdv fulfills a similar role. This machinery is related to the eukaryotic ESCRT-III machinery which, while best known for its role in cell sorting, also has been seen to fulfill a role in separation between divided cell, suggesting an ancestral role in cell division.

Both bacteria and eukaryotes, but not archaea, make spores. Some species of Haloarchaea undergo phenotypic switching and grow as several different cell types, including thick-walled structures that are resistant to osmotic shock and allow the archaea to survive in water at low salt concentrations, but these are not reproductive structures and may instead help them reach new habitats.

Ecology

Habitats

Archaea exist in a broad range of habitats, and as a major part of global ecosystems, may represent about 20% of microbial cells in the oceans. The first-discovered archaeans were extremophiles. Indeed, some archaea survive high temperatures, often above 100 °C (212 °F), as found in geysers, black smokers, and oil wells. Other common habitats include very cold habitats and highly saline, acidic, or alkaline water. However, archaea include mesophiles that grow in mild conditions, in swamps and marshland, sewage, the oceans, the intestinal tract of animals, and soils.

Extremophile archaea are members of four main physiological groups. These are the halophiles, thermophiles, alkaliphiles, and acidophiles. These groups are not comprehensive or phylum-specific, nor are they mutually exclusive, since some archaea belong to several groups. Nonetheless, they are a useful starting point for classification.

Archaea that grow in the hot water of the Morning Glory Hot Spring in Yellowstone National Park produce a bright colour

Halophiles, including the genus *Halobacterium*, live in extremely saline environments such as salt lakes and outnumber their bacterial counterparts at salinities greater than 20–25%. Thermophiles grow best at temperatures above 45 °C (113 °F), in places such as hot springs; *hyperthermophilic* archaea grow optimally at temperatures greater than 80 °C (176 °F). The archaeal *Methanopyrus kandleri* Strain 116 can even reproduce at 122 °C (252 °F), the highest recorded temperature of any organism.

Other archaea exist in very acidic or alkaline conditions. For example, one of the most extreme archaean acidophiles is *Picrophilus torridus*, which grows at pH 0, which is equivalent to thriving in 1.2 molar sulfuric acid.

This resistance to extreme environments has made archaea the focus of speculation about the possible properties of extraterrestrial life. Some extremophile habitats are not dissimilar to those on Mars, leading to the suggestion that viable microbes could be transferred between planets in meteorites.

Recently, several studies have shown that archaea exist not only in mesophilic and thermophilic environments but are also present, sometimes in high numbers, at low temperatures as well. For example, archaea are common in cold oceanic environments such as polar seas. Even more significant are the large numbers of archaea found throughout the world's oceans in non-extreme habitats among the plankton community (as part of the picoplankton). Although these archaea can be present in extremely high numbers (up to 40% of the microbial biomass), almost none of these species have been isolated and studied in pure culture.Consequently, our understanding of the role of archaea in ocean ecology is rudimentary, so their full influence on global biogeochemical cycles remains largely unexplored. Some marine Crenarchaeota are capable of nitrification, suggesting these organisms may affect the oceanic nitrogen cycle, although these oceanic Crenarchaeota may also use other sources of energy. Vast numbers of archaea are also found in the sediments that cover the sea floor, with these organisms making up the majority of living cells at depths over 1 meter below the ocean bottom. It has been demonstrated that in all oceanic surface sediments (from 1000- to 10,000-m water depth), the impact of viral infection is higher on archaea than on bacteria and virus-induced lysis of archaea accounts for up to one-third of the total microbial biomass killed, resulting in the release of ~0.3 to 0.5 gigatons of carbon per year globally.

Role in Chemical cycling

Archaea recycle elements such as carbon, nitrogen and sulfur through their various habitats. Although these activities are vital for normal ecosystem function, archaea can also contribute to human-made changes, and even cause pollution.

Archaea carry out many steps in the nitrogen cycle. This includes both reactions that remove nitrogen from ecosystems (such as nitrate-based respiration and denitrification) as well as processes that introduce nitrogen (such as nitrate assimilation and nitrogen fixation). Researchers recently discovered archaeal involvement in ammonia oxidation reactions. These reactions are particularly important in the oceans. The archaea also appear crucial for ammonia oxidation in soils. They produce nitrite, which other microbes then oxidize to nitrate. Plants and other organisms consume the latter.

In the sulfur cycle, archaea that grow by oxidizing sulfur compounds release this element from rocks, making it available to other organisms. However, the archaea that do this, such as *Sulfolobus*, produce sulfuric acid as a waste product, and the growth of these organisms in abandoned mines can contribute to acid mine drainage and other environmental damage.

In the carbon cycle, methanogen archaea remove hydrogen and play an important role in the decay of organic matter by the populations of microorganisms that act as decomposers in anaerobic ecosystems, such as sediments, marshes and sewage-treatment works.

Global methane levels in 2011 had increased by a factor of 2.5 since pre-industrial times: from 722 ppb to 1800 ppb, the highest value in at least 800,000 years.Methane has an anthropogenic global warming potential (AGWP) of 29, which means that it's 29 times stronger in heat-trapping than carbon dioxide is, over a 100-year time scale.

Interactions with other Organisms

The well-characterized interactions between archaea and other organisms are either mutual or commensal. There are no clear examples of known archaeal pathogens or parasites. However, some species of methanogens have been suggested to be involved in infections in the mouth, and *Nanoarchaeum equitans* may be a parasite of another species of archaea, since it only survives and reproduces within the cells of the Crenarchaeon *Ignicoccus hospitalis*, and appears to offer no benefit to its host. In contrast, Archaeal Richmond Mine Acidophilic Nanoorganisms (ARMAN) occasionally connect with other archaeal cells in acid mine drainage biofilms. The nature of this relationship is unknown. However, it is distinct from that of *Nanarchaeaum–Ignicoccus* in that the ultrasmall ARMAN cells are usually seen independent of the Thermoplasmatales cells.

Methanogenic archaea form a symbiosis with termites.

Mutualism

One well-understood example of mutualism is the interaction between protozoa and methanogenic archaea in the digestive tracts of animals that digest cellulose, such as ruminants and termites. In these anaerobic environments, protozoa break down plant cellulose to obtain energy. This process releases hydrogen as a waste product, but high levels of hydrogen reduce energy production. When methanogens convert hydrogen to methane, protozoa benefit from more energy.

In anaerobic protozoa, such as *Plagiopyla frontata*, archaea reside inside the protozoa and consume hydrogen produced in their hydrogenosomes. Archaea also associate with larger organisms. For example, the marine archaean *Cenarchaeum symbiosum* lives within (is an endosymbiont of) the sponge *Axinella mexicana*.

Commensalism

Archaea can also be commensals, benefiting from an association without helping or harming the other organism. For example, the methanogen *Methanobrevibacter smithii* is by far the most common archaean in the human flora, making up about one in ten of all the prokaryotes in the human gut. In termites and in humans, these methanogens may in fact be mutualists, interacting with other microbes in the gut to aid digestion. Archaean communities also associate with a range of other organisms, such as on the surface of corals, and in the region of soil that surrounds plant roots (the rhizosphere).

Significance in Technology and Industry

Extremophile archaea, particularly those resistant either to heat or to extremes of acidity and alkalinity, are a source of enzymes that function under these harsh conditions. These enzymes have found many uses. For example, thermostable DNA polymerases, such as the Pfu DNA polymerase from *Pyrococcus furiosus*, revolutionized molecular biology by allowing the polymerase chain reaction to be used in research as a simple and rapid technique for cloning DNA. In industry, amylases, galactosidases and pullulanases in other species of *Pyrococcus* that function at over 100 °C (212 °F) allow food processing at high temperatures, such as the production of low lactose milk and whey. Enzymes from these thermophilic archaea also tend to be very stable in organic solvents, allowing their use in environmentally friendly processes in green chemistry that synthesize organic compounds. This stability makes them easier to use in structural biology. Consequently, the counterparts of bacterial or eukaryotic enzymes from extremophile archaea are often used in structural studies.

In contrast to the range of applications of archaean enzymes, the use of the organisms themselves in biotechnology is less developed. Methanogenic archaea are a vital part of sewage treatment, since they are part of the community of microorganisms that carry out anaerobic digestion and produce biogas. In mineral processing, acidophilic archaea display promise for the extraction of metals from ores, including gold, cobalt and copper.

Archaea host a new class of potentially useful antibiotics. A few of these archaeocins have been characterized, but hundreds more are believed to exist, especially within *Haloarchaea* and *Sulfolobus*. These compounds differ in structure from bacterial antibiotics, so they may have novel modes of action. In addition, they may allow the creation of new selectable markers for use in archaeal molecular biology.

EUKARYOTE

A **eukaryote** is any organism whose cells contain a nucleus and other organelles enclosed within membranes. Eukaryotes belong to the taxon **Eukarya** or **Eukaryota**. The defining feature that sets eukaryotic cells apart from prokaryotic cells (Bacteria and Archaea) is that they have membrane-bound organelles, especially the nucleus, which contains the genetic material and is enclosed by the nuclear envelope. The presence of a nucleus gives eukaryotes their name, which comes from the Greek ε (*eu*, "well" or "true") and κάρυον (*karyon*, "nut" or "kernel"). Eukaryotic cells also contain other membrane-bound organelles such as mitochondria and the Golgi apparatus. In addition, plants and algae contain chloroplasts. Eukaryotic organisms may be unicellular or multicellular. Only eukaryotes form multicellular organisms consisting of many kinds of tissue made up of different cell types.

Eukaryotes can reproduce both asexually through mitosis and sexually through meiosis and gamete fusion. In mitosis, one cell divides to produce two genetically identical cells. In meiosis, DNA replication is followed by two rounds of cell division to produce four daughter cells each with half the number of chromosomes as the original parent cell (haploid cells). These act as sex cells (gametes – each gamete has just one complement of chromosomes, each a unique mix of the corresponding pair of parental chromosomes) resulting from genetic recombination during meiosis.

The domain Eukaryota appears to be monophyletic, and so makes up one of the three domains of life. The two other domains, Bacteria and Archaea, are prokaryotes and have none of the above features. Eukaryotes represent a tiny minority of all living things. However, due to their much larger size, eukaryotes' collective worldwide biomass is estimated at about equal to that of prokaryotes. Eukaryotes first developed approximately 1.6–2.1 billion years ago (during the Proterozoic eon).

Cell Features

Eukaryotic cells are typically much larger than those of prokaryotes. They have a variety of internal membrane-bound structures, called organelles, and a cytoskeleton composed of microtubules, microfilaments, and intermediate filaments, which play an important role in defining the cell's organization and shape. Eukaryotic DNA is divided into several linear bundles called chromosomes, which are separated by a microtubular spindle during nuclear division.

Internal Membrane

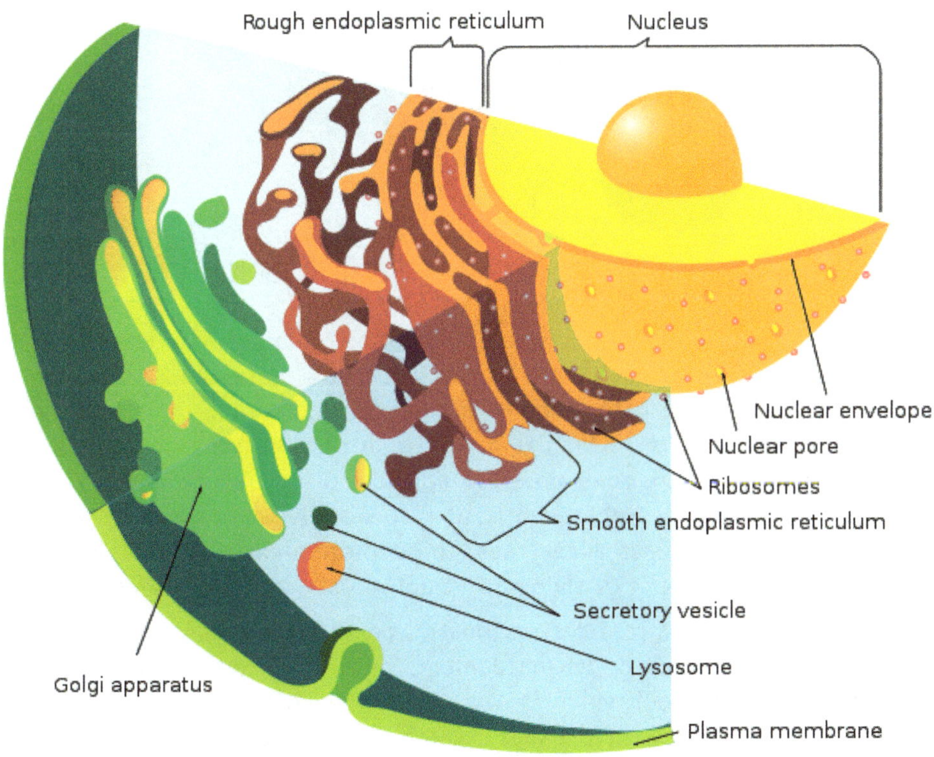

Detail of the endomembrane system and its components

Eukaryote cells include a variety of membrane-bound structures, collectively referred to as the endomembrane system. Simple compartments, called vesicles or vacuoles, can form by budding off other membranes. Many cells ingest food and other materials through a process of endocytosis, where the outer membrane invaginates and then pinches off to form a vesicle. It is probable that most other membrane-bound organelles are ultimately derived from such vesicles. Alternatively some products produced by the cell can leave in a vesicle through exocytosis.

A 3D rendering of an animal cell cut in half.

The nucleus is surrounded by a double membrane (commonly referred to as a nuclear membrane or nuclear envelope), with pores that allow material to move in and out. Various tube- and sheet-like extensions of the nuclear membrane form what is called the endoplasmic reticulum or ER, which is involved in protein transport and maturation. It includes the rough ER where ribosomes are attached to synthesize proteins, which enter the interior space or lumen. Subsequently, they generally enter vesicles, which bud off from the smooth ER. In most eukaryotes, these protein-carrying vesicles are released and further modified in stacks of flattened vesicles, called Golgi bodies or dictyosomes.

Vesicles may be specialized for various purposes. For instance, lysosomes contain digestive enzymes that break down the contents of food vacuoles, and peroxisomes are used to break down peroxide, which is toxic otherwise. Many protozoa have contractile vacuoles, which collect and expel excess water, and extrusomes, which expel material used to deflect predators or capture prey. In higher plants, most of a cell's volume is taken up by a central vacuole, which primarily maintains its osmotic pressure.

Mitochondria and Plastids

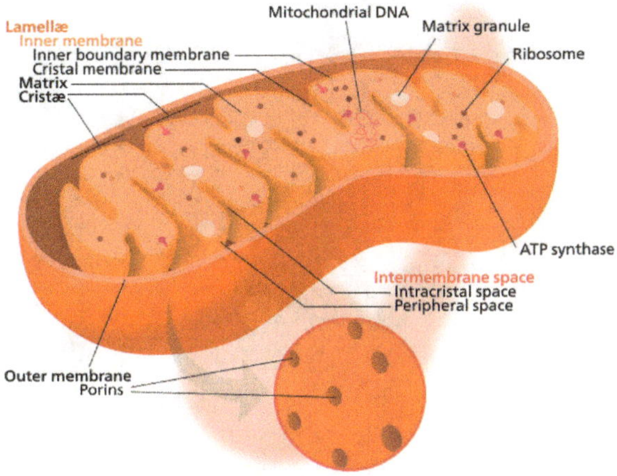

Simplified structure of a mitochondrion

Mitochondria are organelles found in nearly all eukaryotes that provide energy to the cell by converting ingested sugars into ATP. They are surrounded by two membranes (each a phospholipid bi-layer), the inner of which is folded into invaginations called cristae, where aerobic respiration takes place. Mitochondria contain their own DNA. They are now generally held to have developed from endosymbiotic prokaryotes, probably proteobacteria. Protozoa and microbes that lack mitochondria, such as the amoebozoan Pelomyxa and metamonads such as Giardia and Trichomonas, have usually been found to contain mitochondrion-derived organelles, such as hydrogenosomes and mitosomes, and thus probably lost the mitochondria secondarily.

In 2016, *Monocercomonoides*, a metamonad flagellate which resides in the intestines of the chinchilla, has been found to lack mitochondria entirely. *Monocercomonoides* obtains its energy by enzymatic action on nutrients absorbed from the environment. It has also acquired, by lateral gene transfer, a cytosolic sulphur mobilisation system which provides the clusters of iron and sulfur required for protein synthesis. The normal mitochondrial iron-sulphur cluster pathway is considered to have been lost secondarily.

Plants and various groups of algae also have plastids. Plastids have their own DNA and are developed from endosymbionts, in this case cyanobacteria. They usually take the form of chloroplasts, which like cyanobacteria contain chlorophyll and produce organic compounds (such as glucose) through photosynthesis. Others are involved in storing food. Although plastids probably had a single origin, not all plastid-containing groups are closely related. Instead, some eukaryotes have obtained them from others through secondary endosymbiosis or ingestion.

Endosymbiotic origins have also been proposed for the nucleus, for which see below, and for eukaryotic flagella, supposed to have developed from spirochaetes. This is not generally accepted, both from a lack of cytological evidence and difficulty in reconciling this with cellular reproduction.

Cytoskeletal Structures

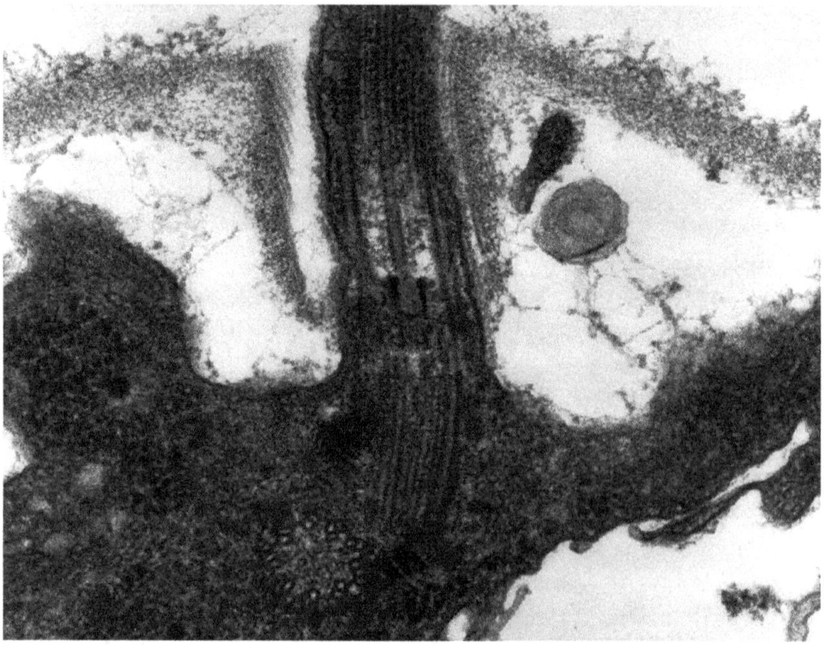

Longitudinal section through the flagellum of *Chlamydomonas reinhardtii*

Many eukaryotes have long slender motile cytoplasmic projections, called flagella, or similar structures called cilia. Flagella and cilia are sometimes referred to as undulipodia, and are variously involved in movement, feeding, and sensation. They are composed mainly of tubulin. These are entirely distinct from prokaryotic flagellae. They are supported by a bundle of microtubules arising from a basal body, also called a kinetosome or centriole, characteristically arranged as nine doublets surrounding two singlets. Flagella also may have hairs, or mastigonemes, and scales connecting membranes and internal rods. Their interior is continuous with the cell's cytoplasm.

Microfilamental structures composed of actin and actin binding proteins, e.g., α-actinin, fimbrin, filamin are present in submembraneous cortical layers and bundles, as well. Motor proteins of microtubules, e.g., dynein or kinesin and actin, e.g., myosins provide dynamic character of the network.

Centrioles are often present even in cells and groups that do not have flagella, but conifers and flowering plants have neither. They generally occur in groups of one or two, called kinetids, that give rise to various microtubular roots. These form a primary component of the cytoskeletal structure, and are often assembled over the course of several cell divisions, with one flagellum retained from the parent and the other derived from it. Centrioles may also be associated in the formation of a spindle during nuclear division.

The significance of cytoskeletal structures is underlined in the determination of shape of the cells, as well as their being essential components of migratory responses like chemotaxis and chemokinesis. Some protists have various other microtubule-supported organelles. These include the radiolaria and heliozoa, which produce axopodia used in flotation or to capture prey, and the haptophytes, which have a peculiar flagellum-like organelle called the haptonema.

Cell Wall

The cells of plants, fungi, and most chromalveolates have a cell wall, a layer outside the cell membrane, providing the cell with structural support, protection, and a filtering mechanism. The cell wall also prevents over-expansion when water enters the cell.

The major polysaccharides making up the primary cell wall of land plants are cellulose, hemicellulose, and pectin. The cellulose microfibrils are linked via hemicellulosic tethers to form the cellulose-hemicellulose network, which is embedded in the pectin matrix. The most common hemicellulose in the primary cell wall is xyloglucan.

Differences among Eukaryotic Cells

There are many different types of eukaryotic cells, though animals and plants are the most familiar eukaryotes, and thus provide an excellent starting point for understanding eukaryotic structure. Fungi and many protists have some substantial differences, however.

Animal Cell

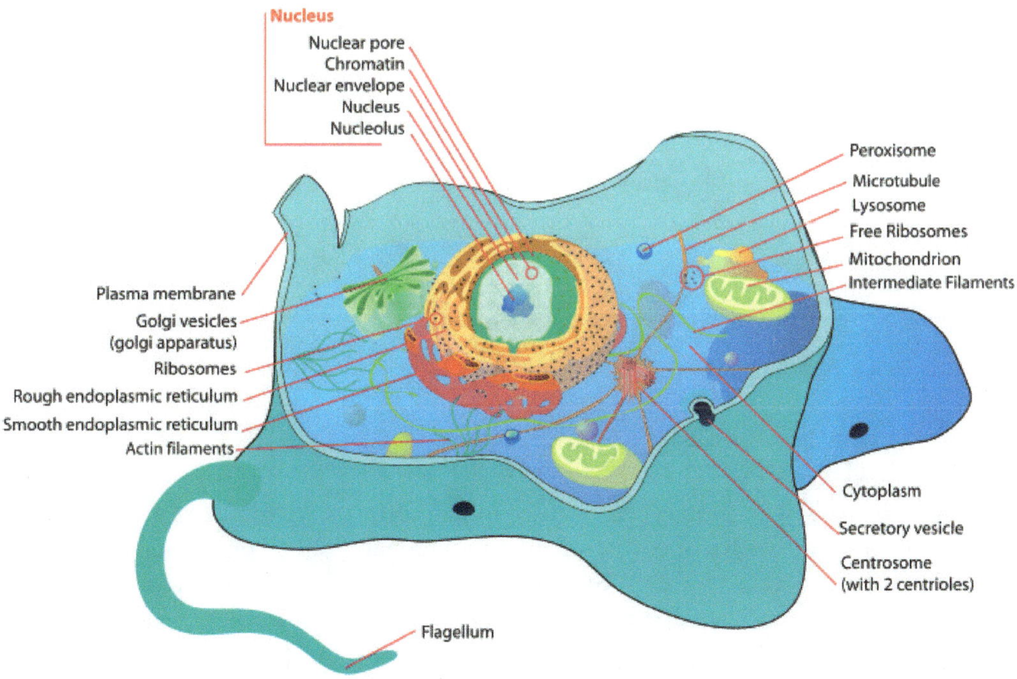

Structure of a typical animal cell

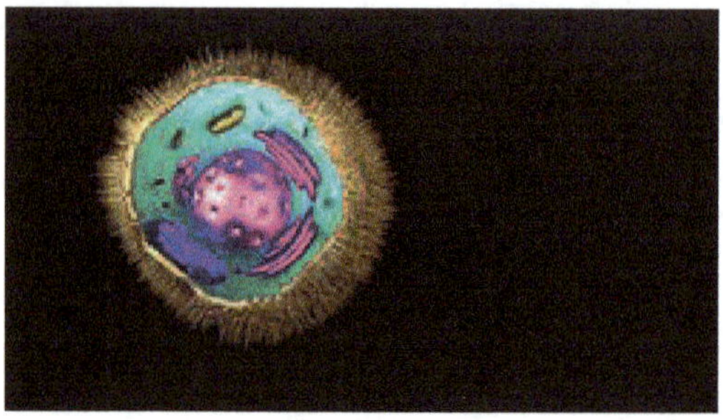

3D Simulation of animal cell

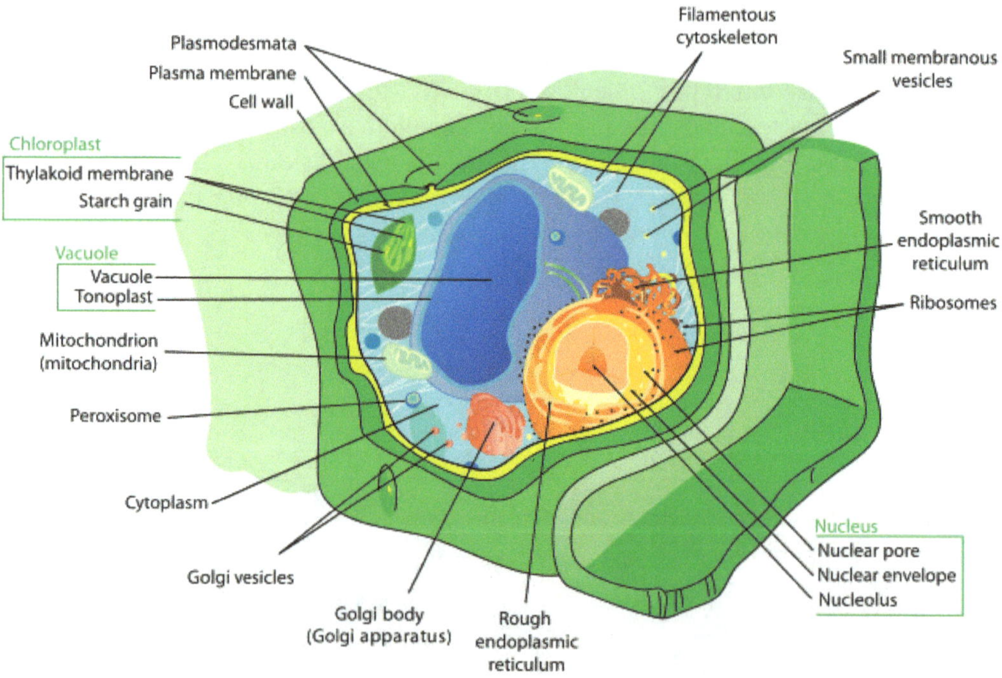

Structure of a typical plant cell

All animals consist of eukaryotic cells. Animal cells are distinct from those of other eukaryotes, most notably plants, as they lack cell walls and chloroplasts and have smaller vacuoles. Due to the lack of a cell wall, animal cells can adopt a variety of shapes. A phagocytic cell can even engulf other structures.

There are many other types of cell. For instance, there are approximately 210 distinct cell types in the adult human body.

Plant Cell

Plant cells are quite different from the cells of the other eukaryotic organisms. Their distinctive features are:

- A large central vacuole (enclosed by a membrane, the tonoplast), which maintains the cell's turgor and controls movement of molecules between the cytosol and sap
- A primary cell wall containing cellulose, hemicellulose and pectin, deposited by the protoplast on the outside of the cell membrane; this contrasts with the cell walls of fungi, which contain chitin, and the cell envelopes of prokaryotes, in which peptidoglycans are the main structural molecules
- The plasmodesmata, linking pores in the cell wall that allow each plant cell to communicate with other adjacent cells; this is different from the functionally analogous system of gap junctions between animal cells.
- Plastids, especially chloroplasts that contain chlorophyll, the pigment that gives plants their green color and allows them to perform photosynthesis
- Bryophytes and seedless vascular plants lack flagellae and centrioles except in the sperm cells. Sperm of cycads and *Ginkgo* are large, complex cells that swim with hundreds to thousands of flagellae.
- Conifers (Pinophyta) and flowering plants (Angiospermae) lack the flagellae and centrioles that are present in animal cells.

Fungal Cell

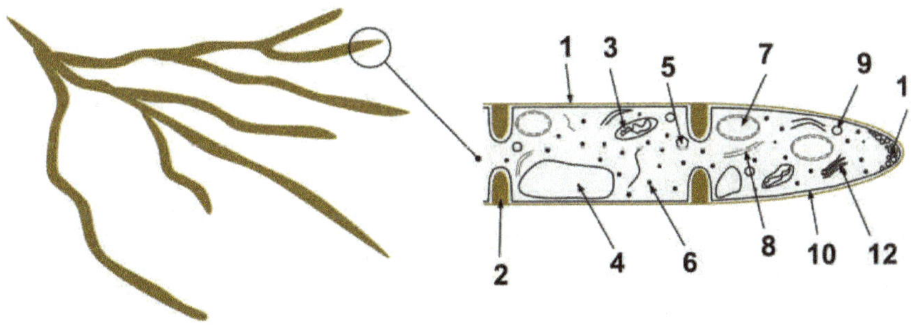

Fungal Hyphae Cells

1- Hyphal wall 2- Septum 3- Mitochondrion 4- Vacuole 5- Ergosterol crystal 6- Ribosome 7- Nucleus 8- Endoplasmic reticulum 9- Lipid body 10- Plasma membrane 11- Spitzenkörper 12- Golgi apparatus

Fungal cells are most similar to animal cells, with the following exceptions:

- A cell wall that contains chitin
- Less definition between cells; the hyphae of higher fungi have porous partitions

called septa, which allow the passage of cytoplasm, organelles, and, sometimes, nuclei. Primitive fungi have few or no septa, so each organism is essentially a giant multinucleate supercell; these fungi are described as coenocytic.

- Only the most primitive fungi, chytrids, have flagella.

Other Eukaryotic Cells

Eukaryotes are a very diverse group, and their cell structures are equally diverse. Many have cell walls; many do not. Many have chloroplasts, derived from primary, secondary, or even tertiary endosymbiosis; and many do not. Some groups have unique structures, such as the cyanelles of the glaucophytes, the haptonema of the haptophytes, or the ejectisomes of the cryptomonads. Other structures, such as pseudopods, are found in various eukaryote groups in different forms, such as the lobose amoebozoans or the reticulose foraminiferans.

Reproduction

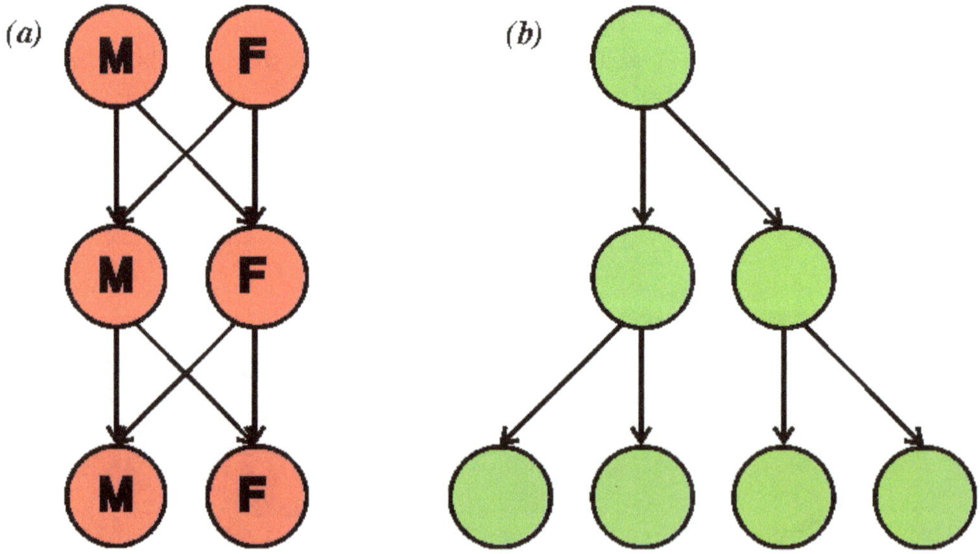

This diagram illustrates the *twofold cost of sex*. If each individual were to contribute to the same number of offspring (two), *(a)* the sexual population remains the same size each generation, where the *(b)* asexual population doubles in size each generation.

Cell division generally takes place asexually by mitosis, a process that allows each daughter nucleus to receive one copy of each chromosome. In most eukaryotes, there is also a process of sexual reproduction, typically involving an alternation between haploid generations, wherein only one copy of each chromosome is present, and diploid generations, wherein two copies of each chromosome are present, occurring through meiosis. There is considerable variation in this pattern.

Eukaryotes have a smaller surface area to volume ratio than prokaryotes, and thus have lower metabolic rates and longer generation times. In some multicellular organisms, cells specialized for metabolism will have enlarged surface areas, such as intestinal vili.

The evolution of sexual reproduction may be a primordial and fundamental characteristic of eukaryotes. Based on a phylogenetic analysis, Dacks and Roger proposed that facultative sex was present in the common ancestor of all eukaryotes. A core set of genes that function in meiosis is present in both *Trichomonas vaginalis* and *Giardia intestinalis*, two organisms previously thought to be asexual. Since these two species are descendants of lineages that diverged early from the eukaryotic evolutionary tree, it was inferred that core meiotic genes, and hence sex, were likely present in a common ancestor of all eukaryotes. Other studies on eukaryotic species once thought to be asexual have revealed evidence for a sexual cycle. For instance, parasitic protozoa of the genus Leishmania have recently been shown to have a sexual cycle. Also, evidence now indicates that amoeba, that were previously regarded as asexual, are anciently sexual and that the majority of present-day asexual groups likely arose recently and independently.

Classification

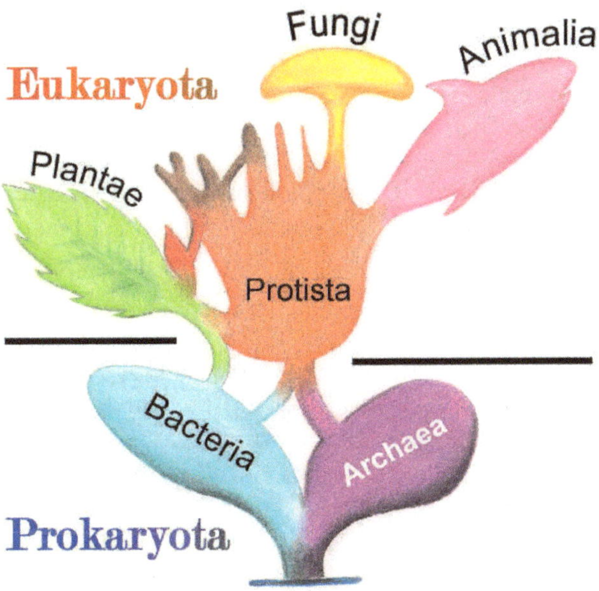

Phylogenetic and symbiogenetic tree of living organisms, showing the origins of eukaryotes & prokaryotes

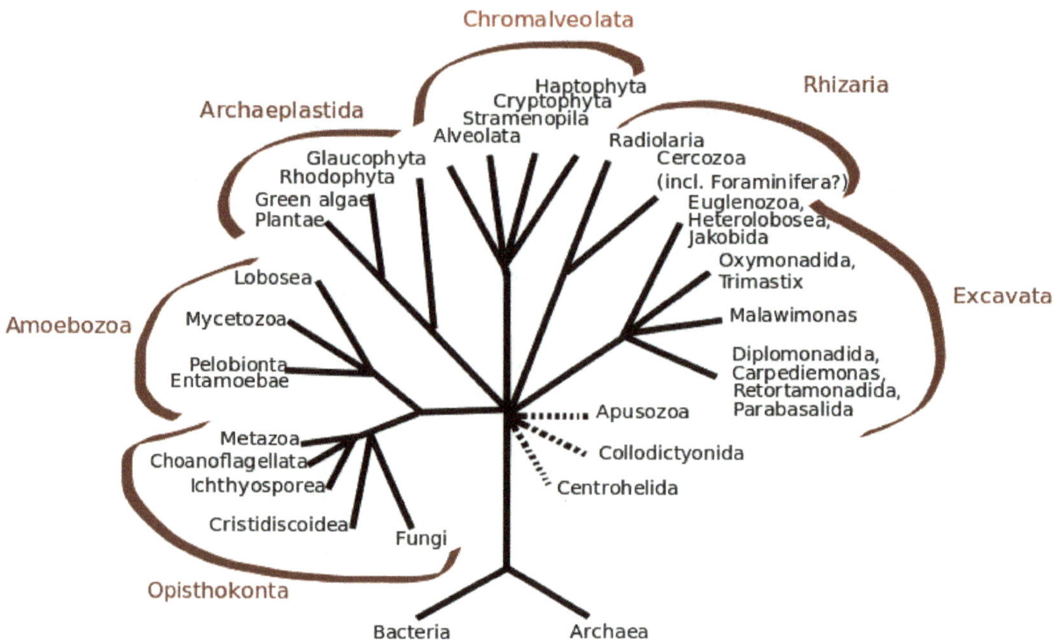

One hypothesis of eukaryotic relationships. The Opisthokonta group includes both animals (Metazoa) and fungi. Plants (Plantae) are placed in Archaeplastida.

In antiquity, the two clades of animals and plants were recognized. They were given the taxonomic rank of Kingdom by Linnaeus. Though he included the fungi with plants with some reservations, it was later realized that they are quite distinct and warrant a separate kingdom, the composition of which was not entirely clear until the 1980s. The various single-cell eukaryotes were originally placed with plants or animals when they became known. In 1830, the German biologist Georg A. Goldfuss coined the word *protozoa* to refer to organisms such as ciliates, and this group was expanded until it encompassed all single-celled eukaryotes, and given their own kingdom, the Protista, by Ernst Haeckel in 1866. The eukaryotes thus came to be composed of four kingdoms:

- Kingdom Protista
- Kingdom Plantae
- Kingdom Fungi
- Kingdom Animalia

The protists were understood to be "primitive forms", and thus an evolutionary grade, united by their primitive unicellular nature. The disentanglement of the deep splits in the tree of life only really started with DNA sequencing, leading to a system of domains rather than kingdoms as top level rank being put forward by Carl Woese, uniting all the eukaryote kingdoms under the eukaryote domain. At the same time, work on the protist tree intensified, and is still actively going on today. Several alternative classifications have been forwarded, though there is no consensus in the field.

A classification produced in 2005 for the International Society of Protistologists, which reflected the consensus of the time, divided the eukaryotes into six supposedly monophyletic `supergroups›. However, in the same year (2005), doubts were expressed as to whether some of these supergroups were monophyletic, particularly the Chromalveolata, and a review in 2006 noted the lack of evidence for several of the supposed six supergroups. A revised classification in 2012 recognizes five supergroups.

Archaeplastida (or Primoplantae)	Land plants, green algae, red algae, and glaucophytes
SAR supergroup	Stramenopiles (brown algae, diatoms, etc.), Alveolata, and Rhizaria (Foraminifera, Radiolaria, and various other amoeboid protozoa).
Excavata	Various flagellate protozoa
Amoebozoa	Most lobose amoeboids and slime molds
Opisthokonta	Animals, fungi, choanoflagellates, etc.

\There are also smaller groups of eukaryotes whose position is uncertain or seems to fall outside the major groups. In particular, Haptophyta, Cryptophyta, Centrohelida, Telonemia, Picozoa, Apusomonadida, Ancyromonadida, Breviatea, and the genus *Collodictyon*. Overall, it seems that, although progress has been made, there are still very significant uncertainties in the evolutionary history and classification of eukaryotes. As Roger & Simpson said in 2009 "with the current pace of change in our understanding of the eukaryote tree of life, we should proceed with caution."

In an article published in *Nature Microbiology* in April 2016 the authors, "reinforced once again that the life we see around us – plants, animals, humans and other so-called eukaryotes – represent a tiny percentage of the world's biodiversity." They classified eukaryote "based on the inheritance of their information systems as opposed to lipid or other cellular structures." Jillian F. Banfield of the University of California, Berkeley and fellow scientists used a super computer to generate a diagram of a new tree of life based on DNA from 3000 species including 2,072 known species and 1,011 newly reported microbial organisms, whose DNA they had gathered from diverse environments. As the capacity to sequence DNA became easier, Banfield and team were able to do metagenomic sequencing—"sequencing whole communities of organisms at once and picking out the individual groups based on their genes alone."

Phylogeny

The rRNA trees constructed during the 1980s and 1990s left most eukaryotes in an unresolved "crown" group (not technically a true crown), which was usually divided by the form of the mitochondrial cristae; see crown eukaryotes. The few groups that lack mitochondria branched separately, and so the absence was believed to be primitive; but this is now considered an artifact of long-branch attraction, and they are known to have lost them secondarily.

As of 2011, there is widespread agreement that the Rhizaria belong with the Stramenopiles and the Alveolata, in a clade dubbed the SAR supergroup, so that Rhizaria is not one of the main eukaryote groups; also that the Amoebozoa and Opisthokonta are each monophyletic and form a clade, often called the unikonts. Beyond this, there does not appear to be a consensus.

It has been estimated that there may be 75 distinct lineages of eukaryotes. Most of these lineages are protists.

The known eukaryote genome sizes vary from 8.2 megabases (Mb) in *Babesia bovis* to 112,000–220,050 Mb in the dinoflagellate *Prorocentrum micans*, suggesting that the genome of the ancestral eukaryote has undergone considerable variation during its evolution. The last common ancestor of all eukaryotes is believed to have been a phagotrophic protist with a nucleus, at least one centriole and cilium, facultatively aerobic mitochondria, sex (meiosis and syngamy), a dormant cyst with a cell wall of chitin and/or cellulose and peroxisomes. Later endosymbiosis led to the spread of plastids in some lineages.

Five Supergroups

A global tree of eukaryotes from a consensus of phylogenetic evidence (in particular, phylogenomics), rare genomic signatures, and morphological characteristics is presented in Adl *et al.* 2012 and Burki 2014.

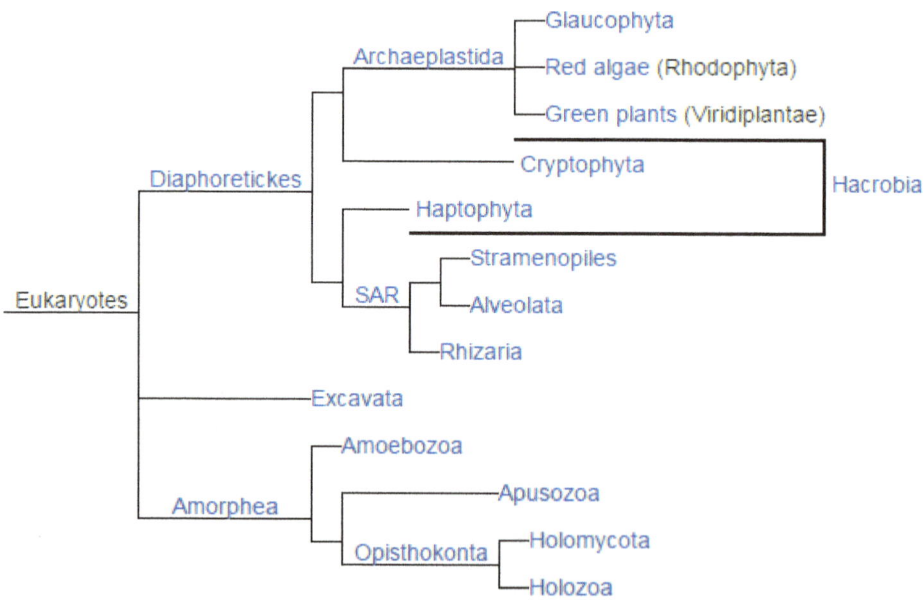

In some analyses, the Hacrobia group (Haptophyta + Cryptophyta) is placed next to Archaeplastida, but in other ones it is nested inside the Archaeplastida.However, several recent studies have concluded that Haptophyta and Cryptophyta do not form a monophyletic group. The former could be a sister group to the SAR group, the latter cluster with the Archaeplastida (plants in the broad sense). As of February 2012, it remains unclear whether the Hacrobia forms a monophyletic group.

The division of the eukaryotes into two primary clades, bikonts (Archaeplastida + SAR + Excavata) and unikonts (Amoebozoa + Opisthokonta), derived from an ancestral biflagellar organism and an ancestral uniflagellar organism, respectively, had been suggested earlier. A 2012 study produced a somewhat similar division, although noting that the terms "unikonts" and "bikonts" were not used in the original sense.

Cavalier-Smith's Tree

Cavalier-Smith 2010, 2013, and 2014 places the eukaryotic tree's root between Excavata (with ventral feeding groove supported by a microtubular root) and the grooveless Euglenozoa:

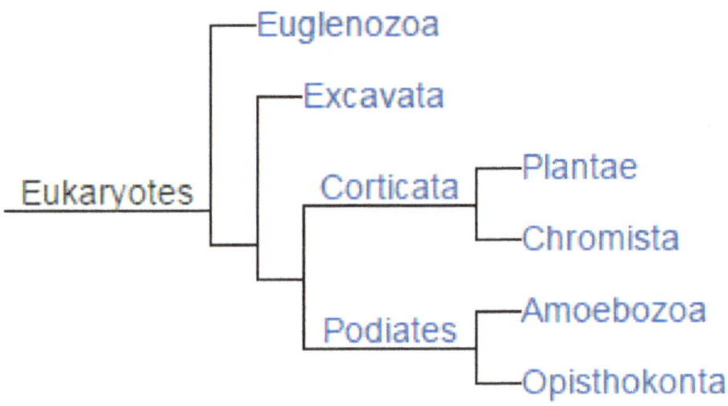

Alternative Views

Other analyses place the SAR supergroup within an expanded Chromalveolata, although they differ on the placement of the resulting five groups. Rogozin *et al.* in 2009 produced the tree shown below, where the primary division is between the Archaeplastida and all other eukaryotes.

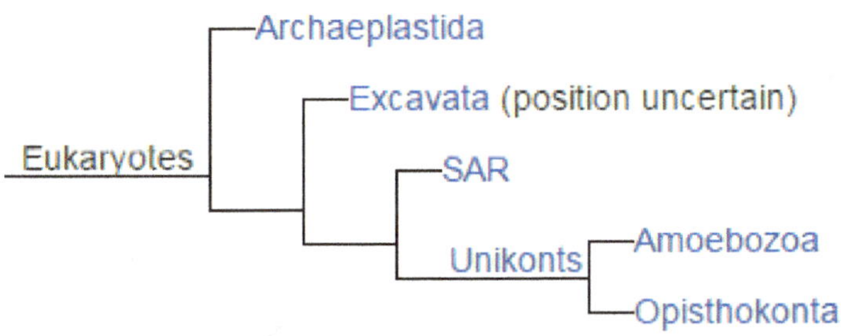

A paper published in 2009, which re-examined the data used in some of the analyses presented above as well as performing new ones, strongly suggested that the Archaeplastida are polyphyletic. The phylogeny finally proposed in the paper is shown below.

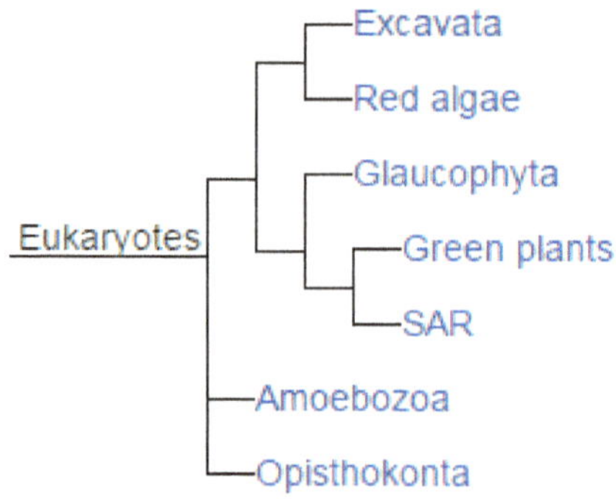

There are also the cladistic analyses of Diana Lipscomb based on classical data which have red algae as basal and Archeoplastida as paraphyletic. In the survey by Parfrey et al. it is recovered in only 42.6% of the molecular analyses that include it, that is, 26 out of 61. It is not recovered in Goloboff et al.'s combined analysis, and is mostly weakly supported in other molecular analyses. Laura Parfrey et al. point out that Archeoplastida support comes primarily from phylogenomic analyses and these may be picking up misleading endosymbiotic gene transfer signal of genes independently transferred from the plastid to the host nucleus in the 3 archeoplastid clades. And Stiller and Harrell emphasize that the group can be explained by "short-branch exclusion" and "subtle and easily overlooked biases can dominate the overall results of molecular phylogenetic analyses of ancient eukaryotic relationships. Sources of potential phylogenetic artifact should be investigated routinely, not just when obvious 'long-branch attraction' is encountered."

Red algae as basal is also supported by molecular evidence. *Cyanidioschyzon*, a red alga, is considered basal (sister group to the rest of eukaryotes) by Nagashima et al. and Seckbach. It has the most primitive chloroplast, only 1 mitochondrion, no vacuoles, no trienoic acids, and the smallest eukaryotic genome at 8 Mbp.

Rhizaria are only moderately supported (65.5% of the studies, i.e., 19 of 29), statistical support for them is inconsistent in multigene genealogies with larger taxon sampling and the group is ambiguously supported in Goloboff et al.

Origin of Eukaryotes

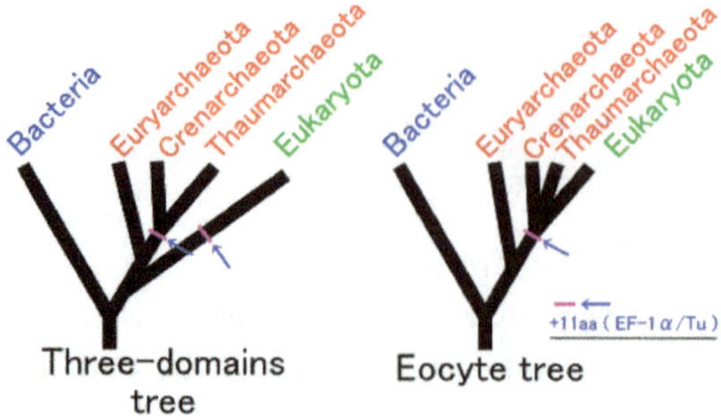

The three-domains tree and the Eocyte hypothesis.

Fossils

The origin of the eukaryotic cell is considered a milestone in the evolution of life, since eukaryotes include all complex cells and almost all multicellular organisms. The timing of this series of events is hard to determine; Knoll (2006) suggests they developed approximately 1.6–2.1 billion years ago. Some acritarchs are known from at least 1.65 billion years ago, and the possible alga *Grypania* has been found as far back as 2.1 billion years ago. The *Geosiphon*-like fossil fungus *Diskagma* has been found in paleosols 2.2 billion years old

Organized living structures have been found in the black shales of the Palaeoproterozoic Francevillian B Formation in Gabon, dated at 2.1 billion years old. Eukaryotic life could have evolved at that time.Fossils that are clearly related to modern groups start appearing an estimated 1.2 billion years ago, in the form of a red alga, though recent work suggests the existence of fossilized filamentous algae in the Vindhya basin dating back perhaps to 1.6 to 1.7 billion years ago.

Biomarkers suggest that at least stem eukaryotes arose even earlier. The presence of steranes in Australian shales indicates that eukaryotes were present in these rocks dated at 2.7 billion years old.

Relationship to Archaea

Eukaryotes are more closely related to Archaea than Bacteria, at least in terms of nuclear DNA and genetic machinery, and one controversial idea is to place them with Archaea in the clade Neomura. However, in other respects, such as membrane composition, eukaryotes are similar to Bacteria. Three main explanations for this have been proposed:

- Eukaryotes resulted from the complete fusion of two or more cells, wherein the cytoplasm formed from a eubacterium, and the nucleus from an archaeon, from a virus, or from a pre-cell.
- Eukaryotes developed from Archaea, and acquired their eubacterial characteristics from the proto-mitochondrion.
- Eukaryotes and Archaea developed separately from a modified eubacterium.

The chronocyte hypothesis for the origin of the eukaryotic cell postulates that a primitive eukaryotic cell was formed by the endosymbiosis of both archaea and bacteria by a third type of cell, termed a chronocyte.

Endomembrane System and Mitochondria

The origins of the endomembrane system and mitochondria are also unclear. The **phagotrophic hypothesis** proposes that eukaryotic-type membranes lacking a cell wall originated first, with the development of endocytosis, whereas mitochondria were acquired by ingestion as endosymbionts. The **syntrophic hypothesis** proposes that the proto-eukaryote relied on the proto-mitochondrion for food, and so ultimately grew to surround it. Here the membranes originated after the engulfment of the mitochondrion, in part thanks to mitochondrial genes (the hydrogen hypothesis is one particular version).

In a study using genomes to construct supertrees, Pisani *et al.* (2007) suggest that, along with evidence that there was never a mitochondrion-less eukaryote, eukaryotes evolved from a syntrophy between an archaea closely related to Thermoplasmatales and an α-proteobacterium, likely a symbiosis driven by sulfur or hydrogen. The mitochondrion and its genome is a remnant of the α-proteobacterial endosymbiont.

Hypotheses for the Origin of Eukaryotes

Different hypotheses have been proposed as to how eukaryotic cells came into existence. These hypotheses can be classified into two distinct classes – autogenous models and chimeric models.

Autogenous Models

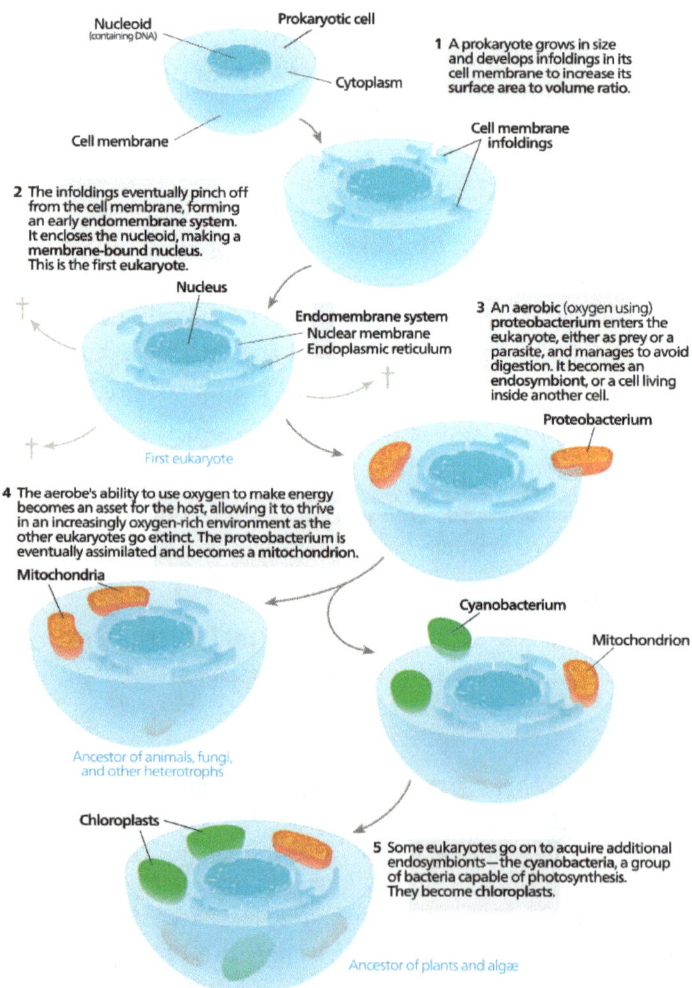

An autogenous model for the origin of eukaryotes.

Autogenous models propose that a proto-eukaryotic cell containing a nucleus existed first, and later acquired mitochondria. According to this model, a large prokaryote developed invaginations in its plasma membrane in order to obtain enough surface area to service its cytoplasmic volume. As the invaginations differentiated in function, some became separate compartments—giving rise to the endomembrane system, including the endoplasmic reticulum, golgi apparatus, nuclear membrane, and single membrane structures such as lysosomes. Mitochondria are proposed to come from the endosymbiosis of an aerobic proteobacterium, and it's assumed that all the eukaryotic lineages that did not acquire mitochondria became extinct. Chloroplasts came about from another endosymbiotic event involving cyanobacteria. Since all eukaryotes have mitochondria, but not all have chloroplasts, mitochondria are thought to have come first. This is the serial endosymbiosis theory.

Some models propose that the origins of double layered organelles, such as mitochondria and chloroplasts, in the proto-eukaryotic cell is due to the compartmentalization of DNA vesicles that were formed from the secondary invaginations or more detailed infoldings of cellular membrane.

Chimeric Models

Chimeric models claim that two prokaryotic cells existed initially – an archaeon and a bacterium. These cells underwent a merging process, either by a physical fusion or by endosymbiosis, thereby leading to the formation of a eukaryotic cell. Within these chimeric models, some studies further claim that mitochondria originated from a bacterial ancestor while others emphasize the role of endosymbiotic processes behind the origin of mitochondria.

Based on the process of mutualistic symbiosis, the hypotheses can be categorized as – the serial endosymbiotic theory (SET), the hydrogen hypothesis (mostly a process of symbiosis where hydrogen transfer takes place among different species), and the syntrophy hypothesis.

According to serial endosymbiotic theory (championed by Dr. Lynn Margulis), a union between a motile anaerobic bacterium (like *Spirochaeta*) and a thermoacidophilic crenarchaeon (like *Thermoplasma* which is sulfidogenic in nature) gave rise to the present day eukaryotes. This union established a motile organism capable of living in the already existing acidic and sulfurous waters. Oxygen is known to cause toxicity to organisms that lack the required metabolic machinery. Thus, the archaeon provided the bacterium with a highly beneficial reduced environment (sulfur and sulfate were reduced to sulfide). In microaerophilic conditions, oxygen was reduced to water thereby creating a mutual benefit platform. The bacterium on the other hand, contributed the necessary fermentation products and electron acceptors along with its motility feature to the archaeon thereby gaining a swimming motility for the organism. From a consortium of bacterial and archaeal DNA originated the nuclear genome of eukaryotic cells. Spirochetes gave rise to the motile features of eukaryotic cells. Endosymbiotic unifications of the ancestors of alpha-proteobacteria and cyanobacteria, led to the origin of mitochondria and plastids respectively. For example, *Thiodendron* has been known to have originated via an ectosymbiotic process based on a similar syntrophy of sulfur existing between the two types of bacteria – *Desulphobacter* and *Spirochaeta*. However, such an association based on motile symbiosis have never been observed practically. Also there is no evidence of archaeans and spirochetes adapting to intense acid-based environments.

In the hydrogen hypothesis, the symbiotic linkage of an anaerobic and autotrophic methanogenic archaeon (host) with an alpha-proteobacterium (the symbiont) gave rise to the eukaryotes. The host utilized hydrogen (H_2) and carbon dioxide (CO_2) to produce methane while the symbiont, capable of aerobic respiration, expelled H_2 and CO_2 as byproducts of anaerobic fermentation process. The host's methanogenic environment worked as a sink for H_2, which resulted in heightened bacterial fermentation. Endosymbiotic gene transfer (EGT) acted as a catalyst for the host to acquire the symbionts' carbohydrate metabolism and turn heterotrophic in nature. Subsequently, the host's methane forming capability was lost. Thus, the origins of the heterotrophic organelle (symbiont) are identical to the origins of the eukaryotic lineage. In this hypothesis, the presence of H_2 represents the selective force that forged eukaryotes out of prokaryotes.

The syntrophy hypothesis was developed in contrast to the hydrogen hypothesis and proposes the existence of two symbiotic events. According to this theory, eukaryogenesis (i.e. origin of eukaryotic cells) occurred based on metabolic symbiosis (syntrophy) between a methanogenic archaeon and a delta-proteobacterium. This syntrophic symbiosis was initially facilitated by H_2 transfer between different species under anaerobic environments. In earlier stages, an alpha-proteobacterium became a member of this integration, and later developed into the mitochondrion. Gene transfer from a delta-proteobacterium to an archaeon led to the methanogenic archaeon developing into a nucleus. The archaeon constituted the genetic apparatus while the delta-proteobacterium contributed towards the cytoplasmic features. This theory incorporates two selective forces that were needed to be considered during the time of nucleus evolution – (a) presence of metabolic partitioning in order to avoid the harmful effects of the co-existence of anabolic and catabolic cellular pathways, and (b) prevention of abnormal biosynthesis of proteins that occur due to a vast spread of introns in the archaeal genes after acquiring the mitochondrion and the loss of methanogenesis.

Thus, the origin of eukaryotes by endosymbiotic processes has been broadly recognized and accepted so far. Mitochondria and plastids have been known to originate from a bacterial ancestor during parallel adaptation to anaerobiosis. However, there still remains a greater need in assessing the question of how much eukaryotic complexity is being originated via an implementation of these symbiogenetic theories.

PROTIST

Protist is an informal term for any eukaryotic organism that is not an animal, plant or fungus. The protists do not form a natural group, or clade, but are often grouped together for convenience, like algae or invertebrates. In some systems of biological classification, such as the popular 5-kingdom scheme proposed by Robert Whittaker in 1969, the protists make up a kingdom called **Protista**, composed of "organisms which are unicellular or unicellular-colonial and which form no tissues."

Besides their relatively simple levels of organization, protists do not necessarily have much in common. When used, the term "protists" is now considered to mean similar-appearing but diverse taxa that are not related through an exclusive common ancestor, and have different life cycles, trophic levels, modes of locomotion, and cellular structures. In the classification system of Lynn Margulis, the term *protist* is reserved for microscopic organisms, while the more inclusive term ***Protoctista*** is applied to a biological kingdom which includes certain large multicellular eukaryotes, such as kelp, red algae and slime molds. Others use the term *protist* more broadly, to encompass both microbial eukaryotes and macroscopic organisms that do not fit into the other traditional kingdoms.

In cladistic systems, there are no equivalents to the taxa Protista or Protoctista, both terms referring to a paraphyletic group which spans the entire eukaryotic tree of life. In cladistic classification, the contents of Protista are distributed among various *supergroups* (SAR, Archaeplastida, Excavata, Opisthokonta, etc.) and "Protista", ''Protoctista" and "Protozoa" are considered obsolete. However, the term "protist" continues to be used informally as a catch-all term for eukaryotic microorganisms. For example, the phrase "protist pathogen" may be used to denote any disease-causing microbe which is not bacteria, virus, viroid or metazoa.

Subdivisions

The term *protista* was first used by Ernst Haeckel in 1866. Protists were traditionally subdivided into several groups based on similarities to the "higher" kingdoms such as:

Protozoa the unicellular "animal-like" (***heterotrophic***/parasitic) protozoa which was further sub-divided based on motility such as (flagellated) *Flagellata*, (ciliated) *Ciliophora*, (phagocytic) *amoeba* and spore-forming *Sporozoans*

Protophyta the "plant-like" (***autotrophic***) protophyta (mostly unicellular algae)

Molds the "fungus-like" (***saprophytic***) slime molds and water molds.

Some protists, sometimes called ambiregnal protists, have been considered to be both protozoa and algae or fungi (e.g., slime molds and flagellated algae), and names for these have been published under either or both of the *ICN* and the *ICZN*. Conflicts, such as these – for example the dual-classification of Euglenids and Dinobryons, which are mixotrophic – is an example of why the kingdom Protista was adopted.

These traditional subdivisions, largely based on superficial commonalities, have been replaced by classifications based on phylogenetics (evolutionary relatedness among organisms). Molecular analyses in modern taxonomy have been used to redistribute former members of this group into diverse and sometimes distantly related phyla. For instance, the water molds are now considered to be closely related to photosynthetic organisms such as Brown algae and Diatoms, the slime molds are grouped mainly under Amoebozoa, and the Amoebozoa itself includes only a subset of "Amoeba" group, and significant number of erstwhile "Amoeboid" genera are distributed among Rhizarians and other Phyla.

However, the older terms are still used as informal names to describe the morphology and ecology of various protists. For example, the term *protozoa* is used to refer to heterotrophic species of protists that do not form filaments.

Classification

Historical Classifications

Among the pioneers in the study of the protists, which were almost ignored by Linnaeus except for some genera (e.g., *Vorticella, Chaos, Volvox, Corallina, Conferva, Ulva, Chara, Fucus*) were Leeuwenhoek, O. F. Müller, C. G. Ehrenberg and Félix Dujardin. The first groups used to classify microscopic organism were the Animalcules and the Infusoria. In 1817, the German naturalist Georg August Goldfuss introduced the word *Protozoa* to refer to organisms such as ciliates and corals. After the cell theory of Schwann and Schleiden (1838–39), this group was modified in 1848 by Carl von Siebold to include only animal-like unicellular organisms, such as foraminifera and amoebae. The formal taxonomic category *Protoctista* was first proposed in the early 1860s by John Hogg, who argued that the protists should include what he saw as primitive unicellular forms of both plants and animals. He defined the Protoctista as a "fourth kingdom of nature", in addition to the then-traditional kingdoms of plants, animals and minerals. The kingdom of minerals was later removed from taxonomy in 1866 by Ernst Haeckel, leaving plants, animals, and the protists (*Protista*), defined as a "kingdom of primitive forms".

In 1938, Herbert Copeland resurrected Hogg's label, arguing that Haeckel's term *Protista* included anucleated microbes such as bacteria, which the term "Protoctista" (literally meaning "first established beings") did not. In contrast, Copeland's term included nucleated eukaryotes such as diatoms, green algae and fungi. This classification was the basis for Whittaker's later definition of Fungi, Animalia, Plantae and Protista as the four kingdoms of life. The kingdom Protista was later modified to separate prokaryotes into the separate kingdom of Monera, leaving the protists as a group of eukaryotic microorganisms. These five kingdoms remained the accepted classification until the development of molecular phylogenetics in the late 20th century, when it became apparent that neither protists nor monera were single groups of related organisms (they were not monophyletic groups).

Modern Classifications

Many systematists today do not treat Protista as a formal taxon, but the term "protist" is still commonly used for convenience in two ways. The most popular contemporary definition is a phylogenetic one, that identifies a paraphyletic group: a protist is any eukaryote that is not an animal, (land) plant, or (true) fungus; this definitionexcludes many unicellular groups, like the Microsporidia (fungi), many Chytridiomycetes (fungi), and yeasts (fungi), and also a non-unicellular group included in Protista in the past, the Myxozoa (animal). Some systematists judge paraphyletic taxa acceptable, and use Protista in this sense as a formal taxon (as found in some secondary textbooks, for pedagogical purpose).

The other definition describes protists primarily by functional or biological criteria: protists are essentially those eukaryotes that are never multicellular, that either exist as independent cells, or if they occur in colonies, do not show differentiation into tissues (but vegetative cell differentiation may occur restricted to sexual reproduction, alternate vegetative morphology, and quiescent or resistant stages, such as cysts); this definition excludes many brown, many red and some green algae, which may have tissues.

The taxonomy of protists is still changing. Newer classifications attempt to present monophyletic groups based on morphological (especially ultrastructural), biochemical (chemotaxonomy) and DNA sequence (molecular research) information. However, there are sometimes discordances between molecular and morphological investigations; these can be categorized as two types: (i) one morphology, multiple lineages (e.g. morphological convergence, cryptic species) and (ii) one lineage, multiple morphologies (e.g. phenotypic plasticity, multiple life-cycle stages).

Because the protists as a whole are paraphyletic, new systems often split up or abandon the kingdom, instead treating the protist groups as separate lines of eukaryotes. The recent scheme by Adl *et al.* (2005) does not recognize formal ranks (phylum, class, etc.) and instead treats groups as clades of phylogenetically related organisms. This is intended to make the classification more stable in the long term and easier to update. Some of the main groups of protists, which may be treated as phyla, are listed in the taxobox, upper right. Many are thought to be monophyletic, though there is still uncertainty. For instance, the excavates are probably not monophyletic and the chromalveolates are probably only monophyletic if the haptophytes and cryptomonads are excluded.

Metabolism

Nutrition can vary according to the type of protist. Most eukaryotic algae are autotrophic, but the pigments were lost in some groups. Other protists are heterotrophic, and may present phagotrophy, osmotrophy, saprotrophy or parasitism. Some are mixotrophic. Some protists that do not have / lost chloroplasts/mitochondria have entered into endosymbiontic relationship with other bacteria/algae to replace the missing functionality. For example *Paramecium bursaria* & *Paulinella* have captured a green algae (*Zoochlorella*) & an cyanobacterium respectively that act as a replacement for chloroplast. Meanwhile, a protist, Mixotricha paradoxa that has lost its mitochondria uses endosymbiontic bacteria as mitochondria and ectosymbiontic hair-like bacteria (*Treponema spirochetes*) for locomotion.

Many protists are flagellate, for example, and filter feeding can take place where the flagella find prey. Other protists can engulf bacteria and other food particles, by extending their cell membrane around them to form a food vacuole and digesting them internally in a process termed phagocytosis.

Nutritional types in protist metabolism

Nutritional type	Source of energy	Source of carbon	Examples
Photoautotrophs	Sunlight	Organic compounds or carbon fixation	Most algae
Chemoheterotrophs	Organic compounds	Organic compounds	Apicomplexa, Trypanosomes or Amoebae

Reproduction

Some protists reproduce sexually using gametes, while others reproduce asexually by binary fission.

Some species, for example *Plasmodium falciparum*, have extremely complex life cycles that involve multiple forms of the organism, some of which reproduce sexually and others asexually. However, it is unclear how frequently sexual reproduction causes genetic exchange between different strains of *Plasmodium* in nature and most populations of parasitic protists may be clonal lines that rarely exchange genes with other members of their species.

Eukaryotes emerged in evolution more than 1.5 billion years ago. The earliest eukaryotes were likely protists. Although sexual reproduction is widespread among extant eukaryotes, it seemed unlikely until recently, that sex could be a primordial and fundamental characteristic of eukaryotes. A principal reason for this view was that sex appeared to be lacking in certain pathogenic protists whose ancestors branched off early from the eukaryotic family tree. However, several of these protists are now known to be capable of, or to recently have had the capability for, meiosis and hence sexual reproduction. For example, the common intestinal parasite *Giardia lamblia* was once considered to be a descendant of a protist lineage that predated the emergence of meiosis and sex. However, *G. lamblia* was recently found to have a core set of genes that function in meiosis and that are widely present among sexual eukaryotes. These results suggested that *G. lamblia* is capable of meiosis and thus sexual reproduction. Furthermore, direct evidence for meiotic recombination, indicative of sex, was also found in *G. lamblia*.

The pathogenic parasitic protists of the genus *Leishmania* have been shown to be capable of a sexual cycle in the invertebrate vector, likened to the meiosis undertaken in the trypanosomes.

Trichomonas vaginalis, a parasitic protist, is not known to undergo meiosis, but when Malik et al. tested for 29 genes that function in meiosis, they found 27 to be present, including 8 of 9 genes specific to meiosis in model eukaryotes. These findings suggest that *T. vaginalis* may be capable of meiosis. Since 21 of the 29 meiotic genes were also present in G. lamblia, it appears that most of these meiotic genes were likely present in a common ancestor of *T. vaginalis* and *G. lamblia*. These two species are descendants of protist lineages that are highly divergent among eukaryotes, leading Malik et al. to suggest that these meiotic genes were likely present in a common ancestor of all eukaryotes.

Based on a phylogenetic analysis, Dacks and Roger proposed that facultative sex was present in the common ancestor of all eukaryotes.

This view was further supported by a study of amoebae by Lahr et al. Amoeba have generally been regarded as asexual protists. However these authors describe evidence that most amoeboid lineages are anciently sexual, and that the majority of asexual groups likely arose recently and independently. It should be noted that early researchers (e.g., Calkins) have interpreted phenomena related to chromidia (chromatin granules free in the cytoplasm) in amoeboid organisms as sexual reproduction.

Protists generally reproduce asexually under favorable environmental conditions, but tend to reproduce sexually under stressful conditions, such as starvation or heat shock.

Oxidative stress, which is associated with the production of reactive oxygen species leading to DNA damage, also appears to be an important factor in the induction of sex in protists.

Ecology

Protists live in almost any environment that contains liquid water. Many protists, such as algae, are photosynthetic and are vital primary producers in ecosystems, particularly in the ocean as part of the plankton. Other protists include pathogenic species such as the kinetoplastid *Trypanosoma brucei*, which causes sleeping sickness and species of the apicomplexan *Plasmodium* which cause malaria.

Parasitism: Role as Pathogens

Some protists are significant parasites of animals (e.g., five species of the parasitic genus *Plasmodium* cause malaria in humans and many others cause similar diseases in other vertebrates), plants (the oomycete *Phytophthora infestans* causes late blight in potatoes) or even of other protists. Protist pathogens share many metabolic pathways with their eukaryotic hosts. This makes therapeutic target development extremely difficult – a drug that harms a protist parasite is also likely to harm its animal/plant host. A more thorough understanding of protist biology may allow these diseases to be treated more efficiently. For example, the apicoplast (a nonphotosynthetic chloroplast but essential to carry out important functions other than photosynthesis) present in apicomplexans provides an attractive target for treating diseases caused by dangerous pathogens such as plasmodium.

Recent papers have proposed the use of viruses to treat infections caused by protozoa.

Researchers from the Agricultural Research Service are taking advantage of protists as pathogens to control red imported fire ant (*Solenopsis invicta*) populations in Argentina. Spore-producing protists such as *Kneallhazia solenopsae* (recognized as a sister clade or the closest relative to the fungus kingdom now) can reduce red fire ant populations by 53–100%. Researchers have also been able to infect phorid fly parasitoids of the ant with the protist without harming the flies. This turns the flies into a vector that can spreads the pathogenic protist between red fire ant colonies.

Fossil Record

Many protists have neither hard parts nor resistant spores, and their fossils are extremely rare or unknown. Examples of such groups include the apicomplexans,most ciliates, some green algae (the Klebsormidiales), choanoflagellates, oomycetes, brown algae, yellow-green algae, excavates (e.g., euglenids). Some of these have been found preserved in amber (fossilized tree resin) or under unusual conditions (e.g., *Paleoleishmania*, a kinetoplastid).

Others are relatively common in the fossil record, as the diatoms, golden algae, haptophytes (coccoliths), silicoflagellates, tintinnids (ciliates), dinoflagellates, green algae, red algae, heliozoans, radiolarians, foraminiferans, ebriids and testate amoebae (euglyphids, arcellaceans). Some are even used as paleoecological indicators to reconstruct ancient environments.

More probable eukaryote fossils begin to appear at about 1.8 billion years ago, the acritarchs, spherical fossils of likely algal protists. Another possible representant of early fossil eukaryotes are the Gabonionta.

FUNGUS

A **fungus** is any member of the group of eukaryotic organisms that includes unicellular microorganisms such as yeasts and molds, as well as multicellular fungi that produce familiar fruiting forms known as mushrooms. These organisms are classified as a kingdom, **Fungi**, which is separate from the other eukaryotic life kingdoms of plants and animals.

A characteristic that places fungi in a different kingdom from plants, bacteria and some protists, is chitin in their cell walls. Similar to animals, fungi are heterotrophs; they acquire their food by absorbing dissolved molecules, typically by secreting digestive enzymes into their environment. Fungi do not photosynthesise. Growth is their means of mobility, except for spores (a few of which are flagellated), which may travel through the air or water. Fungi are the principal decomposers in ecological systems. These and other differences place fungi in a single group of related organisms, named the *Eumycota* (*true fungi* or *Eumycetes*), which share a common ancestor (form a *monophyletic group*), an interpretation that is also strongly supported by molecular phylogenetics. This fungal group is distinct from the structurally similar myxomycetes (slime molds) and oomycetes (water molds). The discipline of biology devoted to the study of fungi is known as mycology (from the Greek μύκης, mukēs, meaning "fungus"). In the past, mycology was regarded as a branch of botany, although it is now known fungi are genetically more closely related to animals than to plants.

Abundant worldwide, most fungi are inconspicuous because of the small size of their structures, and their cryptic lifestyles in soil or on dead matter. Fungi include symbionts of plants, animals, or other fungi and also parasites. They may become noticeable when fruiting, either as mushrooms or as molds. Fungi perform an essential role in the decomposition of organic matter and have fundamental roles in nutrient cycling and exchange in the environment. They have long been used as a direct source of human food, in the form of mushrooms and truffles; as a leavening agent for bread; and in the fermentation of various food products, such as wine, beer, and soy sauce. Since the 1940s, fungi have been used for the production of antibiotics, and, more recently, various enzymes produced by fungi are used industrially and in detergents. Fungi are also used as biological pesticides to control weeds, plant diseases and insect pests. Many species produce bioactive compounds called mycotoxins, such as alkaloids and polyketides, that are toxic to animals including humans. The fruiting structures of a few species contain psychotropic compounds and are consumed recreationally or in traditional spiritual ceremonies. Fungi can break down manufactured materials and buildings, and become significant pathogens of humans and other animals. Losses of crops due to fungal diseases (e.g., rice blast disease) or food spoilage can have a large impact on human food supplies and local economies.

The fungus kingdom encompasses an enormous diversity of taxa with varied ecologies, life cycle strategies, and morphologies ranging from unicellular aquatic chytrids to large mushrooms. However, little is known of the true biodiversity of Kingdom Fungi, which has been estimated at 1.5 million to 5 million species, with about 5% of these having been formally classified. Ever since the pioneering 18th and 19th century taxonomical works of Carl Linnaeus, Christian Hendrik Persoon, and Elias Magnus Fries, fungi have been classified according to their morphology (e.g., characteristics such as spore color or microscopic features) or physiology. Advances in molecular genetics have opened the way for DNA analysis to be incorporated into taxonomy, which has sometimes challenged the historical groupings based on morphology and other traits. Phylogenetic studies published in the last decade have helped reshape the classification within Kingdom Fungi, which is divided into one subkingdom, seven phyla, and ten subphyla.

Etymology

The English word *fungus* is directly adopted from the Latin *fungus* (mushroom), used in the writings of Horace and Pliny. This in turn is derived from the Greek word *sphongos* (σφογγος "sponge"), which refers to the macroscopic structures and morphology of mushrooms and molds; the root is also used in other languages, such as the German *Schwamm* ("sponge") and *Schimmel* ("mold"). The use of the word *mycology*, which is derived from the Greek *mykes* (μύκης "mushroom") and *logos* (λόγος "discourse"), to denote the scientific study of fungi is thought to have originated in 1836 with English naturalist Miles Joseph Berkeley's publication *The English Flora of Sir James Edward Smith, Vol. 5.* A group of all the fungi present in a particular area or geographic region is known as mycobiota (plural noun, no singular), e.g., "the mycobiota of Ireland".

Characteristics

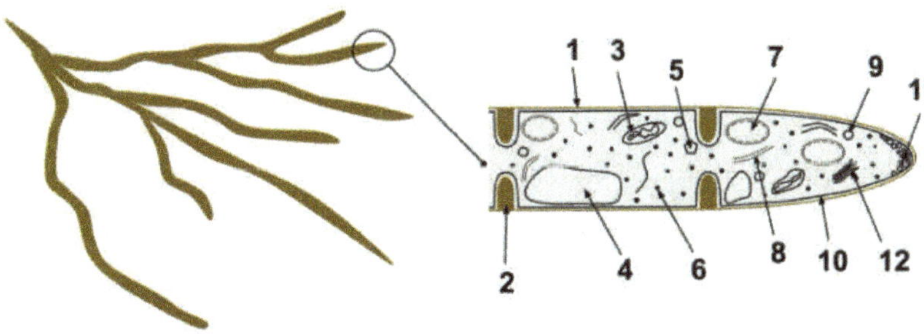

Fungal hyphae cells

- Hyphal wall
- Septum
- Mitochondrion
- Vacuole
- Ergosterol crystal
- Ribosome

- Nucleus
- Endoplasmic reticulum
- Lipid body
- Plasma membrane
- Spitzenkörper
- Golgi apparatus

Before the introduction of molecular methods for phylogenetic analysis, taxonomists considered fungi to be members of the plant kingdom because of similarities in lifestyle: both fungi and plants are mainly immobile, and have similarities in general morphology and growth habitat. Like plants, fungi often grow in soil and, in the case of mushrooms, form conspicuous fruit bodies, which sometimes resemble plants such as mosses. The fungi are now considered a separate kingdom, distinct from both plants and animals, from which they appear to have diverged around one billion years ago. Some morphological, biochemical, and genetic features are shared with other organisms, while others are unique to the fungi, clearly separating them from the other kingdoms:

Shared features:

- With other eukaryotes: Fungal cells contain membrane-bound nuclei with chromosomes that contain DNA with noncoding regions called introns and coding regions called exons. Fungi have membrane-bound cytoplasmic organelles such as mitochondria, sterol-containing membranes, and ribosomes of the 80S type. They have a characteristic range of soluble carbohydrates and storage compounds, including sugar alcohols (e.g., mannitol), disaccharides, (e.g., trehalose), and polysaccharides (e.g., glycogen, which is also found in animals).
- With animals: Fungi lack chloroplasts and are heterotrophic organisms and so require preformed organic compounds as energy sources.
- With plants: Fungi have a cell wall and vacuoles. They reproduce by both sexual and asexual means, and like basal plant groups (such as ferns and mosses) produce spores. Similar to mosses and algae, fungi typically have haploid nuclei.
- With euglenoids and bacteria: Higher fungi, euglenoids, and some bacteria produce the amino acid L-lysine in specific biosynthesis steps, called the α-aminoadipate pathway.
- The cells of most fungi grow as tubular, elongated, and thread-like (filamentous) structures called hyphae, which may contain multiple nuclei and extend by growing at their tips. Each tip contains a set of aggregated vesicles—cellular structures consisting of proteins, lipids, and other organic molecules—called the Spitzenkörper. Both fungi and oomycetes grow as filamentous hyphal cells. In contrast, similar-looking organisms, such as filamentous green algae, grow by repeated cell division within a chain of cells. There are also single-celled fungi (yeasts) that do not form hyphae, and some fungi have both hyphal and yeast forms.
- In common with some plant and animal species, more than 70 fungal species display bioluminescence.

Unique features:

- Some species grow as unicellular yeasts that reproduce by budding or binary fission. Dimorphic fungi can switch between a yeast phase and a hyphal phase in response to environmental conditions.
- The fungal cell wall is composed of glucans and chitin; while glucans are also found in plants and chitin in the exoskeleton of arthropods, fungi are the only organisms that combine these two structural molecules in their cell wall. Unlike those of plants and oomycetes, fungal cell walls do not contain cellulose.

Omphalotus nidiformis, a bioluminescent mushroom

Most fungi lack an efficient system for the long-distance transport of water and nutrients, such as the xylem and phloem in many plants. To overcome this limitation, some fungi, such as *Armillaria*, form rhizomorphs, which resemble and perform functions similar to the roots of plants. As eukaryotes, fungi possess a biosynthetic pathway for producing terpenes that uses mevalonic acid and pyrophosphate as chemical building blocks. Plants and some other organisms have an additional terpene biosynthesis pathway in their chloroplasts, a structure fungi and animals do not have. Fungi produce several secondary metabolites that are similar or identical in structure to those made by plants. Many of the plant and fungal enzymes that make these compounds differ from each other in sequence and other characteristics, which indicates separate origins and convergent evolution of these enzymes in the fungi and plants.

Diversity

Bracket fungi on a tree stump

Fungi have a worldwide distribution, and grow in a wide range of habitats, including extreme environments such as deserts or areas with high salt concentrations or ionizing radiation, as well as in deep sea sediments. Some can survive the intense UV and cosmic radiation encountered during space travel. Most grow in terrestrial environments, though several species live partly or solely in aquatic habitats, such as the chytrid fungus *Batrachochytrium dendrobatidis*, a parasite that has been responsible for a worldwide decline in amphibian populations. This organism spends part of its life cycle as a motile zoospore, enabling it to propel itself through water and enter its amphibian host. Other examples of aquatic fungi include those living in hydrothermal areas of the ocean.

Around 100,000 species of fungi have been formally described by taxonomists, but the global biodiversity of the fungus kingdom is not fully understood. On the basis of observations of the ratio of the number of fungal species to the number of plant species in selected environments, the fungal kingdom has been estimated to contain about 1.5 million species. A recent (2011) estimate suggests there may be over 5 million species. In mycology, species have historically been distinguished by a variety of methods and concepts. Classification based on morphological characteristics, such as the size and shape of spores or fruiting structures, has traditionally dominated fungal taxonomy. Species may also be distinguished by their biochemical and physiological characteristics, such as their ability to metabolize certain biochemicals, or their reaction to chemical tests. The biological species concept discriminates species based on their ability to mate. The application of molecular tools, such as DNA sequencing and phylogenetic analysis, to study diversity has greatly enhanced the resolution and added robustness to estimates of genetic diversity within various taxonomic groups.

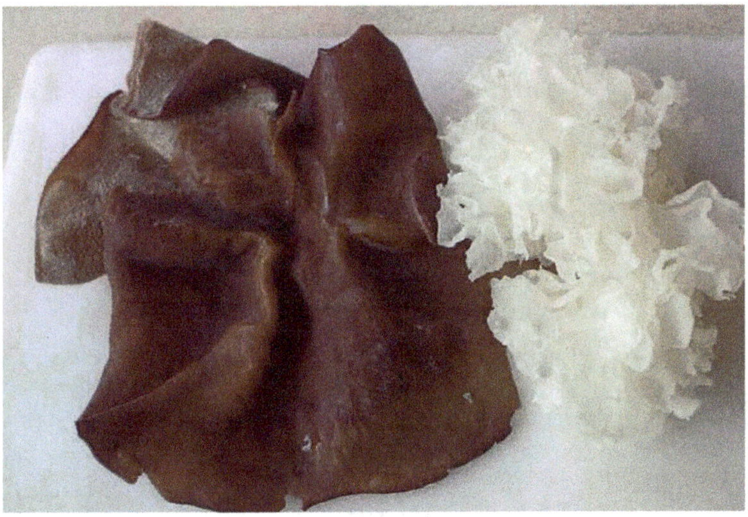

Two types of edible fungi

Mycology

Mycology is the branch of biology concerned with the systematic study of fungi, including their genetic and biochemical properties, their taxonomy, and their use to humans as a source of medicine, food, and psychotropic substances consumed for religious purposes, as well as their dangers, such as poisoning or infection. The field of phytopathology, the study of plant diseases, is closely related because many plant pathogens are fungi.

In 1729, Pier A. Micheli first published descriptions of fungi.

The use of fungi by humans dates back to prehistory; Ötzi the Iceman, a well-preserved mummy of a 5,300-year-old Neolithic man found frozen in the Austrian Alps, carried two species of polypore mushrooms that may have been used as tinder (*Fomes fomentarius*), or for medicinal purposes (*Piptoporus betulinus*). Ancient peoples have used fungi as food sources–often unknowingly–for millennia, in the preparation of leavened bread and fermented juices. Some of the oldest written records contain references to the destruction of crops that were probably caused by pathogenic fungi.

Mycology is a relatively new science that became systematic after the development of the microscope in the 17th century. Although fungal spores were first observed by Giambattista della Porta in 1588, the seminal work in the development of mycology is considered to be the publication of Pier Antonio Micheli's 1729 work *Nova plantarum genera*. Micheli not only observed spores but also showed that, under the proper conditions, they could be induced into growing into the same species of fungi from which they originated. Extending the use of the binomial system of nomenclature introduced by Carl Linnaeus in his *Species plantarum* (1753), the Dutch Christian Hendrik Persoon (1761–1836) established the first classification of mushrooms with such skill so as to be considered a founder of modern mycology. Later, Elias Magnus Fries (1794–1878) further elaborated the classification of fungi, using spore color and various microscopic characteristics, methods still used by taxonomists today. Other notable early contributors to mycology in the 17th–19th and early 20th centuries include Miles Joseph Berkeley, August Carl Joseph Corda, Anton de Bary, the brothers Louis René and Charles Tulasne, Arthur H. R. Buller, Curtis G. Lloyd, and Pier Andrea Saccardo. The 20th century has seen a modernization of mycology that has come from advances in biochemistry, genetics, molecular biology, and biotechnology. The use of DNA sequencing technologies and phylogenetic analysis has provided new insights into fungal relationships and biodiversity, and has challenged traditional morphology-based groupings in fungal taxonomy.

Morphology

Microscopic Structures

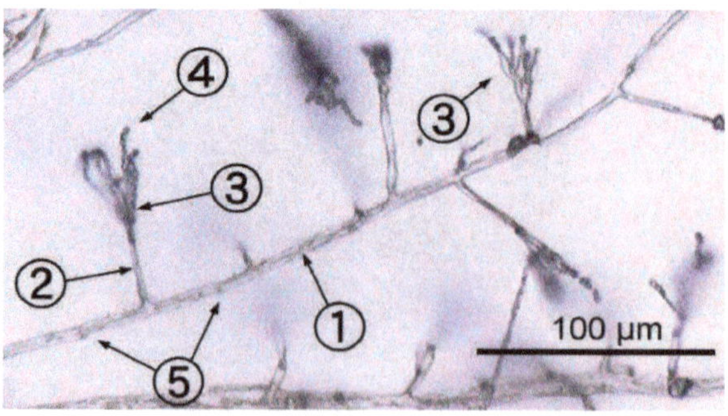

An environmental isolate of *Penicillium*

- hypha
- conidiophore
- phialide
- conidia
- septa

Most fungi grow as hyphae, which are cylindrical, thread-like structures 2–10 μm in diameter and up to several centimeters in length. Hyphae grow at their tips (apices); new hyphae are typically formed by emergence of new tips along existing hyphae by a process called *branching*, or occasionally growing hyphal tips fork, giving rise to two parallel-growing hyphae. Hyphae also sometimes fuse when they come into contact, a process called hyphal fusion (or anastamosis). These growth processes lead to the development of a mycelium, an interconnected network of hyphae. Hyphae can be either septate or coenocytic. Septate hyphae are divided into compartments separated by cross walls (internal cell walls, called septa, that are formed at right angles to the cell wall giving the hypha its shape), with each compartment containing one or more nuclei; coenocytic hyphae are not compartmentalized. Septa have pores that allow cytoplasm, organelles, and sometimes nuclei to pass through; an example is the dolipore septum in fungi of the phylum Basidiomycota. Coenocytic hyphae are in essence multinucleate supercells.

Many species have developed specialized hyphal structures for nutrient uptake from living hosts; examples include haustoria in plant-parasitic species of most fungal phyla, and arbuscules of several mycorrhizal fungi, which penetrate into the host cells to consume nutrients.

Although fungi are opisthokonts—a grouping of evolutionarily related organisms broadly characterized by a single posterior flagellum—all phyla except for the chytrids have lost their posterior flagella. Fungi are unusual among the eukaryotes in having a cell wall that, in addition to glucans (e.g., β-1,3-glucan) and other typical components, also contains the biopolymer chitin.

Macroscopic Structures

Armillaria solidipes

Fungal mycelia can become visible to the naked eye, for example, on various surfaces and substrates, such as damp walls and spoiled food, where they are commonly called molds. Mycelia grown on solid agar media in laboratory petri dishes are usually referred to as colonies. These colonies can exhibit growth shapes and colors (due to spores or pigmentation) that can be used as diagnostic features in the identification of species or groups. Some individual fungal colonies can reach extraordinary dimensions and ages as in the case of a clonal colony of *Armillaria solidipes*, which extends over an area of more than 900 ha (3.5 square miles), with an estimated age of nearly 9,000 years.

The apothecium—a specialized structure important in sexual reproduction in the ascomycetes—is a cup-shaped fruit body that is often macroscopic and holds the hymenium, a layer of tissue containing the spore-bearing cells. The fruit bodies of the basidiomycetes (basidiocarps) and some ascomycetes can sometimes grow very large, and many are well known as mushrooms.

Growth and Physiology

Mold growth covering a decaying peach. The frames were taken approximately 12 hours apart over a period of six days.

The growth of fungi as hyphae on or in solid substrates or as single cells in aquatic environments is adapted for the efficient extraction of nutrients, because these growth forms have high surface area to volume ratios. Hyphae are specifically adapted for growth on solid surfaces, and to invade substrates and tissues. They can exert large penetrative mechanical forces; for example, many plant pathogens, including *Magnaporthe grisea*, form a structure called an appressorium that evolved to puncture plant tissues. The pressure generated by the appressorium, directed against the plant epidermis, can exceed 8 megapascals (1,200 psi). The filamentous fungus *Paecilomyces lilacinus* uses a similar structure to penetrate the eggs of nematodes.

The mechanical pressure exerted by the appressorium is generated from physiological processes that increase intracellular turgor by producing osmolytes such as glycerol. Adaptations such as these are complemented by hydrolytic enzymes secreted into the environment to digest large organic molecules—such as polysaccharides, proteins, and lipids—into smaller molecules that may then be absorbed as nutrients. The vast majority of filamentous fungi grow in a polar fashion (extending in one direction) by elongation at the tip (apex) of the hypha. Other forms of fungal growth include intercalary extension (longitudinal expansion of hyphal compartments that are below the apex) as in the case of some endophytic fungi,or growth by volume expansion during the development of mushroom stipes and other large organs. Growth of fungi as multicellular structures consisting of somatic and reproductive cells—a feature independently evolved in animals and plants—has several functions, including the development of fruit bodies for dissemination of sexual spores (see above) and biofilms for substrate colonization and intercellular communication.

The fungi are traditionally considered heterotrophs, organisms that rely solely on carbon fixed by other organisms for metabolism. Fungi have evolved a high degree of metabolic versatility that allows them to use a diverse range of organic substrates for growth, including simple compounds such as nitrate, ammonia, acetate, or ethanol. In some species the pigment melanin may play a role in extracting energy from ionizing radiation, such as gamma radiation. This form of "radiotrophic" growth has been described for only a few species, the effects on growth rates are small, and the underlying biophysical and biochemical processes are not well known. This process might bear similarity to CO_2 fixation via visible light, but instead uses ionizing radiation as a source of energy.

Reproduction

Polyporus squamosus

Fungal reproduction is complex, reflecting the differences in lifestyles and genetic makeup within this diverse kingdom of organisms. It is estimated that a third of all fungi reproduce using more than one method of propagation; for example, reproduction may occur in two well-differentiated stages within the life cycle of a species, the teleomorph and the anamorph. Environmental conditions trigger genetically determined developmental states that lead to the creation of specialized structures for sexual or asexual reproduction.

These structures aid reproduction by efficiently dispersing spores or spore-containing propagules.

Asexual Reproduction

Asexual reproduction occurs via vegetative spores (conidia) or through mycelial fragmentation. Mycelial fragmentation occurs when a fungal mycelium separates into pieces, and each component grows into a separate mycelium. Mycelial fragmentation and vegetative spores maintain clonal populations adapted to a specific niche, and allow more rapid dispersal than sexual reproduction. The "Fungi imperfecti" (fungi lacking the perfect or sexual stage) or Deuteromycota comprise all the species that lack an observable sexual cycle. Deuteromycota is not an accepted taxonomic clade, and is now taken to mean simply fungi that lack a known sexual stage.

Sexual Reproduction

Sexual reproduction with meiosis has been directly observed in all fungal phyla except Glomeromycota (genetic analysis suggests meiosis in Glomeromycota as well). It differs in many aspects from sexual reproduction in animals or plants. Differences also exist between fungal groups and can be used to discriminate species by morphological differences in sexual structures and reproductive strategies. Mating experiments between fungal isolates may identify species on the basis of biological species concepts. The major fungal groupings have initially been delineated based on the morphology of their sexual structures and spores; for example, the spore-containing structures, asci and basidia, can be used in the identification of ascomycetes and basidiomycetes, respectively. Some species may allow mating only between individuals of opposite mating type, whereas others can mate and sexually reproduce with any other individual or itself. Species of the former mating system are called heterothallic, and of the latter homothallic.

Most fungi have both a haploid and a diploid stage in their life cycles. In sexually reproducing fungi, compatible individuals may combine by fusing their hyphae together into an interconnected network; this process, anastomosis, is required for the initiation of the sexual cycle. Many ascomycetes and basidiomycetes go through a dikaryotic stage, in which the nuclei inherited from the two parents do not combine immediately after cell fusion, but remain separate in the hyphal cells (see heterokaryosis).

In ascomycetes, dikaryotic hyphae of the hymenium (the spore-bearing tissue layer) form a characteristic *hook* at the hyphal septum. During cell division, formation of the hook ensures proper distribution of the newly divided nuclei into the apical and basal hyphal compartments. An ascus (plural *asci*) is then formed, in which karyogamy (nuclear fusion) occurs. Asci are embedded in an ascocarp, or fruiting body. Karyogamy in the asci is followed immediately by meiosis and the production of ascospores. After dispersal, the ascospores may germinate and form a new haploid mycelium.

Sexual reproduction in basidiomycetes is similar to that of the ascomycetes. Compatible haploid hyphae fuse to produce a dikaryotic mycelium. However, the dikaryotic phase is more extensive in the basidiomycetes, often also present in the vegetatively growing mycelium. A specialized anatomical structure, called a clamp connection, is formed at each hyphal septum. As with the structurally similar hook in the ascomycetes, the clamp connection in the basidiomycetes is required for controlled transfer of nuclei during cell division, to maintain the dikaryotic stage with two genetically different nuclei in each hyphal compartment. A basidiocarp is formed in which club-like structures known as basidia generate haploid basidiospores after karyogamy and meiosis. The most commonly known basidiocarps are mushrooms, but they may also take other forms (see Morphology section).

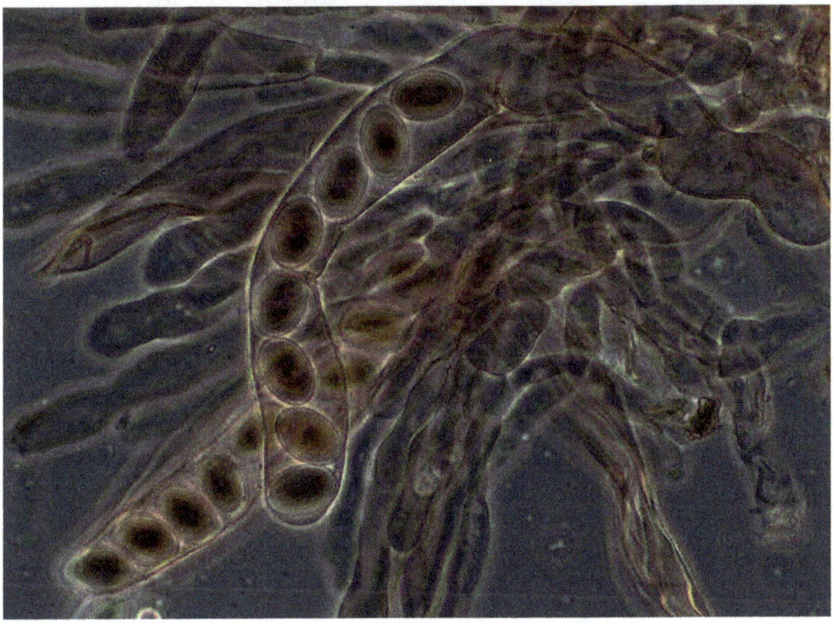

The 8-spore asci of *Morchella elata*, viewed with phase contrast microscopy

In glomeromycetes (formerly zygomycetes), haploid hyphae of two individuals fuse, forming a gametangium, a specialized cell structure that becomes a fertile gamete-producing cell. The gametangium develops into a zygospore, a thick-walled spore formed by the union of gametes. When the zygospore germinates, it undergoes meiosis, generating new haploid hyphae, which may then form asexual sporangiospores. These sporangiospores allow the fungus to rapidly disperse and germinate into new genetically identical haploid fungal mycelia.

Spore Dispersal

Both asexual and sexual spores or sporangiospores are often actively dispersed by forcible ejection from their reproductive structures. This ejection ensures exit of the spores from the reproductive structures as well as traveling through the air over long distances.

The bird's nest fungus *Cyathus stercoreus*

Specialized mechanical and physiological mechanisms, as well as spore surface structures (such as hydrophobins), enable efficient spore ejection. For example, the structure of the spore-bearing cells in some ascomycete species is such that the buildup of substances affecting cell volume and fluid balance enables the explosive discharge of spores into the air. The forcible discharge of single spores termed *ballistospores* involves formation of a small drop of water (Buller's drop), which upon contact with the spore leads to its projectile release with an initial acceleration of more than 10,000 g; the net result is that the spore is ejected 0.01–0.02 cm, sufficient distance for it to fall through the gills or pores into the air below. Other fungi, like the puffballs, rely on alternative mechanisms for spore release, such as external mechanical forces. The bird's nest fungi use the force of falling water drops to liberate the spores from cup-shaped fruiting bodies. Another strategy is seen in the stinkhorns, a group of fungi with lively colors and putrid odor that attract insects to disperse their spores.

Other Sexual Processes

Besides regular sexual reproduction with meiosis, certain fungi, such as those in the genera *Penicillium* and *Aspergillus*, may exchange genetic material via parasexual processes, initiated by anastomosis between hyphae and plasmogamy of fungal cells. The frequency and relative importance of parasexual events is unclear and may be lower than other sexual processes. It is known to play a role in intraspecific hybridization and is likely required for hybridization between species, which has been associated with major events in fungal evolution.

Evolution

In contrast to plants and animals, the early fossil record of the fungi is meager. Factors that likely contribute to the under-representation of fungal species among fossils include the nature of fungal fruiting bodies, which are soft, fleshy, and easily degradable tissues and the microscopic dimensions of most fungal structures, which therefore are not readily evident. Fungal fossils are difficult to distinguish from those of other microbes, and are most easily identified when they resemble extant fungi. Often recovered from a permineralized plant or animal host, these samples are typically studied by making thin-section preparations that can be examined with light microscopy or transmission electron microscopy. Researchers study compression fossils by dissolving the surrounding matrix with acid and then using light or scanning electron microscopy to examine surface details.

The earliest fossils possessing features typical of fungi date to the Proterozoic eon, some 1,430 million years ago (Ma); these multicellular benthic organisms had filamentous structures with septa, and were capable of anastomosis. More recent studies (2009) estimate the arrival of fungal organisms at about 760–1060 Ma on the basis of comparisons of the rate of evolution in closely related groups. For much of the Paleozoic Era (542–251 Ma), the fungi appear to have been aquatic and consisted of organisms similar to the extant chytrids in having flagellum-bearing spores. The evolutionary adaptation from an aquatic to a terrestrial lifestyle necessitated a diversification of ecological strategies for obtaining nutrients, including parasitism, saprobism, and the development of mutualistic relationships such as mycorrhiza and lichenization. Recent (2009) studies suggest that the ancestral ecological state of the Ascomycota was saprobism, and that independent lichenization events have occurred multiple times.

It is presumed that the fungi colonized the land during the Cambrian (542–488.3 Ma), long before land plants. Fossilized hyphae and spores recovered from the Ordovician of Wisconsin (460 Ma) resemble modern-day Glomerales, and existed at a time when the land flora likely consisted of only non-vascular bryophyte-like plants. Prototaxites, which was probably a fungus or lichen, would have been the tallest organism of the late Silurian. Fungal fossils do not become common and uncontroversial until the early Devonian (416–359.2 Ma), when they occur abundantly in the Rhynie chert, mostly as Zygomycota and Chytridiomycota.At about this same time, approximately 400 Ma, the Ascomycota and Basidiomycota diverged, and all modern classes of fungi were present by the Late Carboniferous (Pennsylvanian, 318.1–299 Ma).

Lichen-like fossils have been found in the Doushantuo Formation in southern China dating back to 635–551 Ma. Lichens formed a component of the early terrestrial ecosystems, and the estimated age of the oldest terrestrial lichen fossil is 400 Ma; this date corresponds to the age of the oldest known sporocarp fossil, a *Paleopyrenomycites* species found in the Rhynie Chert. The oldest fossil with microscopic features resembling modern-day basidiomycetes is *Palaeoancistrus*, found permineralized with a fern from the Pennsylvanian. Rare in the fossil record are the Homobasidiomycetes (a taxon roughly equivalent to the mushroom-producing species of the Agaricomycetes). Two amber-preserved specimens provide evidence that the earliest known mushroom-forming fungi (the extinct species *Archaeomarasmius leggetti*) appeared during the late Cretaceous, 90 Ma.

Some time after the Permian–Triassic extinction event (251.4 Ma), a fungal spike (originally thought to be an extraordinary abundance of fungal spores in sediments) formed, suggesting that fungi were the dominant life form at this time, representing nearly 100% of the available fossil record for this period.However, the relative proportion of fungal spores relative to spores formed by algal species is difficult to assess, the spike did not appear worldwide, and in many places it did not fall on the Permian–Triassic boundary.

Taxonomy

Although commonly included in botany curricula and textbooks, fungi are more closely related to animals than to plants and are placed with the animals in the monophyletic group of opisthokonts. Analyses using molecular phylogenetics support a monophyletic origin of the Fungi. The taxonomy of the Fungi is in a state of constant flux, especially due to recent research based on DNA comparisons. These current phylogenetic analyses often overturn classifications based on older and sometimes less discriminative methods based on morphological features and biological species concepts obtained from experimental matings.

There is no unique generally accepted system at the higher taxonomic levels and there are frequent name changes at every level, from species upwards. Efforts among researchers are now underway to establish and encourage usage of a unified and more consistent nomenclature. Fungal species can also have multiple scientific names depending on their life cycle and mode (sexual or asexual) of reproduction. Web sites such as Index Fungorum and ITIS list current names of fungal species (with cross-references to older synonyms).

The 2007 classification of Kingdom Fungi is the result of a large-scale collaborative research effort involving dozens of mycologists and other scientists working on fungal taxonomy. It recognizes seven phyla, two of which—the Ascomycota and the Basidiomycota—are contained within a branch representing subkingdom Dikarya, the most species rich and familiar group, including all the mushrooms, most food-spoilage molds, most plant pathogenic fungi, and the beer, wine, and bread yeasts. The accompanying cladogram depicts the major fungal taxa and their relationship to opisthokont and unikont organisms, based on the work of Philippe Silar and "The Mycota: A Comprehensive Treatise on Fungi as Experimental Systems for Basic and Applied Research". The lengths of the branches are not proportional to evolutionary distances.

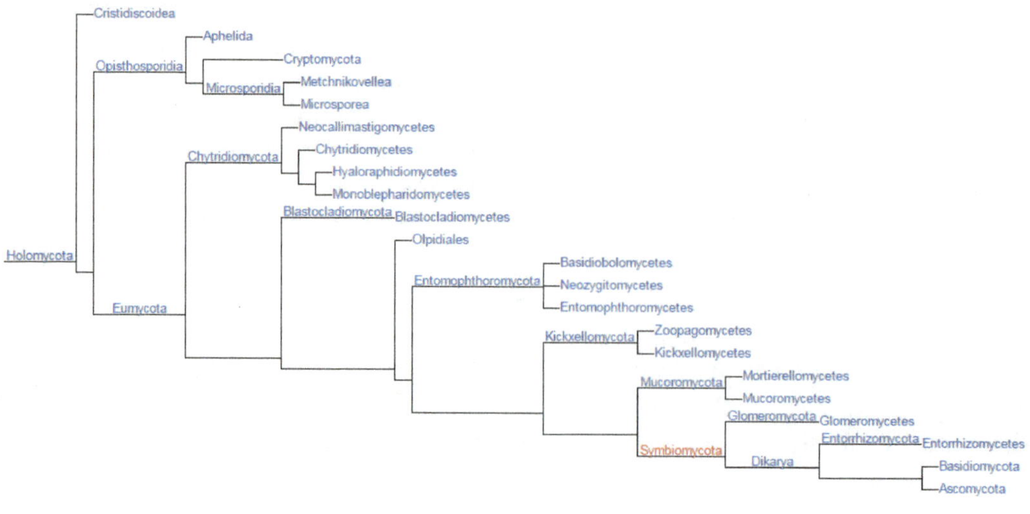

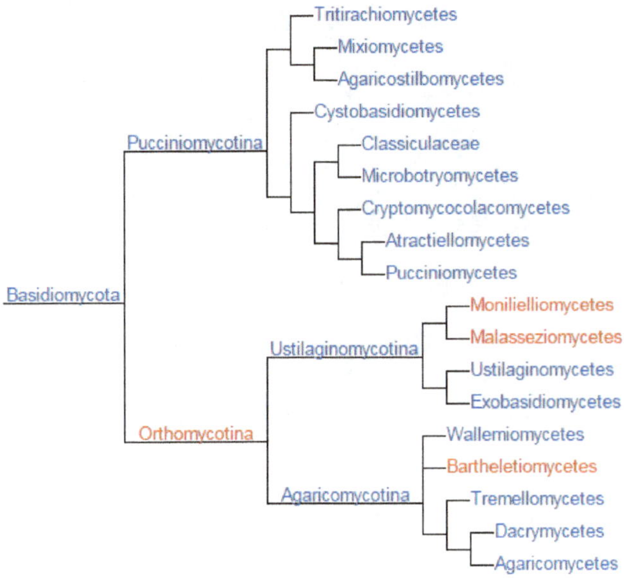

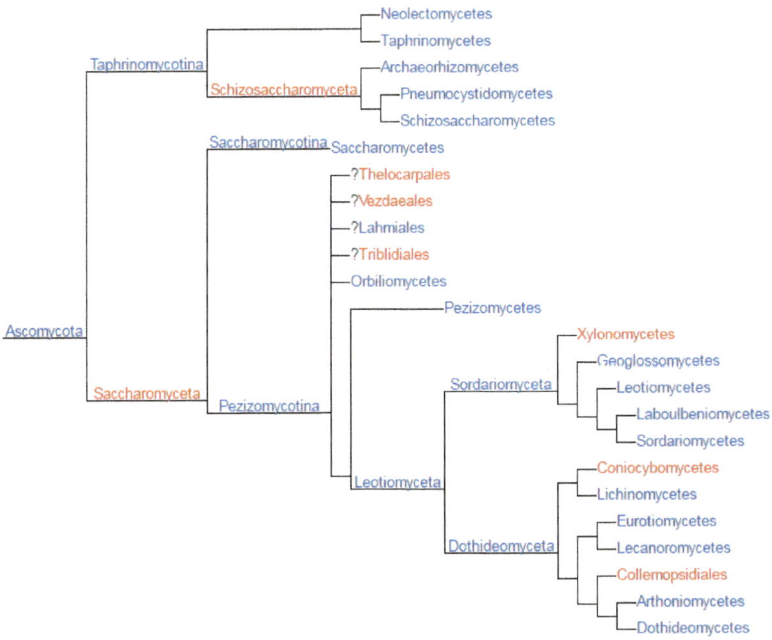

Taxonomic Groups

The major phyla (sometimes called divisions) of fungi have been classified mainly on the basis of characteristics of their sexual reproductive structures. Currently, seven phyla are proposed: Microsporidia, Chytridiomycota, Blastocladiomycota, Neocallimastigomycota, Glomeromycota, Ascomycota, and Basidiomycota.

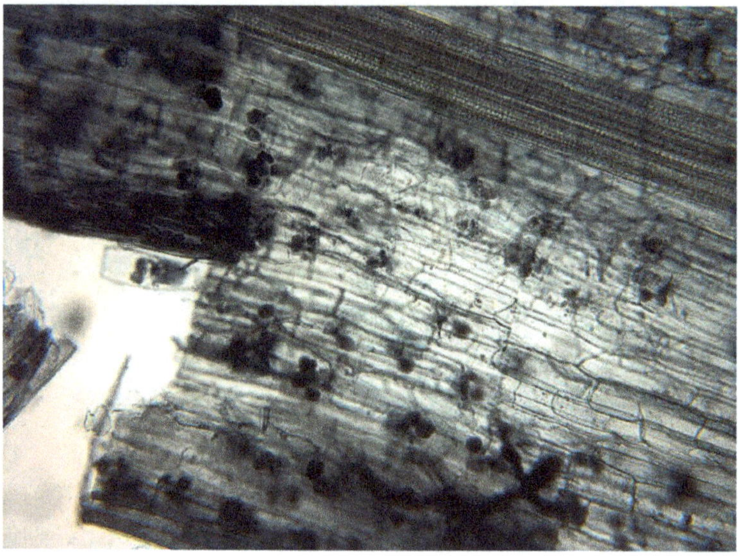

Arbuscular mycorrhiza seen under microscope. Flax root cortical cells containing paired arbuscules.

Phylogenetic analysis has demonstrated that the Microsporidia, unicellular parasites of animals and protists, are fairly recent and highly derived endobiotic fungi (living within the tissue of another species). One 2006 study concludes that the Microsporidia are a sister group to the true fungi; that is, they are each other›s closest evolutionary relative. Hibbett and colleagues suggest that this analysis does not clash with their classification of the Fungi, and although the Microsporidia are elevated to phylum status, it is acknowledged that further analysis is required to clarify evolutionary relationships within this group.

The Chytridiomycota are commonly known as chytrids. These fungi are distributed worldwide. Chytrids and their close relatives Neocallimastigomycota and Blastocladiomycota (below) are the only fungi with active motility, producing zoospores that are capable of active movement through aqueous phases with a single flagellum, leading early taxonomists to classify them as protists. Molecular phylogenies, inferred from rRNA sequences in ribosomes, suggest that the Chytrids are a basal group divergent from the other fungal phyla, consisting of four major clades with suggestive evidence for paraphyly or possibly polyphyly.

The Blastocladiomycota were previously considered a taxonomic clade within the Chytridiomycota. Recent molecular data and ultrastructural characteristics, however, place the Blastocladiomycota as a sister clade to the Zygomycota, Glomeromycota, and Dikarya (Ascomycota and Basidiomycota). The blastocladiomycetes are saprotrophs, feeding on decomposing organic matter, and they are parasites of all eukaryotic groups. Unlike their close relatives, the chytrids, most of which exhibit zygotic meiosis, the blastocladiomycetes undergo sporic meiosis.

The Neocallimastigomycota were earlier placed in the phylum Chytridomycota. Members of this small phylum are anaerobic organisms, living in the digestive system of larger herbivorous mammals and in other terrestrial and aquatic environments enriched in cellulose (e.g., domestic waste landfill sites). They lack mitochondria but contain hydrogenosomes of mitochondrial origin. As in the related chrytrids, neocallimastigomycetes form zoospores that are posteriorly uniflagellate or polyflagellate.

Members of the Glomeromycota form arbuscular mycorrhizae, a form of mutualist symbiosis wherein fungal hyphae invade plant root cells and both species benefit from the resulting increased supply of nutrients. All known Glomeromycota species reproduce asexually. The symbiotic association between the Glomeromycota and plants is ancient, with evidence dating to 400 million years ago. Formerly part of the Zygomycota (commonly known as `sugar› and `pin› molds), the Glomeromycota were elevated to phylum status in 2001 and now replace the older phylum Zygomycota. Fungi that were placed in the Zygomycota are now being reassigned to the Glomeromycota, or the subphyla incertae sedis Mucoromycotina, Kickxellomycotina, the Zoopagomycotina and the Entomophthoromycotina. Some well-known examples of fungi formerly in the Zygomycota include black bread mold (*Rhizopus stolonifer*), and *Pilobolus* species, capable of ejecting spores several meters through the air. Medically relevant genera include *Mucor*, *Rhizomucor*, and *Rhizopus*.

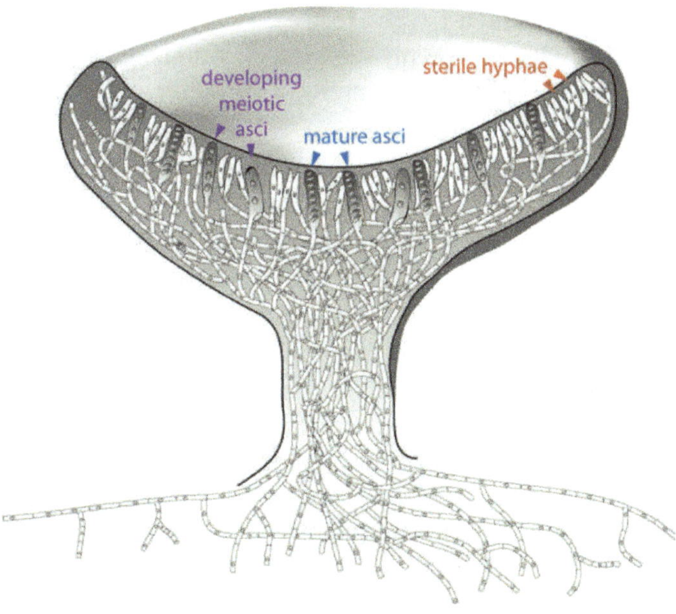

Diagram of an apothecium (the typical cup-like reproductive structure of Ascomycetes) showing sterile tissues as well as developing and mature asci.

The Ascomycota, commonly known as sac fungi or ascomycetes, constitute the largest taxonomic group within the Eumycota. These fungi form meiotic spores called ascospores, which are enclosed in a special sac-like structure called an ascus. This phylum includes morels, a few mushrooms and truffles, unicellular yeasts (e.g., of the genera *Saccharomyces*, *Kluyveromyces*, *Pichia*, and *Candida*), and many filamentous fungi living as saprotrophs, parasites, and mutualistic symbionts (e.g. lichens). Prominent and important genera of filamentous ascomycetes include *Aspergillus*, *Penicillium*, *Fusarium*, and *Claviceps*. Many ascomycete species have only been observed undergoing asexual reproduction (called anamorphic species), but analysis of molecular data has often been able to identify their closest teleomorphs in the Ascomycota. Because the products of meiosis are retained within the sac-like ascus, ascomycetes have been used for elucidating principles of genetics and heredity (e.g., *Neurospora crassa*).

Members of the Basidiomycota, commonly known as the club fungi or basidiomycetes, produce meiospores called basidiospores on club-like stalks called basidia. Most common mushrooms belong to this group, as well as rust and smut fungi, which are major pathogens of grains. Other important basidiomycetes include the maize pathogen *Ustilago maydis*, human commensal species of the genus *Malassezia*, and the opportunistic human pathogen, *Cryptococcus neoformans*.

Fungus-Like Organisms

Because of similarities in morphology and lifestyle, the slime molds (mycetozoans, plasmodiophorids, acrasids, *Fonticula* and labyrinthulids, now in Amoebozoa, Rhizaria, Excavata, Opisthokonta and Stramenopiles, respectively), water molds (oomycetes) and hyphochytrids (both Stramenopiles) were formerly classified in the kingdom Fungi, in groups like Mastigomycotina, Gymnomycota and Phycomycetes. The slime molds were studied also as protozoans, leading to a ambiregnal, duplicated taxonomy.

Unlike true fungi, the cell walls of oomycetes contain cellulose and lack chitin. Hyphochytrids have both chitin and cellulose. Slime molds lack a cell wall during the assimilative phase (except labyrinthulids, which have a wall of scales), and ingest nutrients by ingestion (phagocytosis, except labyrinthulids) rather than absorption (osmotrophy, as fungi, labyrinthulids, oomycetes and hyphochytrids). Neither water molds nor slime molds are closely related to the true fungi, and, therefore, taxonomists no longer group them in the kingdom Fungi. Nonetheless, studies of the oomycetes and myxomycetes are still often included in mycology textbooks and primary research literature.

The Eccrinales and Amoebidiales are opisthokont protists, previously thought to be zygomycete fungi. Other groups now in Opisthokonta (e.g., *Corallochytrium*, Ichthyosporea) were also at given time classified as fungi. The genus *Blastocystis*, now in Stramenopiles, was originally classified as a yeast. *Ellobiopsis*, now in Alveolata, was considered a chytrid. The bacteria were also included in fungi in some classifications, as the group Schizomycetes.

The Rozellida clade, including the "ex-chytrid" *Rozella*, is a genetically disparate group known mostly from environmental DNA sequences that is a sister group to fungi. Members of the group that have been isolated lack the chitinous cell wall that is characteristic of fungi.

The nucleariids, protists currently grouped in the Choanozoa (Opisthokonta), may be the next sister group to the eumycete clade, and as such could be included in an expanded fungal kingdom.

Many Actinomycetales (Actinobacteria), a group with many filamentous bacteria, were also long believed to be fungi.

Ecology

A pin mold decomposing a peach

Although often inconspicuous, fungi occur in every environment on Earth and play very important roles in most ecosystems. Along with bacteria, fungi are the major decomposers in most terrestrial (and some aquatic) ecosystems, and therefore play a critical role in biogeochemical cyclesand in many food webs. As decomposers, they play an essential role in nutrient cycling, especially as saprotrophs and symbionts, degrading organic matter to inorganic molecules, which can then re-enter anabolic metabolic pathways in plants or other organisms.

Symbiosis

Many fungi have important symbiotic relationships with organisms from most if not all Kingdoms. These interactions can be mutualistic or antagonistic in nature, or in the case of commensal fungi are of no apparent benefit or detriment to the host.

With Plants

Mycorrhizal symbiosis between plants and fungi is one of the most well-known plant–fungus associations and is of significant importance for plant growth and persistence in many ecosystems; over 90% of all plant species engage in mycorrhizal relationships with fungi and are dependent upon this relationship for survival.

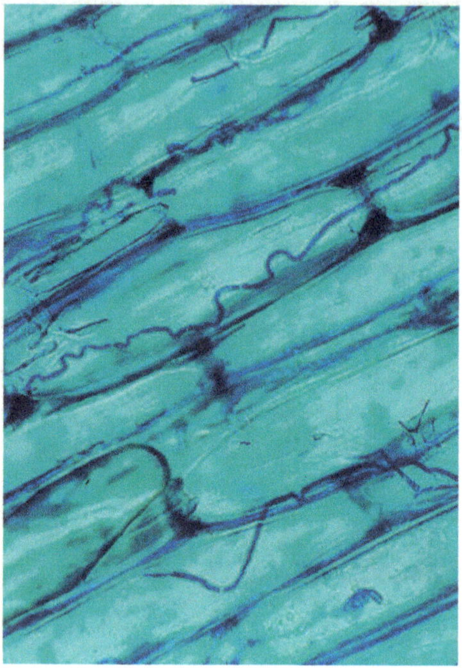

The dark filaments are hyphae of the endophytic fungus *Neotyphodium coenophialum* in the intercellular spaces of tall fescue leaf sheath tissue

The mycorrhizal symbiosis is ancient, dating to at least 400 million years ago. It often increases the plant›s uptake of inorganic compounds, such as nitrate and phosphate from soils having low concentrations of these key plant nutrients. The fungal partners may also mediate plant-to-plant transfer of carbohydrates and other nutrients. Such mycorrhizal communities are called "common mycorrhizal networks". A special case of mycorrhiza is myco-heterotrophy, whereby the plant parasitizes the fungus, obtaining all of its nutrients from its fungal symbiont. Some fungal species inhabit the tissues inside roots, stems, and leaves, in which case they are called endophytes. Similar to mycorrhiza, endophytic colonization by fungi may benefit both symbionts; for example, endophytes of grasses impart to their host increased resistance to herbivores and other environmental stresses and receive food and shelter from the plant in return.

With Algae and Cyanobacteria

Lichens are a symbiotic relationship between fungi and photosynthetic algae or cyanobacteria. The photosynthetic partner in the relationship is referred to in lichen terminology as a "photobiont". The fungal part of the relationship is composed mostly of various species of ascomycetes and a few basidiomycetes. Lichens occur in every ecosystem on all continents, play a key role in soil formation and the initiation of biological succession, and are prominent in some extreme environments, including polar, alpine, and semiarid desert regions. They are able to grow on inhospitable surfaces, including bare soil, rocks, tree bark, wood, shells, barnacles and leaves. As in mycorrhizas, the photobiont provides sugars and other carbohydrates via photosynthesis to the fungus, while the fungus provides minerals and water to the photobiont. The functions of both symbiotic organisms are so closely intertwined that they function almost as a single organism; in most cases the resulting organism differs greatly from the individual components. Lichenization is a common mode of nutrition for fungi; around 20% of fungi—between 17,500 and 20,000 described species—are lichenized. Characteristics common to most lichens include obtaining organic carbon by photosynthesis, slow growth, small size, long life, long-lasting (seasonal) vegetative reproductive structures, mineral nutrition obtained largely from airborne sources, and greater tolerance of desiccation than most other photosynthetic organisms in the same habitat.

The lichen *Lobaria pulmonaria*, a symbiosis of fungal, algal, and cyanobacterial species

With Insects

Many insects also engage in mutualistic relationships with fungi. Several groups of ants cultivate fungi in the order Agaricales as their primary food source, while ambrosia beetles cultivate various species of fungi in the bark of trees that they infest. Likewise, females of several wood wasp species (genus *Sirex*) inject their eggs together with spores of the wood-rotting fungus *Amylostereum areolatum* into the sapwood of pine trees; the growth of the fungus provides ideal nutritional conditions for the development of the wasp larvae. At least one species of stingless bee has a relationship with a fungus in the genus *Monascus*, where the larvae consume and depend on fungus transferred from old to new nests. Termites on the African savannah are also known to cultivate fungi, and yeasts of the genera *Candida* and *Lachancea* inhabit the gut of a wide range of insects, including neuropterans, beetles, and cockroaches; it is not known whether these fungi benefit their hosts. Fungi ingrowing dead wood are essential for xylophagous insects (e.g. woodboring beetles). They deliver nutrients needed by xylophages to nutritionally scarce dead wood. Thanks to this nutritional enrichment the larvae of woodboring insect is able to grow and develop to adulthood. The larvae of many families of fungicolous flies, particularly those within the superfamily Sciaroidea such as the Mycetophilidae and some Keroplatidae feed on fungal fruiting bodies and sterile mycorrhizae.

As Pathogens and Parasites

The plant pathogen *Aecidium magellanicum* causes calafate rust, seen here on a *Berberis* shrub in Chile.

Many fungi are parasites on plants, animals (including humans), and other fungi. Serious pathogens of many cultivated plants causing extensive damage and losses to agriculture and forestry include the rice blast fungus *Magnaporthe oryzae*, tree pathogens such as *Ophiostoma ulmi* and *Ophiostoma novo-ulmi* causing Dutch elm disease and *Cryphonectria parasitica* responsible for chestnut blight, and plant pathogens in the genera *Fusarium*, *Ustilago*, *Alternaria*, and *Cochliobolus*. Some carnivorous fungi, like *Paecilomyces lilacinus*, are predators of nematodes, which they capture using an array of specialized structures such as constricting rings or adhesive nets.

Some fungi can cause serious diseases in humans, several of which may be fatal if untreated. These include aspergillosis, candidiasis, coccidioidomycosis, cryptococcosis, histoplasmosis, mycetomas, and paracoccidioidomycosis. Furthermore, persons with immuno-deficiencies are particularly susceptible to disease by genera such as *Aspergillus*, *Candida*, *Cryptoccocus*, *Histoplasma*, and *Pneumocystis*.Other fungi can attack eyes, nails, hair, and especially skin, the so-called dermatophytic and keratinophilic fungi, and cause local infections such as ringworm and athlete's foot. Fungal spores are also a cause of allergies, and fungi from different taxonomic groups can evoke allergic reactions.

As Targets of Mycoparasites

The organisms which parasitize fungi are known as mycoparasitic organisms. Certain species of the *Pythium* genus, which are oomycetes, have potential as biocontrol agents against certain fungi. Fungi can also act as mycoparasites or antagonists of other fungi, such as *Hypomyces chrysospermus*, which grows on bolete mushrooms.

Fungi can become the target of infection by mycoviruses.

Mycotoxins

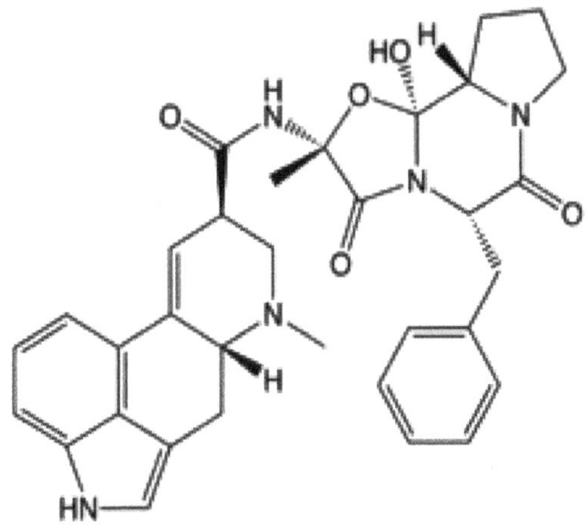

Ergotamine, a major mycotoxin produced by *Claviceps* species, which if ingested can cause gangrene, convulsions, and hallucinations

Many fungi produce biologically active compounds, several of which are toxic to animals or plants and are therefore called mycotoxins. Of particular relevance to humans are mycotoxins produced by molds causing food spoilage, and poisonous mushrooms (see above). Particularly infamous are the lethal amatoxins in some *Amanita* mushrooms, and ergot alkaloids, which have a long history of causing serious epidemics of ergotism (St Anthony's Fire) in people consuming rye or related cereals contaminated with sclerotia of the ergot fungus, *Claviceps purpurea*. Other notable mycotoxins include the aflatoxins, which are insidious liver toxins and highly carcinogenic metabolites produced by certain *Aspergillus* species often growing in or on grains and nuts consumed by humans, ochratoxins, patulin, and trichothecenes (e.g., T-2 mycotoxin) and fumonisins, which have significant impact on human food supplies or animal livestock.

Mycotoxins are secondary metabolites (or natural products), and research has established the existence of biochemical pathways solely for the purpose of producing mycotoxins and other natural products in fungi. Mycotoxins may provide fitness benefits in terms of physiological adaptation, competition with other microbes and fungi, and protection from consumption (fungivory). Many fungal secondary metabolites (or derivatives) are used medically, as described under Human Use below.

Pathogenic Mechanisms

Ustilago maydis is a pathogenic plant fungus that causes smut disease in maize and teosinte. Plants have evolved efficient defense systems against pathogenic microbes such as *U. maydis*. A rapid defense reaction after pathogen attack is the oxidative burst where the plant produces reactive oxygen species at the site of the attempted invasion. *U. maydis* can respond to the oxidative burst with an oxidative stress response, regulated by the gene *YAP1*. The response protects *U. maydis* from the host defense, and is necessary for the pathogen's virulence. Furthermore, *U. maydis* has a well-established recombinational DNA repair system which acts during mitosis and meiosis. The system may assist the pathogen in surviving DNA damage arising from the host plant's oxidative defensive response to infection.

Cryptococcus neoformans is an encapsulated yeast that can live in both plants and animals. *C. neoformans* usually infects the lungs, where it is phagocytosed by alveolar macrophages. Some *C. neoformans* can survive inside macrophages, which appears to be the basis for latency, disseminated disease, and resistance to antifungal agents. One mechanism by which *C. neoformans* survives the hostile macrophage environment is by up-regulating the expression of genes involved in the oxidative stress response. Another mechanism involves meiosis. The majority of *C. neoformans* are mating "type a". Filaments of mating "type a" ordinarily have haploid nuclei, but they can become diploid (perhaps by endoduplication or by stimulated nuclear fusion) to form blastospores. The diploid nuclei of blastospores can undergo meiosis, including recombination, to form haploid basidiospores that can be dispersed. This process is referred to as monokaryotic fruiting. this process requires a gene called *DMC1*, which is a conserved homologue of genes *recA* in bacteria and *RAD51* in eukaryotes, that mediates homologous chromosome pairing during meiosis and repair of DNA double-strand breaks. Thus, *C. neoformans* can undergo a meiosis, monokaryotic fruiting, that promotes recombinational repair in the oxidative, DNA damaging environment of the host macrophage, and the repair capability may contribute to its virulence.

Human Use

The human use of fungi for food preparation or preservation and other purposes is extensive and has a long history. Mushroom farming and mushroom gathering are large industries in many countries. The study of the historical uses and sociological impact of fungi is known as ethnomycology. Because of the capacity of this group to produce an enormous range of natural products with antimicrobial or other biological activities, many species have long been used or are being developed for industrial production of antibiotics, vitamins, and anti-cancer and cholesterol-lowering drugs. More recently, methods have been developed for genetic engineering of fungi, enabling metabolic engineering of fungal species. For example, genetic modification of yeast species—which are easy to grow at fast rates in large fermentation vessels—has opened up ways of pharmaceutical production that are potentially more efficient than production by the original source organisms.

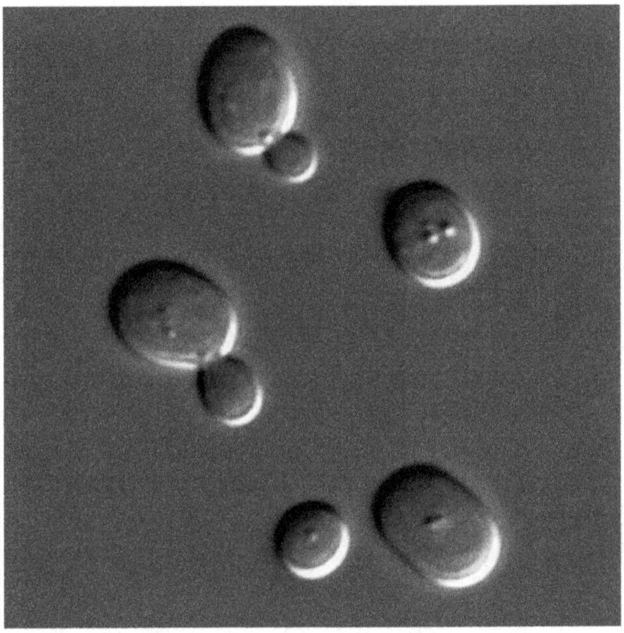

Saccharomyces cerevisiae cells shown with DIC microscopy

Therapeutic Uses

Modern Chemotherapeutics

Many species produce metabolites that are major sources of pharmacologically active drugs. Particularly important are the antibiotics, including the penicillins, a structurally related group of β-lactam antibiotics that are synthesized from small peptides. Although naturally occurring penicillins such as penicillin G (produced by *Penicillium chrysogenum*) have a relatively narrow spectrum of biological activity, a wide range of other penicillins can be produced by chemical modification of the natural penicillins. Modern penicillins are semisynthetic compounds, obtained initially from fermentation cultures, but then structurally altered for specific desirable properties. Other antibiotics produced by fungi include: ciclosporin, commonly used as an immunosuppressant during transplant surgery; and fusidic acid, used to help control infection from methicillin-resistant *Staphylococcus aureus* bacteria. Widespread use of antibiotics for the treatment of bacterial diseases, such as tuberculosis, syphilis, leprosy, and others began in the early 20th century and continues to date. In nature, antibiotics of fungal or bacterial origin appear to play a dual role: at high concentrations they act as chemical defense against competition with other microorganisms in species-rich environments, such as the rhizosphere, and at low concentrations as quorum-sensing molecules for intra- or interspecies signaling. Other drugs produced by fungi include griseofulvin isolated from *Penicillium griseofulvum*, used to treat fungal infections, and statins (HMG-CoA reductase inhibitors), used to inhibit cholesterol synthesis. Examples of statins found in fungi include mevastatin from *Penicillium citrinum* and lovastatin from *Aspergillus terreus* and the oyster mushroom.

Traditional and Folk Medicine

The medicinal fungi *Ganoderma lucidum* (left) and *Ophiocordyceps sinensis* (right)

Certain mushrooms enjoy usage as therapeutics in folk medicines, such as Traditional Chinese medicine. Notable medicinal mushrooms with a well-documented history of use include *Agaricus subrufescens*, *Ganoderma lucidum*, and *Ophiocordyceps sinensis*. Research has identified compounds produced by these and other fungi that have inhibitory biological effects against viruses and cancer cells.Specific metabolites, such as polysaccharide-K, ergotamine, and β-lactam antibiotics, are routinely used in clinical medicine. The shiitake mushroom is a source of lentinan, a clinical drug approved for use in cancer treatments in several countries, including Japan. In Europe and Japan, polysaccharide-K (brand name Krestin), a chemical derived from *Trametes versicolor*, is an approved adjuvant for cancer therapy.

Cultured Foods

Baker's yeast or *Saccharomyces cerevisiae*, a unicellular fungus, is used to make bread and other wheat-based products, such as pizza dough and dumplings. Yeast species of the genus *Saccharomyces* are also used to produce alcoholic beverages through fermentation. Shoyu koji mold (*Aspergillus oryzae*) is an essential ingredient in brewing Shoyu (soy sauce) and sake, and the preparation of miso, while *Rhizopus* species are used for making tempeh. Several of these fungi are domesticated species that were bred or selected according to their capacity to ferment food without producing harmful mycotoxins (see below), which are produced by very closely related *Aspergilli*. Quorn, a meat substitute, is made from *Fusarium venenatum*.

Edible and Poisonous species

Amanita phalloides accounts for the majority of fatal mushroom poisonings worldwide.

Edible mushrooms are well-known examples of fungi. Many are commercially raised, but others must be harvested from the wild. *Agaricus bisporus*, sold as button mushrooms when small or Portobello mushrooms when larger, is a commonly eaten species, used in salads, soups, and many other dishes. Many Asian fungi are commercially grown and have increased in popularity in the West. They are often available fresh in grocery stores and markets, including straw mushrooms (*Volvariella volvacea*), oyster mushrooms (*Pleurotus ostreatus*), shiitakes (*Lentinula edodes*), and enokitake (*Flammulina* spp.).

There are many more mushroom species that are harvested from the wild for personal consumption or commercial sale. Milk mushrooms, morels, chanterelles, truffles, black trumpets, and *porcini* mushrooms (*Boletus edulis*) (also known as king boletes) demand a high price on the market. They are often used in gourmet dishes.

Certain types of cheeses require inoculation of milk curds with fungal species that impart a unique flavor and texture to the cheese. Examples include the blue color in cheeses such as Stilton or Roquefort, which are made by inoculation with *Penicillium roqueforti*. Molds used in cheese production are non-toxic and are thus safe for human consumption; however, mycotoxins (e.g., aflatoxins, roquefortine C, patulin, or others) may accumulate because of growth of other fungi during cheese ripening or storage.

Stilton cheese veined with *Penicillium roqueforti*

Many mushroom species are poisonous to humans, with toxicities ranging from slight digestive problems or allergic reactions as well as hallucinations to severe organ failures and death. Genera with mushrooms containing deadly toxins include *Conocybe*, *Galerina*, *Lepiota*, and, the most infamous, *Amanita*. The latter genus includes the destroying angel *(A. virosa)* and the death cap *(A. phalloides)*, the most common cause of deadly mushroom poisoning. The false morel (*Gyromitra esculenta*) is occasionally considered a delicacy when cooked, yet can be highly toxic when eaten raw. *Tricholoma equestre* was considered edible until it was implicated in serious poisonings causing rhabdomyolysis. Fly agaric mushrooms (*Amanita muscaria*) also cause occasional non-fatal poisonings, mostly as a result of ingestion for its hallucinogenic properties. Historically, fly agaric was used by different peoples in Europe and Asia and its present usage for religious or shamanic purposes is reported from some ethnic groups such as the Koryak people of north-eastern Siberia.

As it is difficult to accurately identify a safe mushroom without proper training and knowledge, it is often advised to assume that a wild mushroom is poisonous and not to consume it.

Pest Control

In agriculture, fungi may be useful if they actively compete for nutrients and space with pathogenic microorganisms such as bacteria or other fungi via the competitive exclusion principle, or if they are parasites of these pathogens. For example, certain species may be used to eliminate or suppress the growth of harmful plant pathogens, such as insects, mites, weeds, nematodes, and other fungi that cause diseases of important crop plants. This has generated strong interest in practical applications that use these fungi in the biological control of these agricultural pests. Entomopathogenic fungi can be used as biopesticides, as they actively kill insects. Examples that have been used as biological insecticides are *Beauveria bassiana*, *Metarhizium* spp, *Hirsutella* spp, *Paecilomyces* (*Isaria*) spp, and *Lecanicillium lecanii*. Endophytic fungi of grasses of the genus *Neotyphodium*, such as *N. coenophialum*, produce alkaloids that are toxic to a range of invertebrate and vertebrate herbivores. These alkaloids protect grass plants from herbivory, but several endophyte alkaloids can poison grazing animals, such as cattle and sheep. Infecting cultivars of pasture or forage grasses with *Neotyphodium* endophytes is one approach being used in grass breeding programs; the fungal strains are selected for producing only alkaloids that increase resistance to herbivores such as insects, while being non-toxic to livestock.

Grasshoppers killed by *Beauveria bassiana*

Bioremediation

Certain fungi, in particular "white rot" fungi, can degrade insecticides, herbicides, pentachlorophenol, creosote, coal tars, and heavy fuels and turn them into carbon dioxide, water, and basic elements. Fungi have been shown to biomineralize uranium oxides, suggesting they may have application in the bioremediation of radioactively polluted sites.

Model Organisms

Several pivotal discoveries in biology were made by researchers using fungi as model organisms, that is, fungi that grow and sexually reproduce rapidly in the laboratory. For example, the one gene-one enzyme hypothesis was formulated by scientists using the bread mold *Neurospora crassa* to test their biochemical theories. Other important model fungi are *Aspergillus nidulans* and the yeasts *Saccaromyces cerevisiae* and *Schizosaccharomyces pombe*, each of which with a long history of use to investigate issues in eukaryotic cell biology and genetics, such as cell cycle regulation, chromatin structure, and gene regulation. Other fungal models have more recently emerged that address specific biological questions relevant to medicine, plant pathology, and industrial uses; examples include *Candida albicans*, a dimorphic, opportunistic human pathogen, *Magnaporthe grisea*, a plant pathogen, and *Pichia pastoris*, a yeast widely used for eukaryotic protein production.

Others

Fungi are used extensively to produce industrial chemicals like citric, gluconic, lactic, and malic acids, and industrial enzymes, such as lipases used in biological detergents,cellulases used in making cellulosic ethanol and stonewashed jeans, and amylases, invertases, proteases and xylanases. Several species, most notably *Psilocybin mushrooms* (colloquially known as *magic mushrooms*), are ingested for their psychedelic properties, both recreationally and religiously.

Habitats and Ecology

Microorganisms are found in almost every habitat present in nature. Even in hostile environments such as the poles, deserts, geysers, rocks, and the deep sea. Some types of microorganisms have adapted to the extreme conditions and sustained colonies; these organisms are known as extremophiles. Extremophiles have been isolated from rocks as much as 7 kilometres below the Earth's surface, and it has been suggested that the amount of living organisms below the Earth's surface is comparable with the amount of life on or above the surface. Extremophiles have been known to survive for a prolonged time in a vacuum, and can be highly resistant to radiation, which may even allow them to survive in space. Many types of microorganisms have intimate symbiotic relationships with other larger organisms; some of which are mutually beneficial (mutualism), while others can be damaging to the host organism (parasitism). If microorganisms can cause disease in a host they are known as pathogens and then they are sometimes referred to as microbes.

Extremophiles

Extremophiles are microorganisms that have adapted so that they can survive and even thrive in conditions that are normally fatal to most life-forms. For example, some species have been found in the following extreme environments:

- Temperature: as high as 130 °C (266 °F), as low as −17 °C (1 °F)
- Acidity/alkalinity: less than pH 0, up to pH 11.5
- Salinity: up to saturation
- Pressure: up to 1,000-2,000 atm, down to 0 atm (e.g. vacuum of space)
- Radiation: up to 5kGy

Extremophiles are significant in different ways. They extend terrestrial life into much of the Earth's hydrosphere, crust and atmosphere, their specific evolutionary adaptation mechanisms to their extreme environment can be exploited in bio-technology, and their very existence under such extreme conditions increases the potential for extraterrestrial life.

Soil Microorganisms

The nitrogen cycle in soils depends on the fixation of atmospheric nitrogen. One way this can occur is in the nodules in the roots of legumes that contain symbiotic bacteria of the genera *Rhizobium*, *Mesorhizobium*, *Sinorhizobium*, *Bradyrhizobium*, and *Azorhizobium*.

Symbiotic Microorganisms

Symbiotic microorganisms such as fungi and algae form an association in lichen. Certain fungi form mycorrhizal symbioses with trees that increase the supply of nutrients to the tree.

Applications

Microorganisms are vital to humans and the environment, as they participate in the carbon and nitrogen cycles, as well as fulfilling other vital roles in virtually all ecosystems, such as recycling other organisms' dead remains and waste products through decomposition. Microorganisms also have an important place in most higher-order multicellular organisms as symbionts. Many blame the failure of Biosphere 2 on an improper balance of microorganisms.

Food Production

Microorganisms are used to make yoghurt, cheese and curd. They are used to leaven bread, and to convert sugars to alcohol in wine and beer.

Soil

Microbes can make nutrients and minerals in the soil available to plants, produce hormones that spur growth, stimulate the plant immune system and trigger or dampen stress responses. In general a more diverse soil microbiome results in fewer plant diseases and higher yield.

Fermentation

Microorganisms are used in brewing, wine making, baking, pickling and other food-making processes.

They are also used to control the fermentation process in the production of cultured dairy products such as yogurt and cheese. The cultures also provide flavor and aroma, and inhibit undesirable organisms.

Hygiene

Hygiene is the avoidance of infection or food spoiling by eliminating microorganisms from the surroundings. As microorganisms, in particular bacteria, are found virtually everywhere, the levels of harmful microorganisms can be reduced to acceptable levels. However, in some cases, it is required that an object or substance be completely sterile, i.e. devoid of all living entities and viruses. A good example of this is a hypodermic needle.

In food preparation microorganisms are reduced by preservation methods (such as the addition of vinegar), clean utensils used in preparation, short storage periods, or by cool temperatures. If complete sterility is needed, the two most common methods are irradiation and the use of an autoclave, which resembles a pressure cooker.

There are several methods for investigating the level of hygiene in a sample of food, drinking water, equipment, etc. Water samples can be filtrated through an extremely fine filter. This filter is then placed in a nutrient medium. Microorganisms on the filter then grow to form a visible colony. Harmful microorganisms can be detected in food by placing a sample in a nutrient broth designed to enrich the organisms in question. Various methods, such as selective media or polymerase chain reaction, can then be used for detection. The hygiene of hard surfaces, such as cooking pots, can be tested by touching them with a solid piece of nutrient medium and then allowing the microorganisms to grow on it.

There are no conditions where all microorganisms would grow, and therefore often several methods are needed. For example, a food sample might be analyzed on three different nutrient mediums designed to indicate the presence of "total" bacteria (conditions where many, but not all, bacteria grow), molds (conditions where the growth of bacteria is prevented by, e.g., antibiotics) and coliform bacteria (these indicate a sewage contamination).

Water Treatment

The majority of all oxidative sewage treatment processes rely on a large range of microorganisms to oxidise organic constituents which are not amenable to sedimentation or flotation. Anaerobic microorganisms are also used to reduce sludge solids producing methane gas (amongst other gases) and a sterile mineralised residue. In potable water treatment, one method, the slow sand filter, employs a complex gelatinous layer composed of a wide range of microorganisms to remove both dissolved and particulate material from raw water.

Energy

Microorganisms are used in fermentation to produce ethanol, and in biogas reactors to produce methane. Scientists are researching the use of algae to produce liquid fuels, and bacteria to convert various forms of agricultural and urban waste into usable fuels.

Chemicals, Enzymes

Microorganisms are used for many commercial and industrial production of chemicals, enzymes and other bioactive molecules. Examples of organic acid produced include

- **Acetic acid**: Produced by the bacterium *Acetobacter aceti* and other acetic acid bacteria (AAB)
- **Butyric acid** (butanoic acid): Produced by the bacterium *Clostridium butyricum*

- **Lactic acid**: *Lactobacillus* and others commonly called as lactic acid bacteria (LAB)
- **Citric acid**: Produced by the fungus *Aspergillus niger*

Microorganisms are used for preparation of bioactive molecules and enzymes.

- Streptokinase produced by the bacterium *Streptococcus* and modified by genetic engineering is used as a clot buster for removing clots from the blood vessels of patients who have undergone myocardial infarctions leading to heart attack.
- Cyclosporin A is a bioactive molecule used as an immunosuppressive agent in organ transplantation
- Statins produced by the yeast *Monascus purpureus* are commercialised as blood cholesterol lowering agents which act by competitively inhibiting the enzyme responsible for synthesis of cholesterol.

Science

Microorganisms are essential tools in biotechnology, biochemistry, genetics, and molecular biology. The yeasts (*Saccharomyces cerevisiae*) and fission yeast (*Schizosaccharomyces pombe*) are important model organisms in science, since they are simple eukaryotes that can be grown rapidly in large numbers and are easily manipulated. They are particularly valuable in genetics, genomics and proteomics. Microorganisms can be harnessed for uses such as creating steroids and treating skin diseases. Scientists are also considering using microorganisms for living fuel cells, and as a solution for pollution.

Warfare

In the Middle Ages, diseased corpses were thrown into castles during sieges using catapults or other siege engines. Individuals near the corpses were exposed to the pathogen and were likely to spread that pathogen to others.

Human Health

Human Digestion

Microorganisms can form an endosymbiotic relationship with other, larger organisms. For example, the bacteria that live within the human digestive system contribute to gut immunity, synthesise vitamins such as folic acid and biotin, and ferment complex indigestible carbohydrates.

Disease

Microorganisms are the causative agents (pathogens) in many infectious diseases. The organisms involved include pathogenic bacteria, causing diseases such as plague, tuberculosis and anthrax; protozoa, causing diseases such as malaria, sleeping sickness, dysentery and toxoplasmosis; and also fungi causing diseases such as ringworm, candidiasis or histoplasmosis. However, other diseases such as influenza, yellow fever or AIDS are caused by pathogenic viruses, which are not usually classified as living organisms and are not, therefore, microorganisms by the strict definition. No clear examples of archaean pathogens are known, although a relationship has been proposed between the presence of some archaean methanogens and human periodontal disease.

Ecology

Microorganisms play critical roles in Earth's biogeochemical cycles as they are responsible for decomposition and play vital parts in carbon and nitrogen fixation, as well as oxygen production.

Etymology and Pronunciation

The word *microorganism* uses combining forms of *micro-* (from the Greek: μικρός, *mikros*, "small") and *organism* from the Greek: ργανισμός, *organismós*, "organism"). It is usually styled solid but is sometimes hyphenated (*micro-organism*), especially in older texts. The word *microbe* comes from μικρός, mikrós, "small" and βίος, bíos, "life".

MICROORGANISMS USED IN FOOD AND BEVERAGE PREPARATION

This is an incomplete list of bacteria and fungi that are used in preparing food.

List of microorganisms used in food and beverage preparation

Microorganism	Type Of Microorganism	Food or Beverage	Reference(s)
Acetobacter aceti	bacterium	chocolate	
Acetobacter aceti	bacterium	vinegar	
Acetobacter cerevisiae	bacterium	Beer	
Acetobacter fabarum	bacterium	chocolate	
Acetobacter fabarum	bacterium	coffee	
Acetobacter lovaniensis	bacterium	vegetables	
Acetobacter malorum	bacterium	vinegar	
Acetobacter orientalis	bacterium	vegetables	
Acetobacter pasteurianus	bacterium	chocolate	
Acetobacter pasteurianus	bacterium	vinegar	
Acetobacter pomorum	bacterium	vinegar	
Acetobacter syzygii	bacterium	chocolate	
Acetobacter syzygii	bacterium	vinegar	
Acetobacter tropicalis	bacterium	chocolate	

Acetobacter tropicalis	bacterium	coffee	
Arthrobacter arilaitensis	bacterium	smear-ripened cheese	
Arthrobacter bergerei	bacterium	smear-ripened cheese	
Arthrobacter globiformis	bacterium	smear-ripened cheese	
Arthrobacter nicotianae	bacterium	surface-ripened cheese	
Arthrobacter nicotianae	bacterium	Tilsit cheese	
Arthrobacter variabilis	bacterium	smear-ripened cheese	
Aspergillus acidus	fungus	tea	
Aspergillus niger	fungus	awamori	
Aspergillus fumigatus	fungus	chocolate	
Aspergillus oryzae	fungus	miso	
Aspergillus oryzae	fungus	sake	
Aspergillus oryzae	fungus	soy sauce	
Aspergillus sojae	fungus	miso	
Aspergillus sojae	fungus	soy sauce	
Bacillus cereus	bacterium	chocolate	
Bacillus coagulans	bacterium	chocolate	
Bacillus licheniformis	bacterium	chocolate	
Bacillus pumilus	bacterium	chocolate	
Bacillus sphaericus	bacterium	stinky tofu	
Bacillus stearothermophilus	bacterium	chocolate	
Bacillus subtilis	bacterium	chocolate	
Bacillus subtilis	bacterium	natto	
Bifidobacterium adolescentis	bacterium	yogurt	
Bifidobacterium animalis	bacterium	daisy	
Bifidobacterium bifidum	bacterium	daisy	
Bifidobacterium breve	bacterium	daisy	
Bifidobacterium breve	bacterium	soy	
Bifidobacterium infantis	bacterium	daisy	
Bifidobacterium lactis	bacterium	daisy	
Bifidobacterium longum	bacterium	daisy	
Bifidobacterium pseudolongum	bacterium	daisy	
Bifidobacterium thermophilum	bacterium	daisy	
Brachybacterium alimentarium	bacterium	Beaufort cheese	
Brachybacterium alimentarium	bacterium	Gruyère cheese	

Brachybacterium tyrofermentans	bacterium	Beaufort cheese	
Brachybacterium tyrofermentans	bacterium	Gruyère cheese	
Brevibacterium aurantiacum	bacterium	cheese	
Brevibacterium casei	bacterium	smear-ripened cheese	
Brevibacterium linens	bacterium	smear-ripened cheese	
Candida colliculosa	fungus	cheese	
Candida colliculosa	fungus	kefir	
Candida exiguus	fungus	sourdough bread	
Candida humicola	fungus	chocolate	
Candida kefyr	fungus	surface-ripened cheese	
Candida krusei	fungus	surface-ripened cheese	
Candida milleri	fungus	sourdough bread	
Candida mycoderma	fungus	Limburger cheese	
Candida pelliculosa	fungus	chocolate	
Candida rugosa	fungus	chocolate	
Candida tropicalis	fungus	chocolate	
Candida utilis	fungus	cheese	
Candida valida	fungus	sourdough	
Candida vini	fungus	Reblochon cheese, wine	
Candida zeylanoides	fungus	Reblochon cheese	
Carnobacterium divergens	bacterium	cheese	
Carnobacterium divergens	bacterium	fish	
Carnobacterium divergens	bacterium	meat	
Carnobacterium maltaromaticum	bacterium	dairy	
Carnobacterium piscicola	bacterium	meat	
Corynebacterium ammoniagenes	bacterium	cheese	
Corynebacterium casei	bacterium	smear-ripened cheese	
Corynebacterium flavescens	bacterium	cheese	
Corynebacterium mooreparkense	bacterium	smear-ripened cheese	
Corynebacterium variabile	bacterium	cheese	
Cyberlindnera mrakii	fungus	wine	
Cystofilobasidium infirmominiatum	fungus	cheese	
Debaryomyces hansenii	fungus	smear-ripened cheese	

Debaryomyces hansenii	fungus	Reblochon cheese	
Debaryomyces kloeckeri	fungus	Limburger cheese	
Enterococcus faecalis	bacterium	cheese	
Enterococcus faecalis	bacterium	cream	
Enterococcus faecalis	bacterium	ham	
Enterococcus faecalis	bacterium	miso	
Enterococcus faecalis	bacterium	pickle	
Enterococcus faecalis	bacterium	sausage	
Enterococcus faecalis	bacterium	soy sauce	
Enterococcus faecium	bacterium	Manchego cheese	
Enterococcus faecium	bacterium	ham	
Enterococcus faecium	bacterium	miso	
Enterococcus faecium	bacterium	pickle	
Enterococcus faecium	bacterium	soy sauce	
Fusarium domesticum	fungus	cheese	
Geotrichum candidum	fungus	cheese	
Gluconacetobacter azotocaptans	bacterium	chocolate	
Gluconacetobacter azotocaptans	bacterium	coffee	
Gluconacetobacter diazotrophicus	bacterium	chocolate	
Gluconacetobacter diazotrophicus	bacterium	coffee	
Gluconacetobacter entanii	bacterium	vinegar	
Gluconacetobacter europaeus	bacterium	vinegar	
Gluconacetobacter hansenii	bacterium	vinegar	
Gluconacetobacter johannae	bacterium	chocolate	
Gluconacetobacter johannae	bacterium	coffee	
Gluconacetobacter oboediens	bacterium	vinegar	
Gluconacetobacter xylinus	bacterium	vinegar	
Gluconobacter oxydans	bacterium	chocolate	
Hafnia alvei	bacterium	cheese	
Halomonas elongata	bacterium	meat	
Issatchenkia orientalis	fungus	kefir	
Kazachstania exigua	fungus	kefir	
Kazachstania unispora	fungus	kefir	
Kloeckera africana	fungus	chocolate	
Kloeckera apis	fungus	chocolate	

Kloeckera javanica	fungus	chocolate	
Kluyveromyces lactis	fungus	cheese	
Kluyveromyces marxianus	fungus	cheese	
Kluyveromyces marxianus	fungus	chocolate	
Kocuria rhizophila	bacterium	cheese	
Kocuria rhizophila	bacterium	meat	
Kocuria varians	bacterium	dairy	
Kocuria varians	bacterium	sausage	
Lactobacillus acetotolerans	bacterium	fruit	
Lactobacillus acetotolerans	bacterium	sourdough bread	
Lactobacillus acetotolerans	bacterium	vegetables	
Lactobacillus acidifarinae	bacterium	sourdough bread	
Lactobacillus acidipiscis	bacterium	dairy	
Lactobacillus acidipiscis	bacterium	fish	
Lactobacillus acidophilus	bacterium	vegetables	
Lactobacillus acidophilus	bacterium	yogurt	
Lactobacillus alimentarius	bacterium	fish	
Lactobacillus alimentarius	bacterium	meat	
Lactobacillus brevis	bacterium	Canestrato Pugliese cheese	
Lactobacillus brevis	bacterium	vegetables	
Lactobacillus brevis ssp. *linens*	bacterium	kefir	
Lactobacillus bucheri	bacterium	bread	
Lactobacillus bucheri	bacterium	wine	
Lactobacillus cacaonum	bacterium	chocolate	
Lactobacillus casei	bacterium	Idiazabal cheese	
Lactobacillus casei	bacterium	Manchego cheese	
Lactobacillus casei	bacterium	Roncal cheese	
Lactobacillus casei	bacterium	yogurt	
Lactobacillus casei ssp. *pseudoplantarum*	bacterium	Grana Padano cheese	
Lactobacillus casei ssp. *pseudoplantarum*	bacterium	Parmigiano-Reggiano cheese	
Lactobacillus cellobiosus	bacterium	chocolate	
Lactobacillus collinoides	bacterium	cider	
Lactobacillus composti	bacterium	*sh ch*	
Lactobacillus coryniformis	bacterium	cheese	

Lactobacillus crispatus	bacterium	sourdough bread	
Lactobacillus curvatus	bacterium	Cacio di Fossa cheese	
Lactobacillus curvatus	bacterium	Canestrato Pugliese cheese	
Lactobacillus curvatus	bacterium	Pecorino Romano cheese	
Lactobacillus curvatus	bacterium	Pecorino Sardo cheese	
Lactobacillus curvatus	bacterium	sausage	
Lactobacillus delbrueckii	bacterium	vegetables	
Lactobacillus delbrueckii ssp. *bulgaricus*	bacterium	cheese	
Lactobacillus delbrueckii ssp. *bulgaricus*	bacterium	yogurt	
Lactobacillus delbrueckii ssp. *lactis*	bacterium	Pecorino Romano cheese	
Lactobacillus dextrinicus	bacterium	meat	
Lactobacillus diolivorans	bacterium	chicha	
Lactobacillus fabifermentans	bacterium	chocolate	
Lactobacillus farciminis	bacterium	fish	
Lactobacillus farciminis	bacterium	soy	
Lactobacillus fermentum	bacterium	chocolate	
Lactobacillus fermentum	bacterium	Pecorino Romano cheese	
Lactobacillus fermentum	bacterium	sourdough bread	
Lactobacillus gasseri	bacterium	dairy	
Lactobacillus gasseri	bacterium	sourdough bread	
Lactobacillus ghanensis	bacterium	chocolate	
Lactobacillus hammesii	bacterium	sourdough bread	
Lactobacillus harbinensis	bacterium	vegetables	
Lactobacillus helveticus	bacterium	cheese	
Lactobacillus helveticus	bacterium	vegetables	
Lactobacillus hilgardii	bacterium	chocolate	
Lactobacillus hilgardii	bacterium	wine	
Lactobacillus homohiochii	bacterium	sake	
Lactobacillus homohiochii	bacterium	sourdough bread	
Lactobacillus jensenii	bacterium	bread	
Lactobacillus johnsonii	bacterium	dairy	
Lactobacillus johnsonii	bacterium	sourdough bread	
Lactobacillus kefiranofaciens	bacterium	kefir	

Lactobacillus kefiri	bacterium	kefir	
Lactobacillus kimchii	bacterium	kimchi	
Lactobacillus kisonensis	bacterium	pickle	
Lactobacillus kunkeei	bacterium	wine	
Lactobacillus mali	bacterium	cider	
Lactobacillus mali	bacterium	rum	
Lactobacillus mali	bacterium	wine	
Lactobacillus manihotivorans	bacterium	cassava	
Lactobacillus mindensis	bacterium	sourdough bread	
Lactobacillus mucosae	bacterium	sourdough bread	
Lactobacillus nagelii	bacterium	wine	
Lactobacillus namuresis	bacterium	sourdough bread	
Lactobacillus nantesis	bacterium	sourdough bread	
Lactobacillus nodensis	bacterium	dairy	
Lactobacillus oeni	bacterium	wine	
Lactobacillus otakiensis	bacterium	pickle	
Lactobacillus panis	bacterium	sourdough bread	
Lactobacillus parabrevis	bacterium	cheese	
Lactobacillus parabrevis	bacterium	kefir	
Lactobacillus parabrevis	bacterium	vegetables	
Lactobacillus parabuchneri	bacterium	sourdough bread	
Lactobacillus paracasei	bacterium	dairy	
Lactobacillus paracasei	bacterium	meat	
Lactobacillus paracasei ssp. *paracasei*	bacterium	Cacio di Fossa cheese	
Lactobacillus paracasei ssp. *paracasei*	bacterium	Canestrato Pugliese cheese	
Lactobacillus paracasei ssp. *paracasei*	bacterium	Pecorino Sardo cheese	
Lactobacillus parakefiri	bacterium	kefir	
Lactobacillus paralimentarius	bacterium	sourdough bread	
Lactobacillus paraplantarum	bacterium	cheese	
Lactobacillus paraplantarum	bacterium	vegetables	
Lactobacillus pentosus	bacterium	Canestrato Pugliese cheese	
Lactobacillus pentosus	bacterium	fish	
Lactobacillus pentosus	bacterium	fruit	
Lactobacillus pentosus	bacterium	wine	

Lactobacillus perolens	bacterium	cheese	
Lactobacillus perolens	bacterium	vegetables	
Lactobacillus plantarum	bacterium	Cacio di Fossa cheese	
Lactobacillus plantarum	bacterium	Canestrato Pugliese cheese	
Lactobacillus plantarum	bacterium	chocolate	
Lactobacillus plantarum	bacterium	Idiazabal cheese	
Lactobacillus plantarum	bacterium	Manchego cheese	
Lactobacillus plantarum	bacterium	Pecorino Romano cheese	
Lactobacillus plantarum	bacterium	Roncal cheese	
Lactobacillus plantarum	bacterium	sausage	
Lactobacillus plantarum	bacterium	vegetables	
Lactobacillus pobuzihii	bacterium	fruit	
Lactobacillus pontis	bacterium	sourdough bread	
Lactobacillus rapi	bacterium	pickle	
Lactobacillus rapi	bacterium	vegetables	
Lactobacillus reuteri	bacterium	sourdough bread	
Lactobacillus rhamnosus	bacterium	Grana Padano cheese	
Lactobacillus rhamnosus	bacterium	Parmigiano-Reggiano cheese	
Lactobacillus rhamnosus	bacterium	meat	
Lactobacillus rhamnosus	bacterium	vegetables	
Lactobacillus rossiae	bacterium	sourdough bread	
Lactobacillus sakei	bacterium	sake	
Lactobacillus sakei	bacterium	sausage	
Lactobacillus salivarius	bacterium	dairy	
Lactobacillus sanfranciscensis	bacterium	sourdough bread	
Lactobacillus satsumensis	bacterium	*sh ch*	
Lactobacillus secaliphilus	bacterium	sourdough bread	
Lactobacillus senmaizukei	bacterium	pickles	
Lactobacillus siliginis	bacterium	sourdough bread	
Lactobacillus similis	bacterium	rum	
Lactobacillus spicheri	bacterium	sourdough bread	
Lactobacillus suebicus	bacterium	fruit	
Lactobacillus spp.	bacterium	butter	
Lactobacillus spp.	bacterium	olive	

Lactobacillus sunkii	bacterium	pickle	
Lactobacillus tucceti	bacterium	dairy	
Lactobacillus tucceti	bacterium	sausage	
Lactobacillus vaccinostercus	bacterium	fruit	
Lactobacillus vaccinostercus	bacterium	vegetables	
Lactobacillus versmoldesis	bacterium	sausage	
Lactobacillus yamanashiensis	bacterium	cider	
Lactobacillus yamanashiensis	bacterium	wine	
Lactococcus lactis	bacterium	buttermilk	
Lactococcus lactis	bacterium	chocolate	
Lactococcus lactis ssp. *cremoris*	bacterium	Cheddar cheese	
Lactococcus lactis ssp. *lactis*	bacterium	cheese	
Lactococcus raffinolactis	bacterium	cheese	
Lactococcus spp.	bacterium	butter	
Lecanicillium lecanii	fungus	cheese	
Leuconostoc carnosum	bacterium	meat	
Leuconostoc citreum	bacterium	cheese	
Leuconostoc citreum	bacterium	fish	
Leuconostoc fallax	bacterium	sauerkraut	
Leuconostoc holzapfelii	bacterium	coffee	
Leuconostoc inhae	bacterium	kimchi	
Leuconostoc kimchii	bacterium	kimchi	
Leuconostoc lactis	bacterium	cheese	
Leuconostoc mesenteroides	bacterium	chocolate	
Leuconostoc mesenteroides	bacterium	vegetables	
Leuconostoc mesenteroides ssp. *cremoris*	bacterium	cheese	
Leuconostoc mesenteroides ssp. *cremoris*	bacterium	vegetables	
Leuconostoc mesenteroides ssp. *dextranicum*	bacterium	butter	
Leuconostoc mesenteroides ssp. *dextranicum*	bacterium	Idiazabal cheese	
Leuconostoc mesenteroides ssp. *dextranicum*	bacterium	pickle	
Leuconostoc mesenteroides ssp. *dextranicum*	bacterium	Roncal cheese	
Leuconostoc mesenteroides ssp. *mesenteroides*	bacterium	Idiazabal cheese	

Leuconostoc mesenteroides ssp. *mesenteroides*	bacterium	Roncal cheese	
Leuconostoc palmae	bacterium	palm wine	
Leuconostoc pseudomesenteroides	bacterium	butter	
Leuconostoc pseudomesenteroides	bacterium	buttermilk	
Leuconostoc pseudomesenteroides	bacterium	sour cream	
Leuconostoc spp.	bacterium	butter	
Leuconostoc spp.	bacterium	olive	
Leuconostoc spp.	bacterium	wine	
Macrococcus caseolyticus	bacterium	cheese	
Macrococcus caseolyticus	bacterium	sausage	
Microbacterium foliorum	bacterium	surface-ripened cheese	
Microbacterium gubbeenense	bacterium	Limburger cheese	
Microbacterium gubbeenense	bacterium	smear-ripened cheese	
Microbacterium gubbeenense	bacterium	Tilsit cheese	
Micrococcus luteus	bacterium	cheese	
Micrococcus lylae	bacterium	sausage	
Mucor hiemalis	fungus	soy bean curd	
Mucor plumbeus	fungus	cheese	
Mucor racemosus	fungus	cheese	
Mucor racemosus	fungus	chocolate	
Neurospora intermedia	fungus	oncom	
Oenococcus oeni	bacterium	wine	
Pediococcus acidilactici	bacterium	sausage	
Pediococcus acidilactici	bacterium	vegetables	
Pediococcus pentosaceus	bacterium	sausage	
Penicillium album	fungus	farmhouse cheeses	
Penicillium camemberti	fungus	cheese	v
Penicillium caseifulvum	fungus	cheese	
Penicillium chrysogenum	fungus	cheese	
Penicillium chrysogenum	fungus	sausage	
Penicillium commune	fungus	surface-ripened cheese	
Penicillium nalgiovense	fungus	cheese	
Penicillium nalgiovense	fungus	ham	
Penicillium nalgiovense	fungus	sausage	
Penicillium roqueforti	fungus	cheese	

Penicillium solitum	fungus	meat	
Pichia fermentans	fungus	dairy	
Pichia fermentans	fungus	kefir	
Pichia fermentans	fungus	wine	
Propionibacterium acidipropionici	bacterium	cheese	
Propionibacterium freudenreichii ssp. *freudenreichii*	bacterium	dairy	
Propionibacterium freudenreichii ssp. *shermanii*	bacterium	Emmental cheese	
Propionibacterium jensenii	bacterium	cheese	
Propionibacterium thoenii	bacterium	cheese	
Proteus vulgaris	bacterium	surface-ripened cheese	
Pseudomonas fluorescens	bacterium	yogurt	
Psychrobacter celer	bacterium	cheese	
Rhizopus microsporus ssp. *oligosporus*	fungus	oncom	
Rhizopus microsporus ssp. *oligosporus*	fungus	tempeh	
Rhodosporidium infirmominiatum	fungus	cheese	
Rhodotorula glutinis	fungus	chocolate	
Rhodotorula minuta	fungus	smear-ripened cheese	
Rhodotorula rubra	fungus	chocolate	
Saccharomyces bayanus	fungus	beer	
Saccharomyces bayanus	fungus	cider	
Saccharomyces bayanus	fungus	wine	
Saccharomyces carlsbergensis	fungus	lager beer	
Saccharomyces cerevisiae	fungus	ale beer	
Saccharomyces cerevisiae	fungus	bread	
Saccharomyces cerevisiae	fungus	cider	
Saccharomyces cerevisiae	fungus	cheese	
Saccharomyces cerevisiae	fungus	chocolate	
Saccharomyces cerevisiae	fungus	wine	
Saccharomyces pastorianus	fungus	lager beer	
Saccharomyces rouzii	fungus	miso	
Saccharomyces uvarum	fungus	lager beer	
Staphylococcus carnosus	bacterium	cheese	

Staphylococcus carnosus ssp. *carnosus*	bacterium	sausage	
Staphylococcus condimenti	bacterium	soy	
Staphylococcus equorum	bacterium	meat	
Staphylococcus equorum ssp. *linens*	bacterium	cheese	
Staphylococcus fleurettii	bacterium	cheese	
Staphylococcus piscifermentans	bacterium	fish	
Staphylococcus saphrophyticus	bacterium	Harzer cheese	
Staphylococcus sciuri ssp. *carnaticus*	bacterium	cheese	
Staphylococcus simulans	bacterium	sausage	
Staphylococcus succinus	bacterium	dairy	
Staphylococcus succinus	bacterium	meat	
Staphylococcus vitulinus	bacterium	cheese	
Staphylococcus vitulinus	bacterium	meat	
Staphylococcus warneri	bacterium	meat	
Staphylococcus xylosus	bacterium	cheese	
Staphylococcus xylosus	bacterium	sausage	
Streptococcus gallolyticus	bacterium	dairy	
Streptococcus salivarius	bacterium	yogurt	
Streptococcus thermophilus	bacterium	cheese	
Streptococcus thermophilus	bacterium	yogurt	
Streptomyces griseus	bacterium	meat	
Tetragenococcus halophilus	bacterium	miso	
Tetragenococcus halophilus	bacterium	soy sauce	
Tetragenococcus koreensis	bacterium	kimchi	
Torulaspora delbrueckii	fungus	smear-ripened cheese	
Torulopsis versatilis	fungus	miso	
Thrichosporon beigelii	fungus	smear-ripened cheese	
Verticillium lecanii	fungus	Tomme cheese	
Weissella beninensis	bacterium	cassava	
Weissella cibaria	bacterium	kimchi	
Weissella fabaria	bacterium	chocolate	
Weissella ghanesis	bacterium	chocolate	
Weissella koreensis	bacterium	kimchi	
Weissella paramesenteroides	bacterium	sausage	

Weissella thailandensis	bacterium	fish	
Yarrowia lipolytica	fungus	Raclette cheese	
Yarrowia lipolytica	fungus	smear-ripened cheese	
Yarrowia lipolytica	fungus	dairy	
Zygotorulaspora florentina	fungus	kefir	
Zymomonas mobilis	bacterium	palm wine	
Zymomonas mobilis	bacterium	pulque	

REFERENCES

1. Madigan M; Martinko J, eds. (2006). Brock Biology of Microorganisms (13th ed.). Pearson Education. p. 1096. ISBN 0-321-73551-X.
2. Rybicki EP (1990). "The classification of organisms at the edge of life, or problems with virus systematics". S Aft J Sci. 86: 182–6. ISSN 0038-2353.
3. Lwoff A (1956). "The concept of virus". J. Gen. Microbiol. 17 (2): 239–53. doi:10.1099/00221287-17-2-239. PMID 13481308.
4. Wade, Nicholas (25 July 2016). "Meet Luca, the Ancestor of All Living Things". New York Times. Retrieved 25 July 2016.
5. University of Georgia (25 August 1998). "First-Ever Scientific Estimate Of Total Bacteria On Earth Shows Far Greater Numbers Than Ever Known Before". Science Daily. Retrieved 10 November 2014.
6. Dose, K; Bieger-Dose, A.; Dillmann, R.; Gill, M.; Kerz, O.; Klein, A.; Meinert, H.; Nawroth, T.; Risi, S.; Stridde, C. (1995). "ERA-experiment "space biochemistry"". Advances in Space Research. 16 (8): 119–129. doi:10.1016/0273-1177(95)00280-R. PMID 11542696.
7. Vaisberg, Horneck G.; Eschweiler, U.; Reitz, G.; Wehner, J.; Willimek, R.; Strauch, K. (1995). "Biological responses to space: results of the experiment "Exobiological Unit" of ERA on EURECA I". Adv Space Res. 16 (8): 105–18. Bibcode:1995AdSpR..16..105V. doi:10.1016/0273-1177(95)00279-N. PMID 11542695.
8. Staff (2012). "The JTM Effect". jtmnutrients.com. Archived from the original on 2012-11-14. Retrieved 10 November 2014.
9. Staff (2014). "The Biosphere". Aspen Global Change Institute. Retrieved 10 November 2014.
10. Choi, Charles Q. (17 March 2013). "Microbes Thrive in Deepest Spot on Earth". LiveScience. Retrieved 17 March 2013.
11. Glud, Ronnie; Wenzhöfer, Frank; Middelboe, Mathias; Oguri, Kazumasa; Turnewitsch, Robert; Canfield, Donald E.; Kitazato, Hiroshi (17 March 2013). "High rates of microbial carbon turnover in sediments in the deepest oceanic trench on Earth". Nature Geoscience. 6 (4): 284–288. Bibcode:2013NatGe...6..284G. doi:10.1038/ngeo1773. Retrieved 17 March 2013.

12. Oskin, Becky (14 March 2013). "Intraterrestrials: Life Thrives in Ocean Floor". LiveScience. Retrieved 17 March 2013.
13. Morelle, Rebecca (15 December 2014). "Microbes discovered by deepest marine drill analysed". BBC News. Retrieved 15 December 2014.
14. Fox, Douglas (20 August 2014). "Lakes under the ice: Antarctica's secret garden". Nature. 512 (7514): 244–246. Bibcode:2014Natur.512..244F. doi:10.1038/512244a. Retrieved 21 August 2014.
15. Mack, Eric (20 August 2014). "Life Confirmed Under Antarctic Ice; Is Space Next?". Forbes. Retrieved 21 August 2014.
16. Christner BC, Morris CE, Foreman CM, Cai R, Sands DC (2008). "Ubiquity of biological ice nucleators in snowfall". Science. 319 (5867): 1214. Bibcode:2008Sci...319.1214C. doi:10.1126/science.1149757. PMID 18309078.
17. 2002 WHO mortality data Accessed 20 January 2007
18. Schopf J (2006). "Fossil evidence of Archaean life". Philos Trans R Soc Lond B Biol Sci. 361 (1470): 869–85. doi:10.1098/rstb.2006.1834. PMC 1578735 . PMID 16754604.
19. Altermann W, Kazmierczak J (2003). "Archean microfossils: a reappraisal of early life on Earth". Res Microbiol. 154 (9): 611–7. doi:10.1016/j.resmic.2003.08.006. PMID 14596897.
20. Cavalier-Smith T (2006). "Cell evolution and Earth history: stasis and revolution". Philos Trans R Soc Lond B Biol Sci. 361 (1470): 969–1006. doi:10.1098/rstb.2006.1842. PMC 1578732 . PMID 16754610.
21. Schopf J (1994). "Disparate rates, differing fates: tempo and mode of evolution changed from the Precambrian to the Phanerozoic". Proc Natl Acad Sci USA. 91 (15): 6735–42. Bibcode:1994PNAS...91.6735S. doi:10.1073/pnas.91.15.6735. PMC 44277 . PMID 8041691.
22. Stanley S (May 1973). "An Ecological Theory for the Sudden Origin of Multicellular Life in the Late Precambrian". Proc Natl Acad Sci USA. 70 (5): 1486–9. Bibcode:1973PNAS...70.1486S. doi:10.1073/pnas.70.5.1486. PMC 433525 . PMID 16592084.
23. DeLong E, Pace N (2001). "Environmental diversity of bacteria and archaea". Syst Biol. 50 (4): 470–8. doi:10.1080/10635150175043504O. PMID 12116647.
24. Schmidt A, Ragazzi E, Coppellotti O, Roghi G (2006). "A microworld in Triassic amber". Nature. 444 (7121): 835. Bibcode:2006Natur.444..835S. doi:10.1038/444835a. PMID 17167469.
25. Schirber, Michael (July 27, 2014). "Microbe's Innovation May Have Started Largest Extinction Event on Earth". Space.com. Astrobiology Magazine. .... That spike in nickel allowed methanogens to take off.
26. Wolska K (2003). "Horizontal DNA transfer between bacteria in the environment". Acta Microbiol Pol. 52 (3): 233–43. PMID 14743976.
27. Enright M, Robinson D, Randle G, Feil E, Grundmann H, Spratt B (May 2002). "The evolutionary history of methicillin-resistant Staphylococcus aureus (MRSA)".

Proc Natl Acad Sci USA. 99 (11): 7687–92. Bibcode:2002PNAS...99.7687E. doi:10.1073/pnas.122108599. PMC 124322 . PMID 12032344.

28. Mahavira is dated 599 BCE - 527 BCE. See. Dundas, Paul; John Hinnels, eds. (2002). The Jains. London: Routledge. ISBN 0-415-26606-8. p. 24
29. Dundas, Paul (2002) p. 88
30. Jaini, Padmanabh (1998). The Jaina Path of Purification. New Delhi: Motilal Banarsidass. ISBN 81-208-1578-5. p. 109
31. Varro On Agriculture 1,xii Loeb
32. Tschanz, David W. "Arab Roots of European Medicine". Heart Views. 4 (2). Archived from the original on 3 May 2011.
33. Colgan, Richard (2009). Advice to the Young Physician: On the Art of Medicine. Springer. p. 33. ISBN 978-1-4419-1033-2.
34. Payne, A.S. The Cleere Observer: A Biography of Antoni Van Leeuwenhoek, p. 13, Macmillan, 1970
35. Leeuwenhoek A (1753). "Part of a Letter from Mr Antony van Leeuwenhoek, concerning the Worms in Sheeps Livers, Gnats, and Animalcula in the Excrements of Frogs". Philosophical Transactions (1683–1775). 22 (260–276): 509–18. doi:10.1098/rstl.1700.0013.
36. Leeuwenhoek A (1753). "Part of a Letter from Mr Antony van Leeuwenhoek, F. R. S. concerning Green Weeds Growing in Water, and Some Animalcula Found about Them". Philosophical Transactions (1683–1775). 23 (277–288): 1304–11. doi:10.1098/rstl.1702.0042.
37. The Nobel Prize in Physiology or Medicine 1905Nobelprize.org Accessed 22 November 2006.
38. O'Brien S, Goedert J (1996). "HIV causes AIDS: Koch's postulates fulfilled". Curr Opin Immunol. 8 (5): 613–18. doi:10.1016/S0952-7915(96)80075-6. PMID 8902385.
39. Borenstein, Seth (13 November 2013). "Oldest fossil found: Meet your microbial mom". Associated Press. Retrieved 15 November 2013.
40. Noffke, Nora; Christian, Christian; Wacey, David; Hazen, Robert M. (8 November 2013). "Microbially Induced Sedimentary Structures Recording an Ancient Ecosystem in the ca. 3.48 Billion-Year-Old Dresser Formation, Pilbara, Western Australia". Astrobiology (journal). 13 (12): 1103–24. doi:10.1089/ast.2013.1030. PMC 3870916 . PMID 24205812. Retrieved 15 November 2013.
41. Ciccarelli FD, Doerks T, von Mering C, Creevey CJ, Snel B, Bork P (2006). "Toward automatic reconstruction of a highly resolved tree of life". Science. 311 (5765): 1283–7. Bibcode:2006Sci...311.1283C. doi:10.1126/science.1123061. PMID 16513982.
42. Gold T (1992). "The deep, hot biosphere". Proc. Natl. Acad. Sci. U.S.A. 89 (13): 6045–9. Bibcode:1992PNAS...89.6045G. doi:10.1073/pnas.89.13.6045. PMC 49434 . PMID 1631089.

43. Whitman W, Coleman D, Wiebe W (1998). "Prokaryotes: The unseen majority". Proc Natl Acad Sci USA. 95 (12): 6578–83. Bibcode:1998PNAS...95.6578W. doi:10.1073/pnas.95.12.6578. PMC 33863 . PMID 9618454.
44. Staff (2 May 2016). "Researchers find that Earth may be home to 1 trillion species". National Science Foundation. Retrieved 6 May 2016.
45. Schulz H, Jorgensen B (2001). "Big bacteria". Annu Rev Microbiol. 55: 105–37. doi:10.1146/annurev.micro.55.1.105. PMID 11544351.
46. Shapiro JA (1998). "Thinking about bacterial populations as multicellular organisms" (PDF). Annu. Rev. Microbiol. 52: 81–104. doi:10.1146/annurev.micro.52.1.81. PMID 9891794. Archived from the original (PDF) on 17 July 2011.
47. Johnsborg O, Eldholm V, Håvarstein LS (December 2007). "Natural genetic transformation: prevalence, mechanisms and function". Res. Microbiol. 158 (10): 767–78. doi:10.1016/j.resmic.2007.09.004. PMID 17997281.
48. Transformation (genetics)
49. Eagon R (1962). "PSEUDOMONAS NATRIEGENS, A MARINE BACTERIUM WITH A GENERATION TIME OF LESS THAN 10 MINUTES". J Bacteriol. 83 (4): 736–7. PMC 279347 . PMID 13888946.
50. Woese C, Kandler O, Wheelis M (1990). "Towards a natural system of organisms: proposal for the domains Archaea, Bacteria, and Eucarya". Proc Natl Acad Sci USA. 87 (12): 4576–9. Bibcode:1990PNAS...87.4576W. doi:10.1073/pnas.87.12.4576. PMC 54159 . PMID 2112744.
51. De Rosa M, Gambacorta A, Gliozzi A (1 March 1986). "Structure, biosynthesis, and physicochemical properties of archaebacterial lipids". Microbiol. Rev. 50 (1): 70–80. PMC 373054 . PMID 3083222.
52. Robertson C, Harris J, Spear J, Pace N (2005). "Phylogenetic diversity and ecology of environmental Archaea". Curr Opin Microbiol. 8 (6): 638–42. doi:10.1016/j.mib.2005.10.003. PMID 16236543.
53. Karner MB, DeLong EF, Karl DM (2001). "Archaeal dominance in the mesopelagic zone of the Pacific Ocean". Nature. 409 (6819): 507–10. doi:10.1038/35054051. PMID 11206545.
54. Sinninghe Damsté JS, Rijpstra WI, Hopmans EC, Prahl FG, Wakeham SG, Schouten S (June 2002). "Distribution of Membrane Lipids of Planktonic Crenarchaeota in the Arabian Sea". Appl. Environ. Microbiol. 68 (6): 2997–3002. doi:10.1128/AEM.68.6.2997-3002.2002. PMC 123986 . PMID 12039760.
55. Leininger, S.; Urich, T.; Schloter, M.; Schwark, L.; Qi, J.; Nicol, G. W.; Prosser, J. I.; Schuster, S. C.; Schleper, C. (2006). "Archaea predominate among ammonia-oxidizing prokaryotes in soils". Nature. 442 (7104): 806–809. Bibcode:2006Natur.442..806L. doi:10.1038/nature04983. PMID 16915287.
56. Eukaryota: More on Morphology. (Retrieved 10 October 2006)
57. Dyall S, Brown M, Johnson P (2004). "Ancient invasions: from endosymbionts to organelles". Science. 304 (5668): 253–7. Bibcode:2004Sci...304..253D. doi:10.1126/science.1094884. PMID 15073369.

58. Bernstein, H; Bernstein, C; Michod, RE (2012). "Chapter 1". In Kimura, Sakura; Shimizu, Sora. DNA repair as the primary adaptive function of sex in bacteria and eukaryotes. DNA Repair: New Research. Hauppauge, N.Y.: Nova Sci. Publ. pp. 1–49. ISBN 978-1-62100-808-8.
59. Cavalier-Smith T (1 December 1993). "Kingdom protozoa and its 18 phyla". Microbiol. Rev. 57 (4): 953–94. PMC 372943 . PMID 8302218.
60. Corliss JO (1992). "Should there be a separate code of nomenclature for the protists?". BioSystems. 28 (1–3): 1–14. doi:10.1016/0303-2647(92)90003-H. PMID 1292654.
61. Devreotes P (1989). "Dictyostelium discoideum: a model system for cell-cell interactions in development". Science. 245 (4922): 1054–8. Bibcode:1989Sci...245.1054D. doi:10.1126/science.2672337. PMID 2672337.
62. Slapeta J, Moreira D, López-García P (2005). "The extent of protist diversity: insights from molecular ecology of freshwater eukaryotes". Proc. Biol. Sci. 272 (1576): 2073–81. doi:10.1098/rspb.2005.3195. PMC 1559898 . PMID 16191619.
63. Moreira D, López-García P (2002). "The molecular ecology of microbial eukaryotes unveils a hidden world" (PDF). Trends Microbiol. 10 (1): 31–8. doi:10.1016/S0966-842X(01)02257-0. PMID 11755083.
64. Lapinski J, Tunnacliffe A (2003). "Anhydrobiosis without trehalose in bdelloid rotifers". FEBS Lett. 553 (3): 387–90. doi:10.1016/S0014-5793(03)01062-7. PMID 14572656.
65. Kumamoto CA, Vinces MD (2005). "Contributions of hyphae and hypha-co-regulated genes to Candida albicans virulence". Cell. Microbiol. 7 (11): 1546–54. doi:10.1111/j.1462-5822.2005.00616.x. PMID 16207242.
66. Thomas, David C. (2002). Seaweeds. London: Natural History Museum. ISBN 0-565-09175-1.
67. Szewzyk U, Szewzyk R, Stenström T (1994). "Thermophilic, anaerobic bacteria isolated from a deep borehole in granite in Sweden". Proc Natl Acad Sci USA. 91 (5): 1810–3. Bibcode:1994PNAS...91.1810S. doi:10.1073/pnas.91.5.1810. PMC 43253 . PMID 11607462.
68. Horneck G (1981). "Survival of microorganisms in space: a review". Adv Space Res. 1 (14): 39–48. doi:10.1016/0273-1177(81)90241-6. PMID 11541716.
69. Strain 121, a hyperthermophilic archaea, has been shown to reproduce at 121 °C (250 °F), and survive at 130 °C (266 °F).
70. Some Psychrophilic bacteria can grow at −17 °C (1 °F), and can survive near absolute zero.
71. Picrophilus can grow at pH -0.06.
72. The alkaliphilic bacteria Bacillus alcalophilus can grow at up to pH 11.5.
73. Dyall-Smith, Mike, HALOARCHAEA, University of Melbourne. Haloarchaea.
74. The piezophilic bacteria Halomonas salaria requires a pressure of 1,000 atm; nanobes, a speculative organism, have been reportedly found in the earth's crust at 2,000 atm.

75. See Deinococcus radiodurans
76. Cavicchioli R (2002). "Extremophiles and the search for extraterrestrial life". Astrobiology. 2 (3): 281–92. Bibcode:2002AsBio...2..281C. doi:10.1089/153110702762027862. PMID 12530238.
77. Barea J, Pozo M, Azcón R, Azcón-Aguilar C (2005). "Microbial co-operation in the rhizosphere". J Exp Bot. 56 (417): 1761–78. doi:10.1093/jxb/eri197. PMID 15911555.
78. Gillen, Alan L. (2007). The Genesis of Germs: The Origin of Diseases and the Coming Plagues. New Leaf Publishing Group. p. 10. ISBN 0-89051-493-3.
79. Vrieze, Jop de (2015-08-14). "The littlest farmhands". Science. 349 (6249): 680–683. doi:10.1126/science.349.6249.680. ISSN 0036-8075. PMID 26273035.
80. "Dairy Microbiology". University of Guelph. Retrieved 9 October 2006.
81. Gray, N.F. (2004). Biology of Wastewater Treatment. Imperial College Press. p. 1164. ISBN 1-86094-332-2.
82. Kitani, Osumu; Carl W. Hall (1989). Biomass Handbook. Taylor & Francis US. p. 256. ISBN 2-88124-269-3.
83. Pimental, David (2007). Food, Energy, and Society. CRC Press. p. 289. ISBN 1-4200-4667-5.
84. Tickell, Joshua; et al. (2000). From the Fryer to the Fuel Tank: The Complete Guide to Using Vegetable Oil as an Alternative Fuel. Biodiesel America. p. 53. ISBN 0-9707227-0-2.
85. Inslee, Jay; et al. (2008). Apollo's Fire: Igniting America's Clean Energy Economy. Island Press. p. 157. ISBN 1-59726-175-0.
86. Biology textbook for class XII. National council of educational research and training. p. 183. ISBN 81-7450-639-X.
87. Castrillo JI, Oliver SG (2004). "Yeast as a touchstone in post-genomic research: strategies for integrative analysis in functional genomics". J. Biochem. Mol. Biol. 37 (1): 93–106. doi:10.5483/BMBRep.2004.37.1.093. PMID 14761307.
88. Suter B, Auerbach D, Stagljar I (2006). "Yeast-based functional genomics and proteomics technologies: the first 15 years and beyond". BioTechniques. 40 (5): 625–44. doi:10.2144/000112151. PMID 16708762.
89. Sunnerhagen P (2002). "Prospects for functional genomics in Schizosaccharomyces pombe". Curr. Genet. 42 (2): 73–84. doi:10.1007/s00294-002-0335-6. PMID 12478386.
90. Soni, S.K. (2007). Microbes: A Source of Energy for 21st Century. New India Publishing. ISBN 81-89422-14-6.
91. Moses, Vivian; et al. (1999). Biotechnology: The Science and the Business. CRC Press. p. 563. ISBN 90-5702-407-1.
92. Langford, Roland E. (2004). Introduction to Weapons of Mass Destruction: Radiological, Chemical, and Biological. Wiley-IEEE. p. 140. ISBN 0-471-46560-7.

93. O'Hara A, Shanahan F (2006). "The gut flora as a forgotten organ". EMBO Rep. 7 (7): 688–93. doi:10.1038/sj.embor.7400731. PMC 1500832 . PMID 16819463.

94. Eckburg P, Lepp P, Relman D (2003). "Archaea and Their Potential Role in Human Disease". Infect Immun. 71 (2): 591–6. doi:10.1128/IAI.71.2.591-596.2003. PMC 145348 . PMID 12540534.

95. Lepp P, Brinig M, Ouverney C, Palm K, Armitage G, Relman D (2004). "Methanogenic Archaea and human periodontal disease". Proc Natl Acad Sci USA. 101 (16): 6176–81. Bibcode:2004PNAS..101.6176L. doi:10.1073/pnas.0308766101. PMC 395942 . PMID 15067114.

CHAPTER 2

Food Contaminant and Foodborne Illness

Chapter Contents

- **Introduction**
- **Agrochemicals**
- **Pesticides and Carcinogens**
- **Hair in Food**
- **Processing Contaminants**
- **Emerging Food Contaminants**
- **Safety and Regulation**
- **Food Contaminant Testing**
- **Foodborne Illness**

A foodborne illness is any illness that results from consuming food contaminated with pathogenic microorganisms, or toxic compounds released by microorganisms. The disease-causing microorganisms are most often bacteria, but can also include fungi, viruses and protozoa. Your food usually travels a long distance from its origin before it gets to your plate. It can be handled by many different people and exposed to many different conditions. This results in many potential points where food can become contaminated.

Some microbes responsible for foodborne illnesses are members of the normal microflora of the food organism. So, for example, the bacteria that are naturally found in the cow intestinal tract can end up contaminating the slabs of beef during butchering. This kind of contamination can occur during slaughter or during any point in the food's processing.

Vegetables and fruits are not immune to contamination with foodborne pathogens, despite the fact that they are picked, not slaughtered. Irrigation water or fertilizers sprayed on crops can be contaminated with animal or human waste containing foodborne pathogens.

INTRODUCTION

Food contamination refers to the presence in food of harmful chemicals and microorganisms which can cause consumer illness. This article addresses the chemical contamination of foods, as opposed to microbiological contamination, which can be found under foodborne illness.

The impact of chemical contaminants on consumer health and well-being is often apparent only after many years of processing.Prolonged exposure at low levels (e.g., cancer). Chemical contaminants present in foods are often unaffected by thermal processing (unlike most microbiological agents). Chemical contaminants can be classified according to the source of contamination and the mechanism by which they enter the food product.

AGROCHEMICALS

Agrochemicals are chemicals used in agricultural practices and animal husbandry with the intent to increase crops and reduce costs. Such agents include pesticides (e.g., insecticides, herbicides, rodenticides), plant growth regulators, veterinary drugs (e.g., nitrofuran, fluoroquinolones, malachite green, chloramphenicol), and bovine somatotropin (rBST).

Environmental contaminants

Environmental contaminants are chemicals that are present in the environment in which the food is grown, harvested, transported, stored, packaged, processed, and consumed. The physical contact of the food with its environment results in its contamination. Possible sources of contamination include:

- **Air**: radionuclides (^{137}Caesium, ^{90}Strontium), polycyclic aromatic hydrocarbons (PAH)
- **Water**: arsenic, mercury
- **Soil**: cadmium, nitrates, perchlorates
- Polychlorinated biphenyls (PCB), dioxins, and polybrominated diphenyl ethers (PBDE) are ubiquitous chemicals
- **Packaging materials**: antimony, tin, lead, perfluorooctanoic acid (PFOA), semicarbazide, benzophenone, isopropyl thioxanthone (ITX), bisphenol A
- **Processing/cooking equipment**: copper, or other metal chips, lubricants, cleaning, and sanitizing agents
- **Naturally occurring toxins**: mycotoxins, phytohaemagglutinin, pyrrolizidine alkaloids, grayanotoxin, mushroom toxins, scombrotoxin (histamine), ciguatera, shellfish toxins (see shellfish poisoning), tetrodotoxin, among many others.

PESTICIDES AND CARCINOGENS

There are many cases of banned pesticides or carcinogens found in foods.

- Greenpeace exposed in 2006 in China that 25% of surveyed supermarkets agricultural products contained banned pesticides. Over 70% of tomatoes that tested were found to have the banned pesticide Lindane, and almost 40%

of the samples had a mix of three or more types of pesticides. Fruits were also tested in this investigation. Tangerines, strawberries, and Kyofung grapes samples were found contaminated by banned pesticides, including the highly toxic Methamidophos. These fruits can also be found in Hong Kong market. Greenpeace says there exists no comprehensive monitoring on fruit produce in the Hong Kong as of 2006.

- In India, soft drinks were found contaminated with high levels of pesticides and insecticides, including lindane, DDT, malathion and chlorpyrifos.
- News of Formaldehyde, a carcinogen was found in Vietnamese national dish, Pho, broke in 2007 Vietnam food scare. Vegetables and fruits were also found to have banned pesticides. "Health agencies have known that Vietnamese soy sauce, the country›s second most popular sauce after fish sauce, has been chock full of cancer agents since at least 2001", thundered the *Thanh Nien* daily. "Why didn't anyone tell us?" The carcinogen in Asian sauces is 3-MCPD and its metabolite 1,3-DCP, which has been an ongoing problem before 2000 affecting multiple continents.
- 2005 Indonesia food scare, carcinogenic formaldehyde was added as a preservative to noodles, tofu, salted fish, and meatballs
- 2008 Chinese milk scandal

HAIR IN FOOD

There is a heavy stigma attached to the presence of hair in food in most societies. There is a risk that it may induce choking and vomiting, and also that it may be contaminated by toxic substances. Views differ as to the level of risk it poses to the inadvertent consumer.

In most countries, people working in the food industry are required to cover their hair because it will contaminate the food. When people are served food which contains hair in restaurants or cafés, it is usual for them to complain to the staff. Despite this, it is not a valid ground on which to sue the restaurant in the United States but in the United Kingdom it breaks the regulations of the UK Food Safety Act 1990, and people can sue over this.

There are a range of possible reasons for the objection to hair in food, ranging from cultural taboos to the simple fact that it is difficult to digest and unpleasant to eat. It may also be interpreted as a sign of more widespread problems with hygiene. The introduction of complete-capture hairnets is believed to have resulted in a decrease in incidents of contamination of this type.

Sometimes protein from human hair is used as a food ingredient, in bread and other such similar products. Such use of human hair in food is forbidden in Islam. Historically, in Judaism, finding hair in food was a sign of bad luck.

PROCESSING CONTAMINANTS

Processing contaminants are generated during the processing of foods (e.g., heating, fermentation). They are absent in the raw materials, and are formed by chemical reactions between natural and/or added food constituents during processing. The presence of these contaminants in processed foods cannot be entirely avoided. Technological processes can be adjusted and/or optimized, however, in order to reduce the levels of formation of processing contaminants. Examples are: nitrosamines, polycyclic aromatic hydrocarbons (PAH), heterocyclic amines, histamine, acrylamide, furan, benzene, trans fat, 3-MCPD, semicarbazide, 4-hydroxynonenal (4-HNE), and ethyl carbamate. There is also the possibility of metal chips from the processing equipment contaminating food. These can be identified using metal detection equipment. In many conveyor lines, the line will be stopped, or when weighing the product with a Check weigher, the item can be rejected for being over- or underweight or because small pieces of metal are detected within it.

EMERGING FOOD CONTAMINANTS

While many food contaminants have been known for decades, the formation and presence of certain chemicals in foods has been discovered relatively recently. These are the so-called emerging food contaminants like acrylamide, furan, benzene, perchlorate, perfluorooctanoic acid (PFOA), 3-monochloropropane-1,3-diol (3-MCPD), 4-hydroxynonenal, and (4-HNE).

SAFETY AND REGULATION

Acceptable Daily Intake (ADI) levels and tolerable concentrations of contaminants in individual foods are determined on the basis of the "No Observed Adverse Effect Level" (NOAEL) in animal experiments, by using a safety factor (usually 100). The maximum concentrations of contaminants allowed by legislation are often well below toxicological tolerance levels, because such levels can often be reasonably achieved by using good agricultural and manufacturing practices.

Regulatory officials, in order to combat the dangers associated with foodborne viruses, are pursuing various possible measures.

- The EFSA published a report in 2011 on "scientific opinion regarding an update of the present knowledge on the occurrence and control of foodborne viruses".
- This year, an expert working group created by the European Committee for Standardization (CEN), is expected to publish a standard method for the detection of norovirus and hepatitis A virus in food.
- The CODEX Committee on Food Hygiene (CCFH) is also working on a guideline which is now ready for final adoption.
- European Commission Regulation (EC) No 2073/2005 of 15 November 2005 indicates that "foodstuffs should not contain micro-organisms or their toxins or metabolites in quantities that present an unacceptable risk for human health", underlining that methods are required for foodborne virus detection.

FOOD CONTAMINANT TESTING

To maintain high quality of food and comply with health, safety and environmental regulatory standards it is best to rely on food contaminant testing through an independent third party such as laboratories, certification companies or similar. For manufacturers the testing for food contaminants can minimize the risk of noncompliance in relation to raw ingredients, semi-manufactured foods and final products. Also, food contaminant testing assures consumers safety and quality of purchased food products and can prevent foodborne diseases, and chemical, microbiological, or physical food hazards.

The establishment of ADIs for certain emerging food contaminants is currently an active area of research and regulatory debate.

FOODBORNE ILLNESS

Foodborne illness (also foodborne disease and colloquially referred to as food poisoning) is any illness resulting from the food spoilage of contaminated food, pathogenic bacteria, viruses, or parasites that contaminate food, as well as chemical or natural toxins such as poisonous mushrooms and various species of beans that have not been boiled for at least 10 minutes.

Symptoms vary depending on the cause, and are described below in this article. A few broad generalizations can be made, e.g.: The incubation period ranges from hours to days, depending on the cause and on how much was consumed. The incubation period tends to cause sufferers to not associate the symptoms with the item consumed, and so to cause sufferers to attribute the symptoms to gastroenteritis for example. Symptoms often include vomiting, fever, and aches, and may include diarrhea. Bouts of vomiting can be repeated with an extended delay in between, because even if infected food was eliminated from the stomach in the first bout, microbes (if applicable) can pass through the stomach into the intestine via cells lining the intestinal walls and begin to multiply. Some types of microbes stay in the intestine, some produce a toxin that is absorbed into the bloodstream, and some can directly invade deeper body tissues.

Causes

Foodborne illness usually arises from improper handling, preparation, or food storage. Good hygiene practices before, during, and after food preparation can reduce the chances of contracting an illness. There is a consensus in the public health community that regular hand-washing is one of the most effective defenses against the spread of foodborne illness. The action of monitoring food to ensure that it will not cause foodborne illness is known as food safety. Foodborne disease can also be caused by a large variety of toxins that affect the environment.

Foodborne illness can also be caused by pesticides or medicines in food and naturally toxic substances such as poisonous mushrooms or reef fish.

Poorly stored food in a refrigerator

Bacteria

Bacteria are a common cause of foodborne illness. In the United Kingdom during 2000, the individual bacteria involved were the following: *Campylobacter jejuni* 77.3%, *Salmonella* 20.9%, *Escherichia coli* O157:H7 1.4%, and all others less than 0.56%. In the past, bacterial infections were thought to be more prevalent because few places had the capability to test for norovirus and no active surveillance was being done for this particular agent. Toxins from bacterial infections are delayed because the bacteria need time to multiply. As a result, symptoms associated with intoxication are usually not seen until 12–72 hours or more after eating contaminated food. However, in some cases, such as Staphylococcal food poisoning, the onset of illness can be as soon as 30 minutes after ingesting contaminated food.

Most common bacterial foodborne pathogens are:

- *Campylobacter jejuni* which can lead to secondary Guillain–Barré syndrome and periodontitis
- *Clostridium perfringens*, the "cafeteria germ"
- *Salmonella* spp. – its *S. typhimurium* infection is caused by consumption of eggs or poultry that are not adequately cooked or by other interactive human-animal pathogens

Salmonella

- *Escherichia coli O157:H7* enterohemorrhagic (EHEC) which can cause hemolytic-uremic syndrome

Other common bacterial foodborne pathogens are:

- *Bacillus cereus*
- *Escherichia coli*, other virulence properties, such as enteroinvasive (EIEC), enteropathogenic (EPEC), enterotoxigenic (ETEC), enteroaggregative (EAEC or EAgEC)
- *Listeria monocytogenes*
- *Shigella* spp.
- *Staphylococcus aureus*
- *Staphylococcal enteritis*
- *Streptococcus*
- *Vibrio cholerae*, including O1 and non-O1
- *Vibrio parahaemolyticus*
- *Vibrio vulnificus*
- *Yersinia enterocolitica* and *Yersinia pseudotuberculosis*

Less common bacterial agents:

- *Brucella* spp.
- *Corynebacterium ulcerans*
- *Coxiella burnetii* or Q fever
- *Plesiomonas shigelloides*

Enterotoxins

In addition to disease caused by direct bacterial infection, some foodborne illnesses are caused by enterotoxins (exotoxins targeting the intestines). Enterotoxins can produce illness even when the microbes that produced them have been killed. Symptom appearance varies with the toxin but may be rapid in onset, as in the case of enterotoxins of *Staphylococcus aureus* in which symptoms appear in one to six hours. This causes intense vomiting including or not including diarrhea (resulting in staphylococcal enteritis), and staphylococcal enterotoxins (most commonly staphylococcal enterotoxin A but also including staphylococcal enterotoxin B) are the most commonly reported enterotoxins although cases of poisoning are likely underestimated. It occurs mainly in cooked and processed foods due to competition with other biota in raw foods, and humans are the main cause of contamination as a substantial percentage of humans are persistent carriers of *S. aureus*. The CDC has estimated about 240,000 cases per year in the United States.

- Clostridium botulinum
- Clostridium perfringens
- Bacillus cereus

The rare but potentially deadly disease botulism occurs when the anaerobic bacterium *Clostridium botulinum* grows in improperly canned low-acid foods and produces botulin, a powerful paralytic toxin.

Pseudoalteromonas tetraodonis, certain species of Pseudomonas and Vibrio, and some other bacteria, produce the lethal tetrodotoxin, which is present in the tissues of some living animal species rather than being a product of decomposition.

Emerging Foodborne Pathogens

Many foodborne illnesses remain poorly understood.

- *Aeromonas hydrophila*, *Aeromonas caviae*, *Aeromonas sobria*

Preventing Bacterial Food Poisoning

Proper storage and refrigeration of food help in the prevention of food poisoning.

Prevention is mainly the role of the state, through the definition of strict rules of hygiene and a public services of veterinary surveying of animal products in the food chain, from farming to the transformation industry and delivery (shops and restaurants). This regulation includes:

- traceability: in a final product, it must be possible to know the origin of the ingredients (originating farm, identification of the harvesting or of the animal) and where and when it was processed; the origin of the illness can thus be tracked and solved (and possibly penalized), and the final products can be removed from the sale if a problem is detected;
- enforcement of hygiene procedures such as HACCP and the "cold chain";
- power of control and of law enforcement of veterinarians.

In August 2006, the United States Food and Drug Administration approved Phage therapy which involves spraying meat with viruses that infect bacteria, and thus preventing infection. This has raised concerns, because without mandatory labelling consumers would not be aware that meat and poultry products have been treated with the spray.

At home, prevention mainly consists of good food safety practices. Many forms of bacterial poisoning can be prevented by cooking it sufficiently, and either eating it quickly or refrigerating it effectively. Many toxins, however, are not destroyed by heat treatment.

Techniques that help prevent food borne illness in the kitchen are hand washing, rinsing produce, preventing cross-contamination, proper storage, and maintaining cooking temperatures. In general, freezing or refrigerating prevents virtually all bacteria from growing, and heating food sufficiently kills parasites, viruses, and most bacteria. Bacteria grow most rapidly at the range of temperatures between 40 °F and 140 °F, called the "danger zone". Storing food below or above the "danger zone" can effectively limit the production of toxins. For storing leftovers, the food must be put in shallow containers for quick cooling and must be refrigerated within two hours. When food is reheated, it must reach an internal temperature of 165 °F (74 °C) or until hot or steaming to kill bacteria.

Mycotoxins and Alimentary Mycotoxicoses

The term alimentary mycotoxicoses refers to the effect of poisoning by Mycotoxins (The term 'mycotoxin' is usually reserved for the toxic chemical products produced by fungi that readily colonize crops) through food consumption. Mycotoxins sometimes have important effects on human and animal health. For example, an outbreak which occurred in the UK in 1960 caused the death of 100,000 turkeys which had consumed aflatoxin-contaminated peanut meal. In the USSR in World War II, 5,000 people died due to Alimentary Toxic Aleukia (ALA). The common foodborne Mycotoxins include:

- Aflatoxins – originated from Aspergillus parasiticus and Aspergillus flavus. They are frequently found in tree nuts, peanuts, maize, sorghum and other oilseeds, including corn and cottonseeds. The pronounced forms of Aflatoxins are those of B1, B2, G1, and G2, amongst which Aflatoxin B1 predominantly targets the liver, which will result in necrosis, cirrhosis, and carcinoma. In the US, the acceptable level of total aflatoxins in foods is less than 20 μg/kg, except for Aflatoxin M1 in milk, which should be less than 0.5 μg/kg. The official document can be found at FDA's website.

- Altertoxins – are those of Alternariol (AOH), Alternariol methyl ether (AME), Altenuene (ALT), Altertoxin-1 (ATX-1), Tenuazonic acid (TeA) and Radicinin (RAD), originated from Alternaria spp. Some of the toxins can be present in sorghum, ragi, wheat and tomatoes. Some research has shown that the toxins can be easily cross-contaminated between grain commodities, suggesting that manufacturing and storage of grain commodities is a critical practice.
- Citrinin
- Citreoviridin
- Cyclopiazonic acid
- Cytochalasins
- Ergot alkaloids / Ergopeptine alkaloids – Ergotamine
- Fumonisins – Crop corn can be easily contaminated by the fungi Fusarium moniliforme, and its Fumonisin B1 will cause Leukoencephalomalacia (LEM) in horses, Pulmonary edema syndrome (PES) in pigs, liver cancer in rats and Esophageal cancer in humans. For human and animal health, both the FDA and the EC have regulated the content levels of toxins in food and animal feed.
- Fusaric acid
- Fusarochromanone
- Kojic acid
- Lolitrem alkaloids
- Moniliformin
- 3-Nitropropionic acid
- Nivalenol
- Ochratoxins – In Australia, The Limit of Reporting (LOR) level for Ochratoxin A (OTA) analyses in 20th Australian Total Diet Survey was 1 μg/kg, whereas the EC restricts the content of OTA to 5 μg/kg in cereal commodities, 3 μg/kg in processed products and 10 μg/kg in dried vine fruits.
- Oosporeine
- Patulin – Currently, this toxin has been advisably regulated on fruit products. The EC and the FDA have limited it to under 50 μg/kg for fruit juice and fruit nectar, while limits of 25 μg/kg for solid-contained fruit products and 10 μg/kg for baby foods were specified by the EC.
- Phomopsins
- Sporidesmin A
- Sterigmatocystin
- Tremorgenic mycotoxins – Five of them have been reported to be associated with molds found in fermented meats. These are Fumitremorgen B, Paxilline, Penitrem A, Verrucosidin, and Verruculogen.
- Trichothecenes – sourced from Cephalosporium, Fusarium, Myrothecium,

Stachybotrys and Trichoderma. The toxins are usually found in molded maize, wheat, corn, peanuts and rice, or animal feed of hay and straw. Four trichothecenes, T-2 toxin, HT-2 toxin, diacetoxyscirpenol (DAS) and deoxynivalenol (DON) have been most commonly encountered by humans and animals. The consequences of oral intake of, or dermal exposure to, the toxins will result in Alimentary toxic aleukia, neutropenia, aplastic anemia, thrombocytopenia and/or skin irritation. In 1993, the FDA issued a document for the content limits of DON in food and animal feed at an advisory level. In 2003, US published a patent that is very promising for farmers to produce a trichothecene-resistant crop.

- Zearalenone
- Zearalenols

Viruses

Viral infections make up perhaps one third of cases of food poisoning in developed countries. In the US, more than 50% of cases are viral and noroviruses are the most common foodborne illness, causing 57% of outbreaks in 2004. Foodborne viral infection are usually of intermediate (1–3 days) incubation period, causing illnesses which are self-limited in otherwise healthy individuals; they are similar to the bacterial forms described above.

- Enterovirus
- Hepatitis A is distinguished from other viral causes by its prolonged (2–6 week) incubation period and its ability to spread beyond the stomach and intestines into the liver. It often results in jaundice, or yellowing of the skin, but rarely leads to chronic liver dysfunction. The virus has been found to cause infection due to the consumption of fresh-cut produce which has fecal contamination.
- Hepatitis E
- Norovirus
- Rotavirus

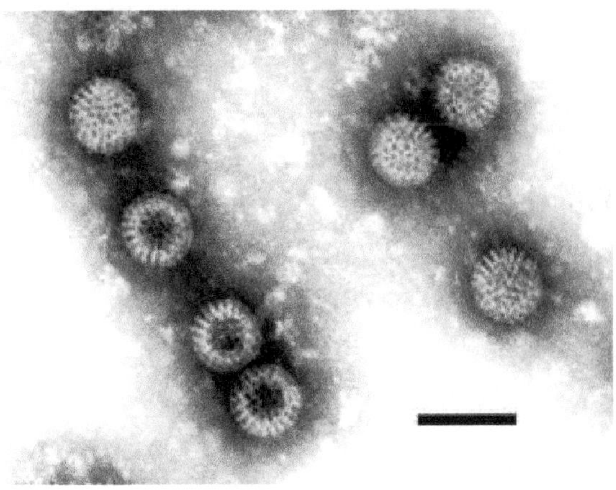

Rotavirus.

Parasites

Most foodborne parasites are zoonoses.

- Platyhelminthes:
  - *Diphyllobothrium* sp.
  - *Nanophyetus* sp.
  - *Taenia saginata*
  - *Taenia solium*
  - *Fasciola hepatica*
- Nematode:
  - *Anisakis* sp.
  - *Ascaris lumbricoides*
  - *Eustrongylides* sp.
  - *Trichinella spiralis*
  - *Trichuris trichiura*
- Protozoa:
  - *Acanthamoeba* and other free-living amoebae
  - *Cryptosporidium parvum*
  - *Cyclospora cayetanensis*
  - *Entamoeba histolytica*
  - *Giardia lamblia*

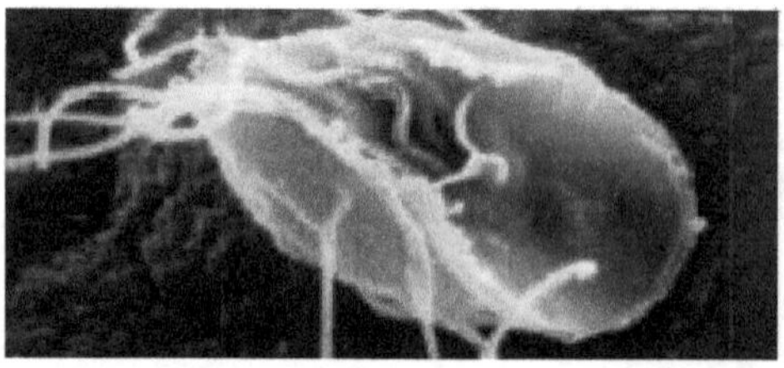

Giardia lamblia

- Sarcocystis hominis
- Sarcocystis suihominis
- Toxoplasma gondii

Natural Toxins

Several foods can naturally contain toxins, many of which are not produced by bacteria. Plants in particular may be toxic; animals which are naturally poisonous to eat are rare. In evolutionary terms, animals can escape being eaten by fleeing; plants can use only passive defenses such as poisons and distasteful substances, for example capsaicin in chili peppers and pungent sulfur compounds in garlic and onions. Most animal poisons are not synthesised by the animal, but acquired by eating poisonous plants to which the animal is immune, or by bacterial action.

- Alkaloids
- Ciguatera poisoning
- Grayanotoxin (honey intoxication)
- Mushroom toxins
- Phytohaemagglutinin (red kidney bean poisoning; destroyed by boiling)
- Pyrrolizidine alkaloids
- Shellfish toxin, including paralytic shellfish poisoning, diarrhetic shellfish poisoning, neurotoxic shellfish poisoning, amnesic shellfish poisoning and ciguatera fish poisoning
- Scombrotoxin
- Tetrodotoxin (fugu fish poisoning)

Some plants contain substances which are toxic in large doses, but have therapeutic properties in appropriate dosages.

- Foxglove contains cardiac glycosides.
- Poisonous hemlock (conium) has medicinal uses.

Other Pathogenic Agents

- Prions, resulting in Creutzfeldt–Jakob disease

"Ptomaine Poisoning"

In 1883, the Italian professor Salmi, of Bologna, introduced the generic name *ptomaine* (from Greek *ptōma*, "fall, fallen body, corpse") for alkaloids found in decaying animal and vegetable matter, especially (as reflected in their names) putrescine and cadaverine. The 1892 *Merck's Bulletin* stated, "We name such products of bacterial origin ptomaines; and the special alkaloid produced by the comma bacillus is variously named Cadaverine, Putrescine, etc." While *The Lancet* stated, "The chemical ferments produced in the system, the...ptomaines which may exercise so disastrous an influence." It is now known that the "disastrous...influence" is due to the direct action of bacteria and only slightly to the alkaloids. Thus, the use of the phrase "ptomaine poisoning" is now obsolete.

Tainted potato salad sickening hundreds at a Communist political convention in Massillon, Ohio, and aboard a Washington DC cruise boat in separate incidents during a single week in 1932 drew national attention to the dangers of so-called "ptomaine poisoning" in the pages of the American news weekly, *Time*. Another newspaper article from 1944 told of more than 150 persons being hospitalized in Chicago with ptomaine poisoning apparently from rice pudding served by a chain of restaurants.

Mechanism

Incubation Period

The delay between consumption of a contaminated food and appearance of the first symptoms of illness is called the incubation period. This ranges from hours to days (and rarely months or even years, such as in the case of listeriosis or bovine spongiform encephalopathy), depending on the agent, and on how much was consumed. If symptoms occur within one to six hours after eating the food, it suggests that it is caused by a bacterial toxin or a chemical rather than live bacteria.

The long incubation period of many foodborne illnesses tends to cause sufferers to attribute their symptoms to gastroenteritis.

During the incubation period, microbes pass through the stomach into the intestine, attach to the cells lining the intestinal walls, and begin to multiply there. Some types of microbes stay in the intestine, some produce a toxin that is absorbed into the bloodstream, and some can directly invade the deeper body tissues. The symptoms produced depend on the type of microbe.

Infectious Dose

The infectious dose is the amount of agent that must be consumed to give rise to symptoms of foodborne illness, and varies according to the agent and the consumer's age and overall health. Pathogens vary in minimum infectious dose; for example, *Shigella sonnei* has a low estimated minimum dose of < 500 colony-forming units (CFU) while *Staphylococcus aureus* has a relatively high estimate.

In the case of *Salmonella* a relatively large inoculum of 1 million to 1 billion organisms is necessary to produce symptoms in healthy human volunteers, as *Salmonellae* are very sensitive to acid. An unusually high stomach pH level (low acidity) greatly reduces the number of bacteria required to cause symptoms by a factor of between 10 and 100.

Epidemiology

Asymptomatic subclinical infection may help spread these diseases, particularly *Staphylococcus aureus*, *Campylobacter, Salmonella, Shigella, V. cholerae,* and *Yersini*. For example, as of 1984 it was estimated that in the United States, 200,000 people were asymptomatic carriers of Salmonella.

United States

In the United States, using FoodNet data from 2000–2007, the CDCP estimated there were 47.8 million foodborne illnesses per year (16,000 cases for 100,000 inhabitants) with 9.4 million of these caused by 31 known identified pathogens.

- 127,839 were hospitalized (43 per 100,000 inhabitants per year).
- 3,037 people died (1.0 per 100,000 inhabitants per year).

Causes of foodborne illness in U. S.			
	Cause	**Annual cases**	**R a t e (per 100,000 inhabitants)**
1	*Norovirus*	5,461,731 cases	X
2	*Salmonella*	1,027,561 cases	X
3	*Clostridium perfringens*	965,958 cases	X
4	Campylobacter	845,024 cases	X

Causes of death by foodborne illness in U. S.			
	Cause	**Annual deaths**	**R a t e (per 100,000 inhabitants)**
1	*Salmonella*	378 cases	0.126
2	*Toxoplasma gondii*	327 cases	0.109
3	*Listeria*	255 cases	0.085
4	*Norovirus*	149 cases	0.050

France

This data pertains to reported medical cases of 23 specific pathogens, as opposed to total population estimates of all food-borne illness for the United States.

In France, for 750,000 cases (1210 per 100,000 inhabitants):

- 70,000 people consulted in the emergency department of an hospital (113 per 100,000 inhabitants.);
- 113,000 people were hospitalized (182 per 100,000 inhabitants);
- 460 people died (0.75 per 100,000 inhabitants).

Causes of foodborne illness in France			
	Cause	**Annual hospitalizations**	**Rate (per 100,000 inhabitants)**
1	***Salmonella***	**~8,000 cases**	**13**
2	***Campylobacter***	**~3,000 cases**	**4.8**
3	**Parasites** **incl. Toxoplasma**	**~500 cases** **~400 cases**	**0.8** **0.65**
4	***Listeria***	**~300 cases**	**0.5**
5	**Hepatitis A**	**~60 cases**	**0.1**

Causes of death by foodborne illness in France			
	Cause	**Annual**	**Rate (per 100,000 inhabitants)**
1	***Salmonella***	**~300 cases**	**0.5**
2	***Listeria***	**~80 cases**	**0.13**
3	**Parasites**	**~37 cases**	**0.06 (95% due to toxoplasma)**
4	***Campylobacter***	**~15 cases**	**0.02**
5	**Hepatitis A**	**~2 cases**	**0.003**

Australia

A study by the Australian National University, published in November 2014, found in 2010 that there were an estimated 4.1 million cases of foodborne gastroenteritis acquired in Australia on average each year, along with 5,140 cases of non-gastrointestinal illness. The study was funded by the Australian Department of Health, Food Standards Australia New Zealand and the NSW Food Authority.

The main causes were Norovirus, pathogenic Escherichia coli, Campylobacter spp. and non-typhoidal Salmonella spp., although the causes of approximately 80% of illnesses were unknown. Approximately 25% (90% CrI: 13%–42%) of the 15.9 million episodes of gastroenteritis that occur in Australia were estimated to be transmitted by contaminated food. This equates to an average of approximately one episode of foodborne gastroenteritis every five years per person. Data on the number of hospitalisations and deaths represent the occurrence of serious foodborne illness. Including gastroenteritis, non-gastroenteritis and sequelae, there were an estimated annual 31,920 (90% CrI: 29,500–35,500) hospitalisations due to foodborne illness and 86 (90% CrI: 70–105) deaths due to foodborne illness circa 2010. This study concludes that these rates are similar to recent estimates in the US and Canada.

A main aim of this study was to compare if foodborne illness incidence had increased over time. In this study, similar methods of assessment were applied to data from circa 2000, which showed that the rate of foodborne gastroenteritis had not changed significantly over time. Two key estimates were the total number of gastroenteritis episodes each year, and the proportion considered foodborne. In circa 2010, it was estimated that 25% of all episodes of gastroenteritis were foodborne. By applying this proportion of episodes due to food to the incidence of gastroenteritis circa 2000, there were an estimated 4.3 million (90% CrI: 2.2–7.3 million) episodes of foodborne gastroenteritis circa 2000, although credible intervals overlap with 2010. Taking into account changes in population size, applying these equivalent methods suggests a 17% decrease in the rate of foodborne gastroenteritis between 2000 and 2010, with considerable overlap of the 90% credible intervals.

This study replaces a previous estimate of 5.4 million cases of food-borne illness in Australia every year, causing:

- 18,000 hospitalizations
- 120 deaths (0.5 deaths per 100,000 inhabitants)
- 2.1 million lost days off work
- 1.2 million doctor consultations
- 300,000 prescriptions for antibiotics.

Most foodborne disease outbreaks in Australia have been linked to raw or minimally cooked eggs or poultry. The Australian Food Safety Information Council estimates that one third of cases of food poisoning occur in the home

Comparison between Countries

Country	Annual deaths per 100,000 inhabitants	Annual hospitalization per 100,000 inhabitants
USA	1.0	43
France	0.75	182
Australia	0.5	82

Outbreaks

The vast majority of reported cases of foodborne illness occur as individual or sporadic cases. The origin of most sporadic cases is undetermined. In the United States, where people eat outside the home frequently, 58% of cases originate from commercial food facilities (2004 FoodNet data). An outbreak is defined as occurring when two or more people experience similar illness after consuming food from a common source.

Often, a combination of events contributes to an outbreak, for example, food might be left at room temperature for many hours, allowing bacteria to multiply which is compounded by inadequate cooking which results in a failure to kill the dangerously elevated bacterial levels.

Outbreaks are usually identified when those affected know each other. However, more and more, outbreaks are identified by public health staff from unexpected increases in laboratory results for certain strains of bacteria. Outbreak detection and investigation in the United States is primarily handled by local health jurisdictions and is inconsistent from district to district. It is estimated that 1–2% of outbreaks are detected.

Society and Culture

United Kingdom

In postwar Aberdeen (1964) a large-scale (>400 cases) outbreak of typhoid occurred, caused by contaminated corned beef which had been imported from Argentina. The corned beef was placed in cans and because the cooling plant had failed, cold river water from the Plate estuary was used to cool the cans. One of the cans had a defect and the meat inside was contaminated. This meat was then sliced using a meat slicer in a shop in Aberdeen, and a lack of cleaning the machinery led to spreading the contamination to other meats cut in the slicer. These meats were then eaten by the people of Aberdeen who then became ill.

Serious outbreaks of foodborne illness since the 1970s prompted key changes in UK food safety law. These included the death of 19 patients in the Stanley Royd Hospital outbreak and the bovine spongiform encephalopathy (BSE, mad cow disease) outbreak identified in the 1980s. The death of 17 people in the 1996 Wishaw outbreak of *E. coli* O157 was a precursor to the establishment of the Food Standards Agency which, according to Tony Blair in the 1998 white paper *A Force for Change* Cm 3830, "would be powerful, open and dedicated to the interests of consumers".

In May 2015, for the second year running, England's Food Standards Agency devoted its annual Food Safety Week to – "The Chicken Challenge". The focus was on the handling of raw chicken in the home and in catering facilities in a drive to reduce the worryingly high levels of food poisoning from the *campylobacter* bacterium. Anne Hardy argues that widespread public education of food hygiene can be useful, particularly through media (T.V cookery programmes) and advertisement. She points to the examples set by Scandinavian societies.

United States

In 2001, the Center for Science in the Public Interest petitioned the United States Department of Agriculture to require meat packers to remove spinal cords before processing cattle carcasses for human consumption, a measure designed to lessen the risk of infection by variant Creutzfeldt–Jakob disease. The petition was supported by the American Public Health Association, the Consumer Federation of America, the Government Accountability Project, the National Consumers League, and Safe Tables Our Priority. This was opposed by the National Cattlemen's Beef Association, the National Renderers Association, the National Meat Association, the Pork Producers Council, sheep raisers, milk producers, the Turkey Federation, and eight other organizations from the animal-derived food industry.

None of the US Department of Health and Human Services targets regarding incidence of foodborne infections were reached in 2007.

Organizations

The World Health Organization Department of Food Safety and Zoonoses (FOS) provides scientific advice for organizations and the public on issues concerning the safety of food. Its mission is to lower the burden of foodborne disease, thereby strengthening the health security and sustainable development of Member States. Foodborne and waterborne diarrhoeal diseases kill an estimated 2.2 million people annually, most of whom are children. WHO works closely with the Food and Agriculture Organization of the United Nations (FAO) to address food safety issues along the entire food production chain—from production to consumption—using new methods of risk analysis. These methods provide efficient, science-based tools to improve food safety, thereby benefiting both public health and economic development.

International Food Safety Authorities Network (INFOSAN)

The International Food Safety Authorities Network (INFOSAN) is a joint program of the WHO and FAO. INFOSAN has been connecting national authorities from around the globe since 2004, with the goal of preventing the international spread of contaminated food and foodborne disease and strengthening food safety systems globally. This is done by:

- Promoting the rapid exchange of information during food safety events;
- Sharing information on important food safety issues of global interest;

- Promoting partnership and collaboration between countries; and
- Helping countries strengthen their capacity to manage food safety risks.

Membership to INFOSAN is voluntary, but is restricted to representatives from national and regional government authorities and requires an official letter of designation. INFOSAN seeks to reflect the multidisciplinary nature of food safety and promote intersectoral collaboration by requesting the designation of Focal Points in each of the respective national authorities with a stake in food safety, and a single Emergency Contact Point in the national authority with the responsibility for coordinating national food safety emergencies; countries choosing to be members of INFOSAN are committed to sharing information between their respective food safety authorities and other INFOSAN members. The operational definition of a food safety authority includes those authorities involved in: food policy; risk assessment; food control and management; food inspection services; foodborne disease surveillance and response; laboratory services for monitoring and surveillance of foods and foodborne diseases; and food safety information, education and communication across the farm-to-table continuum.

Regulatory Steps

Food may be contaminated during all stages of food production and retailing. In order to prevent viral contamination, regulatory authorities in Europe have enacted several measures:

- European Commission Regulation (EC) No 2073/2005 of November 15, 2005
- European Committee for Standardization (CEN): Standard method for the detection of norovirus and hepatitis A virus in food products (shellfish, fruits and vegetables, surfaces and bottled water)
- CODEX Committee on Food Hygiene (CCFH): Guideline for the application of general principles of food hygiene for the control of viruses in food

REFERENCES

1. "food poisoning" at Dorland's Medical Dictionary
2. "Foodborne Illness - Frequently Asked Questions". US Centers for Disease Control and Prevention. Archived from the original on March 3, 2011. Retrieved 3 July 2016.
3. For foodborne illness caused by chemicals, see Food contaminants.
4. "Reducing the risk from E. coli 0157 – controlling cross-contamination". Food Standards Agency, United Kingdom. February 2011. Archived from the original on April 16, 2014. Retrieved 14 August 2016.
5. "Staphylococcal Food Poisoning". U.S. Centers for Disease Control and Prevention. Retrieved 3 July 2016.
6. Humphrey T, O'Brien S, Madsen M (2007). "Campylobacters as zoonotic pathogens: a food production perspective". International Journal of Food Microbiology. 117 (3): 237–57. doi:10.1016/j.ijfoodmicro.2007.01.006. PMID 17368847.

7. "Foodborne Illness: What Consumers Need to Know". USDA.gov. Retrieved 14 August 2016.
8. Tribe, Ingrid G; Cowell, David; Cameron, Peter; Cameron, Scott (2002). "An outbreak of Salmonella Typhimurium phage type 135 infection linked to the consumption of raw shell eggs in an aged care facility". Communicable Diseases Intelligence. 26 (1): 38–9. PMID 11950200. Archived from the original on February 17, 2014.
9. "Salmonella Infection (salmonellosis) and Animals". Centers for Disease Control and Prevention. Retrieved August 12, 2007.
10. Doyle, M. P.; Erickson, M. C. (2006). "Reducing the carriage of foodborne pathogens in livestock and poultry". Poultry science. 85 (6): 960–73. doi:10.1093/ps/85.6.960. PMID 16776463.
11. Food poisoning: Causes. Mayo Clinic.
12. Argudín M, Mendoza MC, Rodicio MR (2010). "Food poisoning and Staphylococcus aureus enterotoxins". Toxins (Basel). 2 (7): 1751–73. doi:10.3390/toxins2071751. PMC 3153270 . PMID 22069659.
13. Staphylococcus aureus: A Problem When Food Is Left Out Too Long. Ohio State University Extension HYG-5564-11].
14. "FDA Says Viruses Safe for Treating Meat". Archived from the original on August 25, 2006. Retrieved September 2, 2014.
15. DeRusha, Jason. "Good Question: Does Washing Fruit Do Anything?". Retrieved 2016-09-18.
16. How Temperatures Affect Food. U.D. Department of Agriculture
17. Mount, Michael. "Fungi and Mycotoxins" (PDF). vetmed.ucdavis.edu. Retrieved August 11, 2007.
18. "Aflatoxins". Center for Food Safety & Applied Nutrition. Retrieved August 12, 2007.
19. "GASGA Technical Leaflet – 3 Mycotoxins in Grain". Food and Agriculture Organization of the United Nations. Retrieved August 12, 2007.
20. "Chapter 2 Foodborne Hazards in Basic Food Safety for Health Workers" (PDF). World Health Organization. Retrieved August 12, 2007.
21. "Sec. 683.100 Action Levels for Aflatoxins in Animal Feeds (CPG 7126.33)". Food and Drug Administration. Retrieved August 13, 2007.
22. Henry, Michael H. "Mycotoxins in Feeds: CVM's Perspective". FDA.gov. Retrieved January 1, 2012.
23. Webley DJ, Jackson KL, Mullins JD, Hocking AD, Pitt JI (1997). "Alternaria toxins in weather-damaged wheat and sorghum in the 1995–1996 Australian harvest". Australian Journal of Agricultural Research. 48 (8): 1249–56. doi:10.1071/A97005.

24. Li F, Yoshizawa T (2000). "Alternaria mycotoxins in weathered wheat from China". J. Agric. Food Chem. 48 (7): 2920–4. doi:10.1021/jf0000171. PMID 10898645.
25. Motta SD, Valente Soares LM (2001). "Survey of Brazilian tomato products for alternariol, alternariol monomethyl ether, tenuazonic acid and cyclopiazonic acid". Food Addit Contam. 18 (7): 630–4. doi:10.1080/02652030117707. PMID 11469319.
26. Li FQ, Toyazaki N, Yoshizawa T (2001). "Production of alternaria mycotoxins by Alternaria alternata isolated from weather-damaged wheat". J. Food Prot. 64 (4): 567–71. PMID 11307900.
27. Marasas WF (1995). "Fumonisins: their implications for human and animal health". Nat. Toxins. 3 (4): 193–8; discussion 221. doi:10.1002/nt.2620030405. PMID 7582616.
28. Soriano, J.M. (2004). "Occurrence of fumonisins in foods". Food Research International. 37 (10): 985–1000. doi:10.1016/j.foodres.2004.06.009.
29. "CVM and Fumonisins". Food and Drug Administration. Archived from the original on August 12, 2007. Retrieved August 13, 2007.
30. "More contaminated maize meal products withdrawn from sale". Food Standards Agency. Retrieved August 12, 2007.
31. "20th Australian Total Diet Survey – Part B". Food Standards Australia New Zealand. Retrieved August 13, 2007.
32. "Worldwide regulations for mycotoxins in food and feed in 2003". FAO FOOD AND NUTRITION PAPER 81. Retrieved August 13, 2007.
33. "Patulin in Apple Juice, Apple Juice Concentrates and Apple Juice Products". Food and Drug Administration. Retrieved August 16, 2007.
34. Sabater-Vilar M, Nijmeijer S, Fink-Gremmels J (2003). "Genotoxicity assessment of five tremorgenic mycotoxins (fumitremorgen B, paxilline, penitrem A, verruculogen, and verrucosidin) produced by molds isolated from fermented meats". J. Food Prot. 66 (11): 2123–9. PMID 14627292.
35. Adejumo TO, Hettwer U, Karlovsky P (May 2007). "Occurrence of Fusarium species and trichothecenes in Nigerian maize". Int. J. Food Microbiol. 116 (3): 350–7. doi:10.1016/j.ijfoodmicro.2007.02.009. PMID 17412440.
36. Mazur LJ, Kim J (2006). "Spectrum of noninfectious health effects from molds". Pediatrics. 118 (6): e1909–26. doi:10.1542/peds.2006-2829. PMID 17142508.
37. Froquet R, Sibiril Y, Parent-Massin D (2001). "Trichothecene toxicity on human megakaryocyte progenitors (CFU-MK)". Hum Exp Toxicol. 20 (2): 84–9. doi:10.1191/096032701677428611. PMID 11327514.
38. Joffe AZ, Yagen B (1977). "Comparative study of the yield of T-2 toxic produced by Fusarium poae, F. sporotrichioides and F. sporotrichioides var. tricinctum strains from different sources". Mycopathologia. 60 (2): 93–7. doi:10.1007/bf00490378. PMID 846559.

39. Hay, R. J. (2007). "Fusarium infections of the skin". Current Opinion in Infectious Diseases. 20 (2): 115. doi:10.1097/QCO.0b013e328014392d.
40. "Guidance for Industry and FDA – Letter to State Agricultural Directors, State Feed Control Officials, and Food, Feed, and Grain Trade Organizations". Food and Drug Administration. Archived from the original on June 9, 2007. Retrieved August 13, 2007.
41. Hohn, Thomas M. "Trichothecene-resistant transgenic plants". U.S. Patent 6,646,184. Priority date March 31, 1999.
42. Dubois E, Hennechart C, Deboosère N, et al. (April 2006). "Intra-laboratory validation of a concentration method adapted for the enumeration of infectious F-specific RNA coliphage, enterovirus, and hepatitis A virus from inoculated leaves of salad vegetables". Int. J. Food Microbiol. 108 (2): 164–71. doi:10.1016/j.ijfoodmicro.2005.11.007. PMID 16387377.
43. Schmidt, Heather Martin. "Improving the microbilological quality and safety of fresh-cut tomatoes by low dose dlectron beam irradiation – Master thesis" (PDF). Retrieved August 11, 2007.
44. Oxford English Dictionary (First ed.). London: Oxford University Press. 1933. p. 1557.
45. Merck's Bulletin, Volume 5, William Henry Porter, 1892.
46. Lancet, 3 Oct 1891, page 752
47. "Report Next Week," Evening Independent (Massillon, OH), July 22, 1932, pg. 6
48. "Medicine: Potato Salad," Time, August 1, 1932.
49. "Poisoning Hits 150 in Chicago". Kenosha Evening News. May 27, 1944.
50. "Food-Related Diseases". US Centers for Disease Control and Prevention.
51. Greig, Judy D. "Infective Doses and Pathogen Carriage". Public Health Agency of Canada.
52. Owens, Michael D (January 2014) Salmonella Infection in Emergency Medicine. emedicine.com
53. Scallan E, Griffin PM, Angulo FJ, Tauxe RV, Hoekstra RM (2011). "Foodborne illness acquired in the United States—unspecified agents". Emerging Infectious Diseases. 17 (1): 16–22. doi:10.3201/eid1701.P21101. PMC 3204615 . PMID 21192849.
54. Scallan E, Hoekstra RM, Angulo FJ, Tauxe RV, Widdowson MA, Roy SL, et al. (2011). "Foodborne illness acquired in the United States—major pathogens". Emerging Infectious Diseases. 17 (1): 7–15. doi:10.3201/eid1701.P11101. PMC 3375761 . PMID 21192848.
55. "Obama Proposes Single Overseer for Food Safety". New York Times. February 20, 2015. Retrieved 2015-02-22. According to the C.D.C., an estimated 87 million Americans are sickened each year by contaminated food, 371,000 are hospitalized with food-related illness and 5,700 die from food-related disease
56. Sabrina Tavernise (July 26, 2013). "F.D.A. Says Importers Must Audit Food

Safety". New York Times. Retrieved July 27, 2013. One in every six Americans becomes ill from eating contaminated food each year, Dr. Margaret A. Hamburg, F.D.A. commissioner, estimated. About 130,000 are hospitalized and 3,000 die.

57. Stephanie Strom (January 4, 2013). "F.D.A. Offers Sweeping Rules to Fight Food Contamination". New York Times. Retrieved January 5, 2013. One in six Americans becomes ill from eating contaminated food each year, the government estimates; of those, roughly 130,000 are hospitalized and 3,000 die.
58. "Report of the French sanitary agencies" (PDF) (in French). INVS/Afssa.
59. "Summary of Report of the French sanitary agencies"(PDF) (in French). INVS/Afssa.
60. Kirk, Martyn; et al. "Foodborne illness in Australia: Annual incidence circa 2010 pp7-9". Australia Department of Health. Australian National University. Retrieved 13 September 2015.
61. "Food borne illness in Australia" (PDF). OzFoodNet.
62. Astridge, K; McPherson, M; Kirk, M; et al. (2011). "Foodborne disease outbreaks in Australia 2001-2009". Food Australia. 63 (12): 44–50.
63. "Food Hygiene". BUPA. Retrieved 13 September 2015.
64. Smith, David F.; Diack, H. Lesley and Pennington, T. Hugh (2005) Food Poisoning, Policy and Politics : Corned Beef and Typhoid in Britain in the 1960s. Boydell Press. ISBN 1-84383-138-4
65. Brian Deer (February 24, 1985). "Food poison deaths probe may reveal NHS flaws". The Sunday Times.
66. Cowden JM, Ahmed S, Donaghy M, Riley A (2001). "Epidemiological investigation of the Central Scotland outbreak of Escherichia coli O157 infection, November to December 1996" (PDF). Epidemiol. Infect. 126: 335–341. doi:10.1017/S0950268801005520. PMC 2869700 . PMID 11467789.
67. "The Food Standards Agency: A Force for Change". gov.uk. Food Standards Agency, United Kingdom. January 14, 1998. Retrieved 14 August 2016.
68. Hardy, Anne (13 January 2016). "Food Poisoning: An On-going Saga". History and Policy. History and Policy. Retrieved 4 July 2016.
69. "Meat Produced by Advanced Meat/Bone Separation Machinery and Meat Recovery (AMR) Systems". Federal Register. U.S. National Archive and Records Administration. Retrieved 3 July 2016.
70. "Livestock Industry Opposed to Stricter Laws on Mad Cow Prevention". January 2003. Retrieved 3 July 2016.
71. Healthy People 2010 Home Page
72. "Preliminary FoodNet data on the incidence of infection with pathogens transmitted commonly through food—10 states, 2007". MMWR Morb. Mortal. Wkly. Rep. 57 (14): 366–70. April 2008. PMID 18401330.
73. "Codex Committee on Food Hygiene (CCFH)", European Commission, Retrieved April 7, 2015

CHAPTER 3

Drinking Water and Water Borne Diseases

Chapter Contents

- **Introduction**
- **Drinking**
- **Methods**
- **Hydration and Dehydration**
- **Alcoholic Beverages**
- **Approach**
- **Methodologies**
- **Pathogen Analysis**
- **Types of Nutrient Media Used in Analysis**

Waterborne diseases are caused by drinking contaminated or dirty water. Contaminated water can cause many types of diarrheal diseases, including Cholera, and other serious illnesses such as Guinea worm disease, Typhoid, and Dysentery. Water related diseases cause 3.4 million deaths each year. LifeStraw water filters remove the bacteria and pathogens that contaminate water to decrease the incidence of waterborne diseases.

INTRODUCTION

Bacteriological water analysis is a method of analysing water to estimate the numbers of bacteria present and, if needed, to find out what sort of bacteria they are. It represents one aspect of water quality. It is a microbiological analytical procedure which uses samples of water and from these samples determines the concentration of bacteria. It is then possible to draw inferences about the suitability of the water for use from these concentrations. This process is used, for example, to routinely confirm that water is safe for human consumption or that bathing and recreational waters are safe to use.

The interpretation and the action trigger levels for different waters vary depending on the use made of the water. Whilst very stringent levels apply to drinking water, more relaxed levels apply to marine bathing waters, where much lower volumes of water are expected to be ingested by users.

DRINKING

Drinking is the act of ingesting water or other liquids into the body through the mouth. Water is required for many of life's physiological processes. Both excessive and inadequate water intake are associated with health problems.

The act of drinking portrayed in statuary—the figure employs a traditional waterskin.

METHODS

A Canada goose scooping water into its bill.

In Humans

When a liquid is poured into an open human mouth, the swallowing process is completed by peristalsis which delivers the liquid to the stomach; much of the activity is abetted by gravity. The liquid may be poured from the hands or drinkware may be used as vessels. Drinking can also be performed by acts of inhalation, typically when imbibing hot liquids or drinking from a spoon. Infants employ a method of suction wherein the lips are pressed tight around a source, as in breastfeeding: a combination of breath and tongue movement creates a vacuum which draws in liquid.

In Aquatic Animals

Amphibians and aquatic animals which live in freshwater do not need to drink: they absorb water steadily through the skin by osmosis. Saltwater fish, however, drink through the mouth as they swim, and purge the excess salt through the gills.

In Land Animals

By necessity, terrestrial animals in captivity become accustomed to drinking water, but most free-roaming animals stay hydrated through the fluids and moisture in fresh food. When conditions impel them to drink from bodies of water, the methods and motions differ greatly among species. Many desert animals do not drink even if water becomes available, but rely on eating succulent plants.

Cats, canines, and ruminants all lower the neck and lap in water with their powerful tongues. Cats and canines lap up water with the tongue in a spoon-like shape. Ruminants and most other herbivores partially submerge the tip of the mouth in order to draw in water by means of a plunging action with the tongue held straight. Cats drink at a significantly slower pace than ruminants, who face greater natural predation hazards. Uniquely, elephants draw water into their trunks and squirt it into their mouths.

Most birds scoop or draw water into the buccal areas of their bills, raising and tilting their heads back to drink. An exception is the common pigeon which can suck in water directly by inhalation.

HYDRATION AND DEHYDRATION

Like nearly all other life forms, humans require water for tissue hydration. Lack of hydration causes thirst, a desire to drink which is regulated by the hypothalamus in response to subtle changes in the body's electrolyte levels and blood volume. A decline in total body water is called dehydration and will eventually lead to death by hypernatremia. Methods used in the management of dehydration include assisted drinking or oral rehydration therapy.

An overconsumption of water can lead to water intoxication, which can dangerously dilute the concentration of salts in the body. Overhydration sometimes occurs among athletes and outdoor laborers, but it can also be a sign of disease or damage to the hypothalamus. A persistent desire to drink inordinate quantities of water is a psychological condition termed polydipsia. It is often accompanied by polyuria and may itself be a symptom of Diabetes mellitus or Diabetes insipidus.

Drinking can be observed in many insect species.

Human Water Requirements

A daily intake of water is required for the normal physiological functioning of the human body. The USDA recommends a daily intake of total water: not necessarily by drinking but by consumption of water contained in other beverages and foods. The recommended intake is 3.7 liters (appx. 1 gallon) per day for an adult male, and 2.7 liters (appx. 0.75 gallon) for an adult female. Other sources, however, claim that a high intake of fresh drinking water, separate and distinct from other sources of moisture, is necessary for good health – eight servings per day of eight fluid ounces (1.8 liters, or 0.5 gallon) is the amount recommended by many nutritionists, although there is no scientific evidence supporting this recommendation.

ALCOHOLIC BEVERAGES

The term "drinking" is often used metonymically for the consumption of alcoholic beverages. Most cultures throughout history have incorporated some number of the wide variety of "strong drinks" into their meals, celebrations, ceremonies, toasts and other occasions. Evidence of fermented drinks in human culture goes back as early as the Neolithic Period, and the first pictorial evidence can be found in Egypt around 4,000 BC.

The drinking of alcohol has developed into a variety of well-established drinking cultures around the world. Despite its popularity, alcohol consumption poses significant health risks. Alcohol abuse and the addiction of alcoholism are common maladies in developed countries worldwide. A high rate of consumption can also lead to cirrhosis, gastritis, gout, pancreatitis, hypertension, various forms of cancer, and numerous other illnesses.

APPROACH

The common feature of all these routine screening procedures is that the primary analysis is for indicator organisms rather than the pathogens that might cause concern. Indicator organisms are bacteria such as non-specific coliforms, *Escherichia coli* and *Pseudomonas aeruginosa* that are very commonly found in the human or animal gut and which, if detected, may suggest the presence of sewage. Indicator organisms are used because even when a person is infected with a more pathogenic bacteria, they will still be excreting many millions times more indicator organisms than pathogens. It is therefore reasonable to surmise that if indicator organism levels are low, then pathogen levels will be very much lower or absent. Judgements as to suitability of water for use are based on very extensive precedents and relate to the probability of any sample population of bacteria being able to be infective at a reasonable statistical level of confidence.

Analysis is usually performed using culture, biochemical and sometimes optical methods. When indicator organisms levels exceed pre-set triggers, specific analysis for pathogens may then be undertaken and these can be quickly detected (where suspected) using specific culture methods or molecular biology.

METHODOLOGIES

Because the analysis is always based on a very small sample taken from a very large volume of water, all methods rely on statistical principles.

Multiple tube method

One of the oldest methods is called the multiple tube method. In this method a measured sub-sample (perhaps 10 ml) is diluted with 100 ml of sterile growth medium and an aliquot of 10 ml is then decanted into each of ten tubes. The remaining 10 ml is then diluted again and the process repeated. At the end of 5 dilutions this produces 50 tubes covering the dilution range of 1:10 through to 1:10000.

The tubes are then incubated at a pre-set temperature for a specified time and at the end of the process the number of tubes with growth in is counted for each dilution. Statistical tables are then used to derive the concentration of organisms in the original sample. This method can be enhanced by using indicator medium which changes colour when acid forming species are present and by including a tiny inverted tube called a Durham tube in each sample tube. The Durham inverted tube catches any gas produced.

The production of gas at 37 degrees Celsius is a strong indication of the presence of *Escherichia coli.*

ATP Testing

An ATP test is the process of rapidly measuring active microorganisms in water through detection adenosine triphosphate (ATP). ATP is a molecule found only in and around living cells, and as such it gives a direct measure of biological concentration and health. ATP is quantified by measuring the light produced through its reaction with the naturally occurring enzyme firefly luciferase using a luminometer. The amount of light produced is directly proportional to the amount of biological energy present in the sample.

Second generation ATP tests are specifically designed for water, wastewater and industrial applications where, for the most part, samples contain a variety of components that can interfere with the ATP assay.

Plate Count

The plate count method relies on bacteria growing a colony on a nutrient medium so that the colony becomes visible to the naked eye and the number of colonies on a plate can be counted. To be effective, the dilution of the original sample must be arranged so that on average between 30 and 300 colonies of the target bacterium are grown. Fewer than 30 colonies makes the interpretation statistically unsound whilst greater than 300 colonies often results in overlapping colonies and imprecision in the count. To ensure that an appropriate number of colonies will be generated several dilutions are normally cultured. This approach is widely utilised for the evaluation of the effectiveness of water treatment by the inactivation of representative microbial contaminants such as "E. coli" following ASTM D5465.

The laboratory procedure involves making serial dilutions of the sample (1:10, 1:100, 1:1000, etc.) in sterile water and cultivating these on nutrient agar in a dish that is sealed and incubated. Typical media include plate count agar for a general count or MacConkey agar to count Gram-negative bacteria such as *E. coli*. Typically one set of plates is incubated at 22°C and for 24 hours and a second set at 37°C for 24 hours. The composition of the nutrient usually includes reagents that resist the growth of non-target organisms and make the target organism easily identified, often by a colour change in the medium. Some recent methods include a fluorescent agent so that counting of the colonies can be automated. At the end of the incubation period the colonies are counted by eye, a procedure that takes a few moments and does not require a microscope as the colonies are typically a few millimetres across.

Membrane Filtration

Most modern laboratories use a refinement of total plate count in which serial dilutions of the sample are vacuum filtered through purpose made membrane filters and these filters are themselves laid on nutrient medium within sealed plates. The methodology is otherwise similar to conventional total plate counts. Membranes have a printed millimetre grid printed on and can be reliably used to count the number of colonies under a binocular microscope.

Pour Plate Method

When the analysis is looking for bacterial species that grow poorly in air, the initial analysis is done by mixing serial dilutions of the sample in liquid nutrient agar which is then poured into bottles which are then sealed and laid on their sides to produce a sloping agar surface. Colonies that develop in the body of the medium can be counted by eye after incubation.

The total number of colonies is referred to as the Total Viable Count (TVC). The unit of measurement is cfu/ml (or colony forming units per millilitre) and relates to the original sample. Calculation of this is a multiple of the counted number of colonies multiplied by the dilution used.

PATHOGEN ANALYSIS

When samples show elevated levels of indicator bacteria, further analysis is often undertaken to look for specific pathogenic bacteria. Species commonly investigated in the temperate zone include *Salmonella typhi* and *Salmonella Typhimurium*. Depending on the likely source of contamination investigation may also extend to organisms such as *Cryptosporidium spp*. In tropical areas analysis of *Vibrio cholerae* is also routinely undertaken.

TYPES OF NUTRIENT MEDIA USED IN ANALYSIS

MacConkey agar is culture medium designed to grow Gram-negative bacteria and stain them for lactose fermentation. It contains bile salts (to inhibit most Gram-positive bacteria), crystal violet dye (which also inhibits certain Gram-positive bacteria), neutral red dye (which stains microbes fermenting lactose), lactose and peptone. Alfred Theodore MacConkey developed it while working as a bacteriologist for the Royal Commission on Sewage Disposal in the United Kingdom.

Endo agar contains peptone, lactose, dipotassium phosphate, agar, sodium sulfite, basic fuchsin and was originally developed for the isolation of *Salmonella typhi*, but is now commonly used in water analysis. As in MacConkey agar, coliform organisms ferment the lactose, and the colonies become red. Non-lactose-fermenting organisms produce clear, colourless colonies against the faint pink background of the medium.

mFC medium is used in membrane filtration and contains selective and differential agents. These include Rosolic acid to inhibit bacterial growth in general, except for faecal coliforms, Bile salts inhibit non-enteric bacteria and Aniline blue indicates the ability of faecal coliforms to ferment lactose to acid that causes a pH change in the medium.

TYEA medium contains tryptone, yeast extract, common salt and L-arabinose per liter of glass distilled water and is a non selective medium usually cultivated at two temperatures (22 and 36°C) to determine a general level of contamination (a.k.a. colony count).

REFERENCES

1. U.S. Environmental Protection Agency (EPA), Washington, D.C. (October 2002). "Method 1680: Fecal Coliforms in Biosolids by Multiple-Tube Fermentation Procedures." Draft. Document no. EPA-821-R-02-026.
2. Hanaor, Dorian A. H.; Sorrell, Charles C. (2014). "Sand Supported Mixed-Phase TiO_2 Photocatalysts for Water Decontamination Applications". Advanced Engineering Materials. 16 (2): 248–254. doi:10.1002/adem.201300259.
3. Hanaor, D.; Michelazzi, M.; Leonelli, C.; Sorrell, C.C. (2011). "The effects of firing conditions on the properties of electrophoretically deposited titanium dioxide films on graphite substrates". Journal of the European Ceramic Society. 31 (15): 2877–2885. doi:10.1016/j.jeurceramsoc.2011.07.007.
4. EPA (2002). «Method 1106.1: Enterococci in Water by Membrane Filtration Using membrane-Enterococcus-Esculin Iron Agar (mE-EIA)." Document no. EPA 821-R-02-021.
5. Neogen Corporation, Lansing, MI (2011). "m-Endo Agar (7724)." Product information sheet no. PI 7724, Rev 1.
6. U.S. Geological Survey. Ohio Water Microbiology Laboratory, Columbus, OH. (January 2007). "mFC agar method for fecal coliforms." Analytical Methods.

CHAPTER 4

Oxygenation and Air Borne Diseases

Chapter Contents

Airborne disease can spread when an infected person coughs, sneezes, or talks, spewing nasal and throat secretions into the air. Certain viruses or bacteria take flight and hang in the air or land on other people or surfaces.

When you breathe airborne pathogenic organisms in, they take up residence inside you. You can also pick up germs when you touch an infected surface, and then touch your own eyes, nose, or mouth.

INTRODUCTION

Environmental **oxygenation** can be important to the sustainability of a particular ecosystem. Insufficient oxygen (environmental hypoxia) may occur in bodies of water such as ponds and rivers, tending to suppress the presence of aerobic organisms such as fish. Deoxygenation increases the relative population of anaerobic organisms such as plants and some bacteria, resulting in fish kills and other adverse events. The net effect is to alter the balance of nature by increasing the concentration of anaerobic over aerobic species.

Oxygenation by water aeration can be part of the environmental remediation of a usually stagnant body of water. For example, Bubbly Creek in Chicago, Illinois, was hypoxic (deficient in oxygen) due to its use as an open sewer by Chicago's meat packing industry but has been oxygenated by introducing compressed air into its waters, increasing the fish population. A similar technique has previously been used in the Thames.

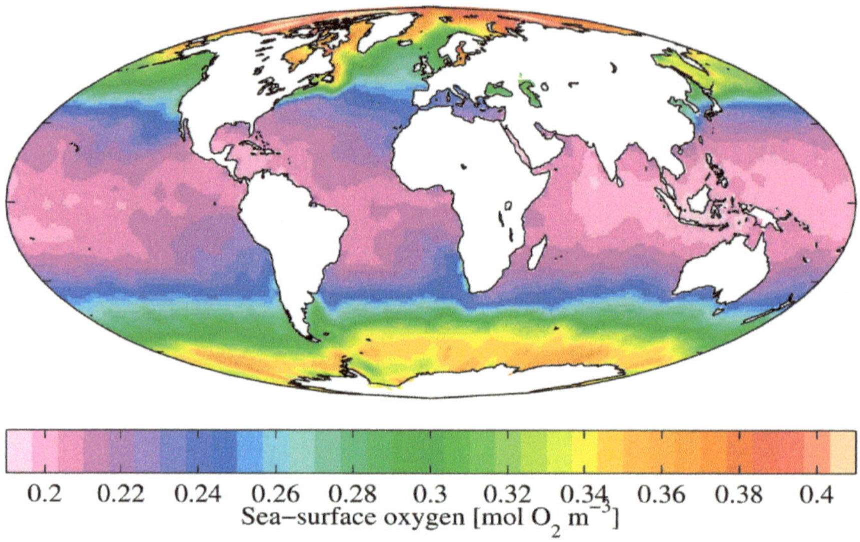

Annual mean sea surface dissolved **oxygen** for the World Ocean. Data from the World Ocean Atlas 2009.

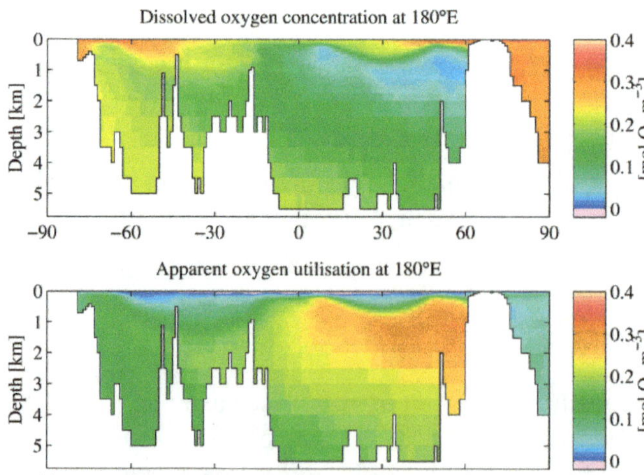

Pacific Ocean sections of dissolved oxygen and apparent oxygen utilisation. Data from the World Ocean Atlas 2009.[1]

Dissolved oxygen (DO) is measured in standard solution units such as millilitres O_2 per liter (ml/L), millimoles O_2 per liter (mmol/L), milligrams O_2 per liter (mg/L) and moles O_2 per cubic meter (mol/mm³). For example, in freshwater under atmospheric pressure at 20 °C, O_2 saturation is 9.1 mg/L.

In aquatic environments, **oxygen saturation** is a relative measure of the amount of oxygen (O_2) dissolved in the water compared to equilibrium conditions.

Supersaturation of oxygen (i.e. saturation levels above 100%) can occur naturally. The most common cause is oxygen production by photosynthetically active species such as plants and algae. According to Henry›s law, the equilibrium oxygen concentration is proportional to the partial pressure of oxygen gas. As air contains about 21% oxygen the equilibrium concentration of pure oxygen gas corresponds to nearly 500% air saturation. The other reason is that the oxygen concentration can be slow to adjust to changes in the environment. A rapid increase in temperature can reduce the equilibrium concentration of oxygen to a value below the actual concentration in the water, giving raise to more than 100% saturation until the system has had time to equilibrate through diffusion.[2] Supersaturation can sometimes be harmful for organisms and cause decompression sickness.

Solubility tables (based upon temperature) and corrections for different salinities and pressures can be found at the USGS web site.[3] Tables such as these of *DO* in milliliters per liter (ml/L) are based upon empirical equations that have been worked out and tested:[4]

$$ln(DO) = A1 + A2 * 100/T + A3 * ln(T/100) + A4 * T/100 + S * [B1 + B2 * T/100 + B3 * (T/100)^2]$$

Where *ln* is the natural logarithm and the other variables take the following values:

A1	=	-173.4292		*B1*	=	-0.033096
A2	=	249.6339		*B2*	=	0.014259
A3	=	143.3483		*B3*	=	-0.001700
A4	=	-21.8492				
T	=	temperature (kelvin)		*S*	=	salinity (g/kg)

To convert the calculated *DO* above from ml/L to mg/L, multiply the answer by (P/T)*0.55130, P=mmHg, T=Kelvin

MEASUREMENT

DO levels are typically measured using "rugged dissolved oxygen" (RDO) equipment which measures luminescence quenching ability of a sample. Increased oxygen levels result in increased quenching which is well characterised and allows accurate measurements to be made with a probe which requires minimal maintenance. Prior to the development of RDO technology membrane redox technology was used which measured oxygen levels using a clark electrode. Electrochemical equipment requires considerable maintenance to remove fouling and prevent degradation of the membrane. Redox methods may also display some cross-sensitivity to other gases such as H_2S.

For small or low concentration (<2ppm) samples RDO equipment is significantly better as it does not consume oxygen in the sample (and therefore does not require stirring) or struggle to measure zero-levels.

Wet chemistry methods such as the Winkler test for dissolved oxygen can also be used for DO measurement but as with all wet chemistry measurements these require a skilled technician to obtain accurate results.

AIRBORNE DISEASE

Airborne diseases can be spread via respiratory droplets expelled from the mouth and nose.

An **airborne disease** is any disease that is caused by pathogens and transmitted through the air. Such diseases include many that are of considerable importance both in human and veterinary medicine. The relevant pathogens may be viruses, bacteria, or fungi, and they may be spread through coughing, sneezing, raising of dust, spraying of liquids, or similar activities likely to generate aerosol particles or droplets. Strictly speaking airborne diseases do not include conditions caused simply by air pollution such as dusts and poisons, though their study and prevention may be related.

Airborne diseases include any that are caused and transmitted through the air. Some are of great medical importance. The pathogens transmitted may be any kind of microbe, and they may be spread in aerosols, dust or liquids. The aerosols might be generated from sources of infection such as the bodily secretions of an infected animal or person, or biological wastes such as accumulate in lofts, caves, garbage and the like. Such infected aerosols may stay suspended in air currents long enough to travel for considerable distances, though the rate of infection decreases sharply with the distance between the source and the organism infected.

Airborne pathogens or allergens often cause inflammation in the nose, throat, sinuses and the lungs. This is caused by the inhalation of these pathogens that affect a person's respiratory system or even the rest of the body. Sinus congestion, coughing and sore throats are examples of inflammation of the upper respiratory air way due to these airborne agents. Air pollution plays a significant role in airborne diseases which is linked to asthma. Pollutants are said to influence lung function by increasing air way inflammation.[1]

Many common infections can spread by airborne transmission at least in some cases, including: Anthrax (inhalational), Chickenpox, Influenza, Measles, Smallpox, Cryptococcosis, and Tuberculosis.

Airborne diseases can also affect non-humans. For example, Newcastle disease is an avian disease that affects many types of domestic poultry worldwide which is transmitted via airborne contamination.[2] Often, airborne pathogens or allergens cause inflammation in the nose, throat, sinuses, and the upper airway lungs. Upper airway inflammation causes coughing congestion,and sore throat. This is caused by the inhalation of these pathogens that affect a person's respiratory system or even the rest of the body. Sinus congestion, coughing and sore throats are examples of inflammation of the upper respiratory air way due to these airborne agents.

Causes

An airborne disease can be caused by exposure to a source: an infected patient or animal, by being transferred from the infected person or animal's mouth, nose, cut, or needle puncture. People receive the disease through a portal of entry: mouth, nose, cut, or needle puncture.

Transmission

Airborne transmission of disease depends on several physical variables endemic to the infectious particle. Environmental factors influence the efficacy of airborne disease transmission; the most evident environmental conditions are temperature and relative humidity. The sum of all the factors that influence temperature and humidity, either meteorological (outdoor) or human (indoor), as well as other circumstances influencing the spread of the droplets containing the infectious particles, as winds, or human behavior, sum up the factors influencing the transmission of airborne diseases.

- Climate and living area. Rainfall (number of rainy daysbeing more important than total precipitation), mean of sunshine daily hours, latitude, altitude are characteristic agents to take in account when assessing the possibility of spread of any airborne infection. Furthermore, some infrequent or exceptional extreme events also influence the dissemination of airborne diseases, as tropical storms, hurricanes, typhoons, or monsoons. Climate conditions determine temperature, winds and relative humidity in any territory, either all year around or at isolated moments (days or weeks). Those are the main factors affecting the spread, duration and infectiousness of droplets containing infectious particles. For instance, influenza virus, is spread easily in northern countries (north hemisphere), because of climate conditions which favour the infectiousness of the virus but on the other hand, in those countries, lots of bacterial infections cannot spread outdoor most of the year, keeping in a latent stage.

UV is harmful to both viruses and bacteria. UV incidence can determine the survival of the infectious particles, so that in those territories with a higher average of sunshine daily hours, and closer to the equator, some particles lose their infectious ability. Infectious particles show and increased survival in the presence of UV light at higher relative humidity levels. It is thought to be due to the protective effect of larger particle sizes, as evaporation would be less at these higher RH levels, showing a protective effect of a thicker water coat.[8]

After isolated events, as tropical storms, has been determined that firstly the quantity of fungal spores is decreased, but a few days later, an exponentially increased number of spores is found, compared to normal conditions.

- Socioeconomics and living conditions. They have a minor role in airborne diseases transmission, but they also have to be taken in consideration. Dwelling is an important aspect. In cities the spread of diseases is faster than in rural areas and outskirts. Normally, cities enclose quarters of buildings, in which the transmission of the viral and bacterial diseases among the neighborhoods is uncomplicated. However, suburban areas are generally more favourable for higher airborne fungal spores

Nearness to large sources of water as rivers and lakes can be a cause of some outbreaks of airborne diseases, after changes in local watershed. Poor sewage systems are usually found in poor countries, especially in the rural areas, and can determine the proliferation of infectious bacteria, that once infecting animal or humans can be transmitted throughout the air.

Working conditions, can also settle infectious airborne diseases. At indoor environments, temperature and relative humidity are mainly affected by HVAC systems (heating, ventilation and air conditioning).Inadequate ventilation is implicated in the airborne transmission of respiratory viruses.Poor maintenance or defects on those systems can foster the conditions for airborne infections.For instance, a poor maintenance of air conditioning systems, can lead to an outbreak of Legionella (mainly Legionella pneumophila), that will spread among the population of the building (workers), before the finding of the focal point. In hospitals, isolation of patients sick of infectious diseases has to be added as a factor, which is noticeable in poor regions, where lack of resources facilitates the spread of infectious diseases.

Prevention

Some ways to prevent airborne diseases include washing hands, using appropriate hand disinfection, getting regular immunizations against diseases believed to be locally present, wearing a respirator and limiting time spent in the presence of any patient likely to be a source of infection.[13] Exposure to a patient or animal with an airborne disease does not guarantee receiving the disease. Because of the changes in host immunity and how much the host was exposed to the particles in the air makes a difference to how the disease affects the body.

With few exceptions, antibiotics are not prescribed for patients to control viral infections. They may however be prescribed to a flu patient for instance, to control or prevent bacterial secondary infections. They also may be used in dealing with air-borne bacterial primary infections, such as pneumonic plague.

Additionally the Centers for Disease Control and Prevention (CDC) has told consumers about vaccination and following careful hygiene and sanitation protocols for airborne disease prevention. Consumers also have access to preventive measures like UV Air purification devices that FDA and EPA-certified laboratory test data has verified as effective in inactivating a broad array of airborne infectious diseases.

REFERENCES

1. Airborne diseases". Retrieved 21 May 2013.
2. Mitchell, Bailey W.; King, Daniel J. (October–December 1994).Effect of Negative Air Ionization on Airborne Transmission of Newcastle Disease Virus". Avian Diseases. 38 (4). JSTOR 1592107. Retrieved 21 May 2013.
3. Pica N, Bouvier NM (2012).Environmental Factors Affecting the Transmission of Respiratory Viruses". Curr Opin Virol. 2 (1): 90–5. doi:10.1016/j.coviro.2011.12.003. PMC 3311988. PMID 22440971.
4. Rodríguez-Rajo FJ, Iglesias I, Jato V. "Variation assessment of airborne Alternaria and Cladosporium spores at different bioclimatical conditions.". Mycol Res. 109: 497–507. doi:10.1017/s0953756204001777. PMID 15912938.
5. Peternel R, Culig J, Hrga I.Atmospheric concentrations of Cladosporium spp. and Alternaria spp. spores in Zagreb (Croatia) and effects of some meteorological factors.". Ann Agric Environ Med. 11: 303–7. PMID 15627341.
6. Sabariego S, Díaz , de la Guardia C, Alba F.The effect of meteorological factors on the daily variation of airborne fungal spores in Granada (southern Spain).". Int J Biometeorol. 44: 1–5. doi:10.1007/s004840050131. PMID 10879421.
7. Hedlund C, Blomstedt Y, Schumann B (2014). "Association of climatic factors with infectious diseases in the Arctic and subarctic region – a systematic review". Glob Health Action. 7: 24161. doi:10.3402/gha.v7.24161. PMC 4079933. PMID 24990685.
8. Tang JW (2009). "The effect of environmental parameters on the survival of airborne infectious agents". J R Soc Interface. 6 Suppl 6: S737–46. doi:10.1098/rsif.2009.0227.focus. PMC 2843949. PMID 19773291.
9. Khan NN, Wilson BL.An environmental assessment of mold concentrations and potential mycotoxin exposures in the greater Southeast Texas area.". J Environ Sci Health A Tox Hazard Subst Environ Eng. 38: 2759–72. PMID 14672314.
10. Aerobiology and Its Role in the Transmission of Infectious Diseases". Retrieved 12 April 2015.

11. Aerosolization`s Roll in Transmission of Healthcare Acquired Conditions". Retrieved 12 April 2015.
12. Legionnaire disease". Retrieved 12 April 2015.
13. American Academy of Orthopaedic Surgeons (AAOS) (2011). Bloodborne and Airborne Pathogens. Jones & Barlett Publishers. p. 2. Retrieved 21 May 2013.
14. Laura Ester Ziady; Nico Small (2006). Prevent and Control Infection: Application Made Easy. Juta and Company Ltd. pp. 119–120. Retrieved 21 May 2013.
15. http://www.cdc.gov/vaccines/vpd-vac/
16. Chamber Test Analysis on Eco-RX Inc. Model 400 Air Purifier" (PDF). Retrieved 4 May 2007.

CHAPTER 5

Nitrogen and Nitrogenous Compounds in the Environment

Chapter Contents

Nitrogen is a chemical element with symbol **N** and atomic number 7. It was first discovered and isolated by Scottish physician Daniel Rutherford in 1772. Although Carl Wilhelm Scheele and Henry Cavendish had independently done so at about the same time, Rutherford is generally accorded the credit because his work was published first. The name nitrogen was suggested by Jean-Antoine-Claude Chaptal in 1790, when it was found that nitrogen was present in nitric acid and nitrates; this name derives from the Greek roots ντρον "nitre" and -γεννν "to form". Antoine Lavoisier suggested instead the name azote, from the Greek ἀζωτικός "no life", as it is an asphyxiant gas; his name is instead used in many languages, such as French, Russian, and Turkish, and appears in the English names of some nitrogen compounds such as hydrazine, azides and azo compounds.

INTRODUCTION

Nitrogen is the lightest member of group 15 of the periodic table, often called the pnictogens. The name comes from the Greek πνίγειν "to choke", directly referencing nitrogen›s asphyxiating properties. It is a common element in the universe, estimated at about seventh in total abundance in the Milky Way and the Solar System. At standard temperature and pressure, two atoms of the element bind to form dinitrogen, a colourless and odorless diatomic gas with the formula N_2. Dinitrogen forms about 78% of Earth's atmosphere, making it the most abundant uncombined element. Nitrogen occurs in all organisms, primarily in amino acids (and thus proteins), in the nucleic acids (DNA and RNA) and in the energy transfer molecule adenosine triphosphate. The human body contains about 3% nitrogen by mass, the fourth most abundant element in the body after oxygen, carbon, and hydrogen. The nitrogen cycle describes movement of the element from the air, into the biosphere and organic compounds, then back into the atmosphere.

Many industrially important compounds, such as ammonia, nitric acid, organic nitrates (propellants and explosives), and cyanides, contain nitrogen. The extremely strong triple bond in elemental nitrogen (N≡N), the second strongest bond in any diatomic molecule, dominates nitrogen chemistry. This causes difficulty for both organisms and industry in converting N_2 into useful compounds, but at the same time means that burning, exploding, or decomposing nitrogen compounds to form nitrogen gas releases large amounts of often useful energy. Synthetically produced ammonia and nitrates are key industrial fertilisers, and fertiliser nitrates are key pollutants in the eutrophication of water systems.

Apart from its use in fertilisers and energy-stores, nitrogen is a constituent of organic compounds as diverse as Kevlar used in high-strength fabric and cyanoacrylate used in superglue. Nitrogen is a constituent of every major pharmacological drug class, including antibiotics. Many drugs are mimics or prodrugs of natural nitrogen-containing signal molecules: for example, the organic nitrates nitroglycerin and nitroprusside control blood pressure by metabolizing into nitric oxide. Many notable nitrogen-containing drugs, such as the natural caffeine and morphine or the synthetic amphetamines, act on receptors of animal neurotransmitters.

HISTORY

Daniel Rutherford, discoverer of nitrogen.

Nitrogen compounds have a very long history, ammonium chloride having been known to Herodotus. They were well known by the Middle Ages. Alchemists knew nitric acid as aqua fortis (strong water), as well as other nitrogen compounds such as ammonium salts and nitrate salts. The mixture of nitric and hydrochloric acids was known as aqua regia (royal water), celebrated for its ability to dissolve gold, the king of metals.

The discovery of nitrogen is attributed to the Scottish physician Daniel Rutherford in 1772, who called it noxious air. Though he did not recognise it as an entirely different chemical substance, he clearly distinguished it from Joseph Black's "fixed air", or carbon dioxide. The fact that there was a component of air that does not support combustion was clear to Rutherford, although he was not aware that it was an element. Nitrogen was also studied at about the same time by Carl Wilhelm Scheele, Henry Cavendish, and Joseph Priestley, who referred to it as burnt air or phlogisticated air. Nitrogen gas was inert enough that Antoine Lavoisier referred to it as "mephitic air" or azote, from the Greek word άζωτικός (azotikos), "no life". In an atmosphere of pure nitrogen, animals died and flames were extinguished. Though Lavoisier›s name was not accepted in English, since it was pointed out that almost all gases (indeed, with the sole exception of oxygen) are mephitic, it is used in many languages (French, Italian, Portuguese, Polish, Russian, Albanian, Turkish, etc.; the German Stickstoff similarly refers to the same characteristic, viz. sticken "to choke or suffocate") and still remains in English in the common names of many nitrogen compounds, such as hydrazine and compounds of the azide ion. Finally, it led to the name "pnictogens" for the group headed by nitrogen, from the Greek πνίγειν "to choke".

The English word nitrogen (1794) entered the language from the French nitrogène, coined in 1790 by French chemist Jean-Antoine Chaptal (1756–1832), from the French nitre (potassium nitrate, also called saltpeter) and the French -gène, "producing", from the Greek -γενής (-genes, "producer, begetter"). Chaptal›s meaning was that nitrogen gas is the essential part of nitric acid, which in turn was produced from niter. In earlier times, niter had been confused with Egyptian "natron" (sodium carbonate) — called νίτρον (nitron) in Greek — which, despite the name, contained no nitrate.

The earliest military, industrial, and agricultural applications of nitrogen compounds used saltpeter (sodium nitrate or potassium nitrate), most notably in gunpowder, and later as fertiliser. In 1910, Lord Rayleigh discovered that an electrical discharge in nitrogen gas produced "active nitrogen", a monatomic allotrope of nitrogen. The "whirling cloud of brilliant yellow light" produced by his apparatus reacted with mercury to produce explosive mercury nitride.

For a long time, sources of nitrogen compounds were limited. Natural sources originated either from biology or deposits of nitrates produced by atmospheric reactions. Nitrogen fixation by industrial processes like the Frank–Caro process (1895–1899) and Haber–Bosch process (1908–1913) eased this shortage of nitrogen compounds, to the extent that half of global food production (see Applications) now relies on synthetic nitrogen fertilisers. At the same time, use of the Ostwald process (1902) to produce nitrates from industrial nitrogen fixation allowed the large-scale industrial production of nitrates as feedstock in the manufacture of explosives in the World Wars of the 20th century.

PROPERTIES

Atomic

The shapes of the five orbitals occupied in nitrogen. The two colours show the phase or sign of the wave function in each region. From left to right: 1s, 2s (cutaway to show internal structure), $2p_x$, $2p_y$, $2p_z$.

A nitrogen atom has seven electrons. In the ground state, they are arranged in the electron configuration $1s^2 2s^2 2p_x^1 2p_y^1 2p_z^1$ z. It therefore has five valence electrons in the 2s and 2p orbitals, three of which (the p-electrons) are unpaired. It has one of the highest electronegativities among the elements (3.04 on the Pauling scale), exceeded only by chlorine (3.16), oxygen (3.44), and fluorine (3.98). Following periodic trends, its single-bond covalent radius of 71 pm is smaller than those of boron (84 pm) and carbon (76 pm), while it is larger than those of oxygen (66 pm) and fluorine (57 pm). The nitride anion, N^{3-}, is much larger at 146 pm, similar to that of the oxide (O^{2-}: 140 pm) and fluoride (F^-: 133 pm) anions. The first three ionisation energies of nitrogen are 1.402, 2.856, and 4.577 $MJ \cdot mol^{-1}$, and the sum of the fourth and fifth is 16.920 $MJ \cdot mol^{-1}$. Due to these very high figures, nitrogen has no simple cationic chemistry.

The lack of radial nodes in the 2p subshell is directly responsible for many of the anomalous properties of the first row of the p-block, especially in nitrogen, oxygen, and fluorine. The 2p subshell is very small and has a very similar radius to the 2s shell, facilitating orbital hybridisation. It also results in very large electrostatic forces of attraction between the nucleus and the valence electrons in the 2s and 2p shells, resulting in very high electronegativities. Hypervalency is almost unknown in these elements for the same reason, because the high electronegativity makes it difficult for a small nitrogen atom to be a central atom in an electron-rich three-center four-electron bond since it would tend to attract the electrons strongly to itself. Thus, despite nitrogen's position at the head of group 15 in the periodic table, its chemistry shows huge differences from that of its heavier congeners phosphorus, arsenic, antimony, and bismuth.

Nitrogen may be usefully compared to its horizontal neighbours carbon and oxygen as well as its vertical neighbours in the pnictogen column (phosphorus, arsenic, antimony, and bismuth). Although each period 2 element from lithium to nitrogen shows some similarities to the period 3 element in the next group from magnesium to sulfur (known as the diagonal relationships), their degree drops off quite abruptly past the boron–silicon pair, so that the similarities of nitrogen to sulfur are mostly limited to sulfur nitride ring compounds when both elements are the only ones present. Nitrogen resembles oxygen far more than it does carbon with its high electronegativity and concomitant capability for hydrogen bonding and the ability to form coordination complexes by donating its lone pairs of electrons. It does not share carbon's proclivity for catenation, with the longest chain of nitrogen yet discovered being composed of only eight nitrogen atoms (PhN=N–N(Ph)–N=N–N(Ph)–N=NPh). One property nitrogen does share with both its horizontal neighbours is its preferentially forming multiple bonds, typically with carbon, nitrogen, or oxygen atoms, through p_π–p_π interactions. This is not possible for its vertical neighbours; thus, the nitrogen oxides, nitrites, nitrates, nitro-, nitroso-, azo-, and diazo-compounds, azides, cyanates, thiocyanates, and imino-derivatives find no echo with phosphorus, arsenic, antimony, or bismuth. By the same token, however, the complexity of the phosphorus oxoacids finds no echo with nitrogen.

Isotopes

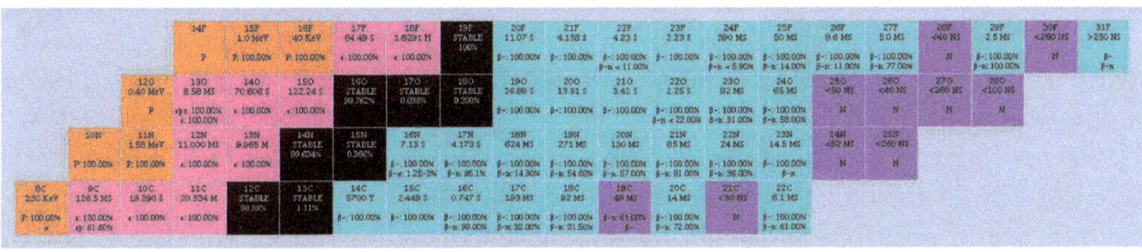

Table of nuclides (Segrè chart) from carbon to fluorine (including nitrogen). Orange indicates proton emission (nuclides outside the proton drip line); pink for positron emission (inverse beta decay); black for stable nuclides; blue for electron emission (beta decay); and violet for neutron emission (nuclides outside the neutron drip line). Proton number increases going up the vertical axis and neutron number going to the right on the horizontal axis.

Nitrogen has two stable isotopes: ^{14}N and ^{15}N. The first is much more common, making up 99.634% of natural nitrogen, and the second (which is slightly heavier) makes up the remaining 0.366%. This leads to an atomic weight of around 14.007 u. Both of these stable isotopes are produced in the CNO cycle in stars, but ^{14}N is more common as its neutron capture is the rate-limiting step. ^{14}N is one of the five stable odd–odd nuclides (a nuclide having an odd number of protons and neutrons); the other four are ^{2}H, ^{6}Li, ^{10}B, and ^{180m}Ta.

The relative abundance of ^{14}N and ^{15}N is practically constant in the atmosphere but can vary elsewhere, due to natural isotopic fractionation from biological redox reactions and the evaporation of natural ammonia or nitric acid. Biologically mediated reactions (e.g., assimilation, nitrification, and denitrification) strongly control nitrogen dynamics in the soil. These reactions typically result in ^{15}N enrichment of the substrate and depletion of the product.

The heavy isotope ^{15}N was first discovered by S. M. Naudé in 1929, soon after heavy isotopes of the neighbouring elements oxygen and carbon were discovered. It presents one of the lowest thermal neutron capture cross-sections of all isotopes. It is frequently used in nuclear magnetic resonance (NMR) spectroscopy to determine the structures of nitrogen-containing molecules, due to its fractional nuclear spin of one-half, which offers advantages for NMR such as narrower line width. ^{14}N, though also theoretically usable, has an integer nuclear spin of one and thus has a quadrupole moment that leads to wider and less useful spectra. ^{15}N NMR nevertheless has complications not encountered in the more ^{1}H and ^{13}C NMR spectroscopy. The low natural abundance of ^{15}N (0.36%) significantly reduces sensitivity, a problem which is only exacerbated by its low gyromagnetic ratio, (only 10.14% that of ^{1}H). As a result, the signal-to-noise ratio for ^{1}H is about 300 times as much as that for ^{15}N at the same magnetic field strength. This may be somewhat alleviated by isotopic enrichment of ^{15}N by chemical exchange or fractional distillation. ^{15}N-enriched compounds have the advantage that under standard conditions, they do not undergo chemical exchange of their nitrogen atoms with atmospheric nitrogen, unlike compounds with labelled hydrogen, carbon, and oxygen isotopes that must be kept away from the atmosphere. The ^{15}N:^{14}N ratio is commonly used in stable isotope analysis in the fields of geochemistry, hydrology, paleoclimatology and paleoceanography, where it is called $\delta^{15}N$.

Of the ten other isotopes produced synthetically, ranging from ^{12}N to ^{23}N, ^{13}N has a half-life of ten minutes and the remaining isotopes have half-lives on the order of seconds (^{16}N and ^{17}N) or even milliseconds. No other nitrogen isotopes are possible as they would fall outside the nuclear drip lines, leaking out a proton or neutron. Given the half-life difference, ^{13}N is the most important nitrogen radioisotope, being relatively long-lived enough to use in positron emission tomography (PET), although its half-life is still short and thus it must be produced at the venue of the PET, for example in a cyclotron via proton bombardment of ^{16}O producing ^{13}N and an alpha particle.

The radioisotope ^{16}N is the dominant radionuclide in the coolant of pressurised water reactors or boiling water reactors during normal operation, and thus it is a sensitive and immediate indicator of leaks from the primary coolant system to the secondary steam cycle, and is the primary means of detection for such leaks. It is produced from ^{16}O (in water) via an (n,p) reaction in which the ^{16}O atom captures a neutron and expels a proton. It has a short half-life of about 7.1 s, but during its decay back to ^{16}O produces high-energy gamma radiation (5 to 7 MeV). Because of this, access to the primary coolant piping in a pressurised water reactor must be restricted during reactor power operation.

CHEMISTRY AND COMPOUNDS

Allotropes

Atomic nitrogen, also known as active nitrogen, is highly reactive, being a triradical with three unpaired electrons. Free nitrogen atoms easily react with most elements to form nitrides, and even when two free nitrogen atoms collide to produce an excited N_2 molecule, they may release so much energy on collision with even such stable molecules as carbon dioxide and water to cause homolytic fission into radicals such as CO and O or OH and H. Atomic nitrogen is prepared by passing an electric discharge through nitrogen gas at 0.1–2 mmHg, which produces atomic nitrogen along with a peach-yellow emission that fades slowly as an afterglow for several minutes even after the discharge terminates.

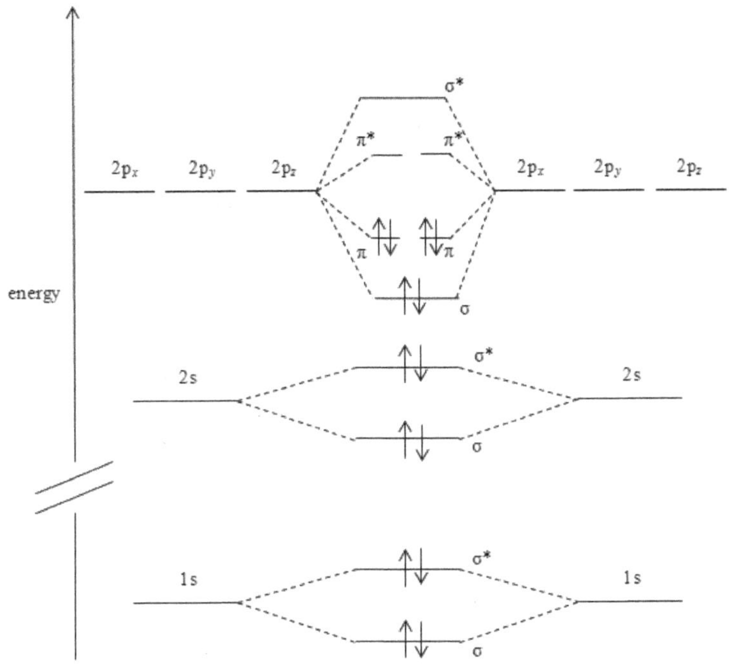

Molecular orbital diagram of dinitrogen molecule, N_2. There are five bonding orbitals and two antibonding orbitals (marked with an asterisk), giving a total bond order of three.

Given the great reactivity of atomic nitrogen, elemental nitrogen usually occurs as molecular N_2, dinitrogen. This molecule is a colourless, odourless, and tasteless diamagnetic gas at standard conditions: it melts at −210 °C and boils at −196 °C. Dinitrogen is mostly unreactive at room temperature, but it will nevertheless react with lithium metal and some transition metal complexes. This is due to its bonding, which is unique among the diatomic elements at standard conditions in that it has an N≡N triple bond. Triple bonds have short bond lengths (in this case, 109.76 pm) and high dissociation energies (in this case, 945.41 kJ/mol), and are thus very strong, explaining dinitrogen›s chemical inertness.

There are some theoretical indications that other nitrogen oligomers and polymers may be possible. If they could be synthesised, they may have potential applications as materials with a very high energy density, that could be used as powerful propellants or explosives. This is because they should all decompose to dinitrogen, whose N≡N triple bond (bond energy 946 kJ·mol^{-1}) is much stronger than those of the N=N double bond (418 kJ·mol^{-1}) or the N–N single bond (160 kJ·mol^{-1}): indeed, the triple bond has more than thrice the energy of the single bond. (The opposite is true for the heavier pnictogens, which prefer polyatomic allotropes.) A great disadvantage is that most neutral polynitrogens are not expected to have a large barrier towards decomposition, and that the few exceptions would be even more challenging to synthesise than the long-sought but still unknown tetrahedrane. This stands in contrast to the well-characterised cationic and anionic polynitrogens azide (N^{3-}) pentazenium (N^{+5}), and pentazolide (cyclic aromatic N−5). Under extremely high pressures (1.1 million atm) and high temperatures (2000 K), as produced in a diamond anvil cell, nitrogen polymerises into the single-bonded cubic gauche crystal structure. This structure is similar to that of diamond, and both have extremely strong covalent bonds, resulting in its nickname "nitrogen diamond".

At atmospheric pressure, molecular nitrogen condenses (liquefies) at 77 K (−195.79 °C) and freezes at 63 K (−210.01 °C) into the beta hexagonal close-packed crystal allotropic form. Below 35.4 K (−237.6 °C) nitrogen assumes the cubic crystal allotropic form (called the alpha phase). Liquid nitrogen, a colourless fluid resembling water in appearance, but with 80.8% of the density (the density of liquid nitrogen at its boiling point is 0.808 g/mL), is a common cryogen. Solid nitrogen has many crystalline modifications. It forms a significant dynamic surface coverage on Pluto and outer moons of the Solar System such as Triton. Even at the low temperatures of solid nitrogen it is fairly volatile and can sublime to form an atmosphere, or condense back into nitrogen frost. It is very weak and flows in the form of glaciers and on Triton geysers of nitrogen gas come from the polar ice cap region.

Dinitrogen Complexes

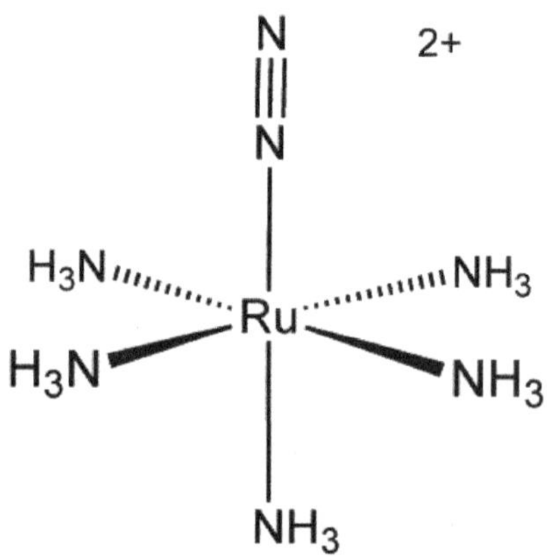

Structure of $[Ru(NH_3)_5(N_2)]^{2+}$(pentaamine(dinitrogen)ruthenium(II)), the first dinitrogen complex to be discovered

The first example of a dinitrogen complex to be discovered was $[Ru(NH_3)_5(N_2)]^{2+}$ (see figure at right), and soon many other such complexes were discovered. These complexes, in which a nitrogen molecule donates at least one lone pair of electrons to a central metal cation, illustrate how N_2 might bind to the metal(s) in nitrogenase and the catalyst for the Haber process: these processes involving dinitrogen activation are vitally important in biology and in the production of fertilisers.

Dinitrogen is able to coordinate to metals in five different ways. The more well-characterised ways are the end-on M←N≡N (η^1) and M←N≡N→M (μ, bis-η^1), in which the lone pairs on the nitrogen atoms are donated to the metal cation. The less well-characterised ways involve dinitrogen donating electron pairs from the triple bond, either as a bridging ligand to two metal cations (μ, bis-η^2) or to just one (η^2). The fifth and unique method involves triple-coordination as a bridging ligand, donating all three electron pairs from the triple bond (μ_3-N_2). A few complexes feature multiple N_2 ligands and some feature N_2 bonded in multiple ways. Since N_2 is isoelectronic with carbon monoxide (CO) and acetylene (C_2H_2), the bonding in dinitrogen complexes is closely allied to that in carbonyl compounds, although N_2 is a weaker σ-donor and π-acceptor than CO. Theoretical studies show that σ donation is a more important factor allowing the formation of the M–N bond than π back-donation, which mostly only weakens the N–N bond, and end-on (η^1) donation is more readily accomplished than side-on (η^2) donation.

Today, dinitrogen complexes are known for almost all the transition metals, accounting for several hundred compounds. They are normally prepared by three methods:

- Replacing labile ligands such as H_2O, H^-, or CO directly by nitrogen: these are often reversible reactions that proceed at mild conditions.
- Reducing metal complexes in the presence of a suitable coligand in excess under nitrogen gas. A common choice include replacing chloride ligands by dimethylphenylphosphine (PMe_2Ph) to make up for the smaller number of nitrogen ligands attached than the original chlorine ligands.
- Converting a ligand with N–N bonds, such as hydrazine or azide, directly into a dinitrogen ligand.

Occasionally the N≡N bond may be formed directly within a metal complex, for example by directly reacting coordinated ammonia (NH_3) with nitrous acid (HNO_2), but this is not generally applicable. Most dinitrogen complexes have colours within the range white-yellow-orange-red-brown; a few exceptions are known, such as the blue $[\{Ti(\eta^5\text{-}C_5H_5)_2\}_2\text{-}(N_2)]$.

Nitrides, Azides, and Nitrido Complexes

Nitrogen bonds to almost all the elements in the periodic table except the first three noble gases, helium, neon, and argon, and some of the very short-lived elements after bismuth, creating an immense variety of binary compounds with varying properties and applications. Many binary compounds are known: with the exception of the nitrogen hydrides, oxides, and fluorides, these are typically called nitrides. Many stoichiometric phases are usually present for most elements (e.g. MnN, Mn_6N_5, Mn_3N_2, Mn_2N, Mn_4N, and Mn_xN for $9.2 < x < 25.3$). They may be classified as "salt-like" (mostly ionic), covalent, "diamond-like", and metallic (or interstitial), although this classification has limitations generally stemming from the continuity of bonding types instead of the discrete and separate types that it implies. They are normally prepared by directly reacting a metal with nitrogen or ammonia (sometimes after heating), or by thermal decomposition of metal amides:

$3\ Ca + N_2 \rightarrow Ca_3N_2$

$3\ Mg + 2\ NH_3 \rightarrow Mg_3N_2 + 3\ H_2$ (at 900 °C)

$3\ Zn(NH_2)_2 \rightarrow Zn_3N_2 + 4\ NH_3$

Many variants on these processes are possible.The most ionic of these nitrides are those of the alkali metals and alkaline earth metals, Li_3N (Na, K, Rb, and Cs do not form stable nitrides for steric reasons) and M_3N_2 (M = Be, Mg, Ca, Sr, Ba). These can formally be thought of as salts of the N^{3-} anion, although charge separation is not actually complete. However, the alkali metal azides NaN_3 and KN_3, featuring the linear N−3 anion, are well-known, as are $Sr(N_3)_2$ and $Ba(N_3)_2$. Azides of the B-subgroup metals (those in groups 11 through 16) are much less ionic, have more complicated structures, and detonate readily when shocked.

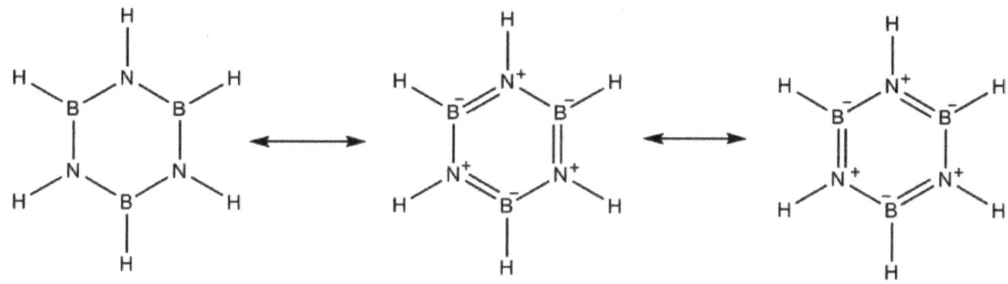

Mesomeric structures of borazine, $(–BH–NH–)_3$

Many covalent binary nitrides are known. Examples include cyanogen ($(CN)_2$), triphosphorus pentanitride (P_3N_5), disulfur dinitride (S_2N_2), and tetrasulfur tetranitride (S_4N_4). The essentially covalent silicon nitride (Si_3N_4) and germanium nitride (Ge_3N_4) are also known: silicon nitride in particular would make a promising ceramic if not for the difficulty of working with and sintering it. In particular, the group 13 nitrides, most of which are promising semiconductors, are isoelectronic with graphite, diamond, and silicon carbide and have similar structures: their bonding changes from covalent to partially ionic to metallic as the group is descended. In particular, since the B–N unit is isoelectronic to C–C, and carbon is essentially intermediate in size between boron and nitrogen, much of organic chemistry finds an echo in boron–nitrogen chemistry, such as in borazine ("inorganic benzene"). Nevertheless, the analogy is not exact due to the ease of nucleophilic attack at boron due to its deficiency in electrons, which is not possible in a wholly carbon-containing ring.

The largest category of nitrides are the interstitial nitrides of formulae MN, M_2N, and M_4N (although variable composition is perfectly possible), where the small nitrogen atoms are positioned in the gaps in a metallic cubic or hexagonal close-packed lattice. They are opaque, very hard, and chemically inert, melting only at very high temperatures (generally over 2500 °C). They have a metallic lustre and conduct electricity as do metals. They hydrolyse only very slowly to give ammonia or nitrogen.

The nitride anion (N^{3-}) is the strongest π donor known amongst ligands (the second-strongest is O^{2-}). Nitrido complexes are generally made by thermal decomposition of azides or by deprotonating ammonia, and they usually involve a terminal $\{\equiv N\}^{3-}$ group. The linear azide anion (N^{3-}), being isoelectronic with nitrous oxide, carbon dioxide, and cyanate, forms many coordination complexes. Further catenation is rare, although N4−

4 (isoelectronic with carbonate and nitrate) is known.

Hydrides

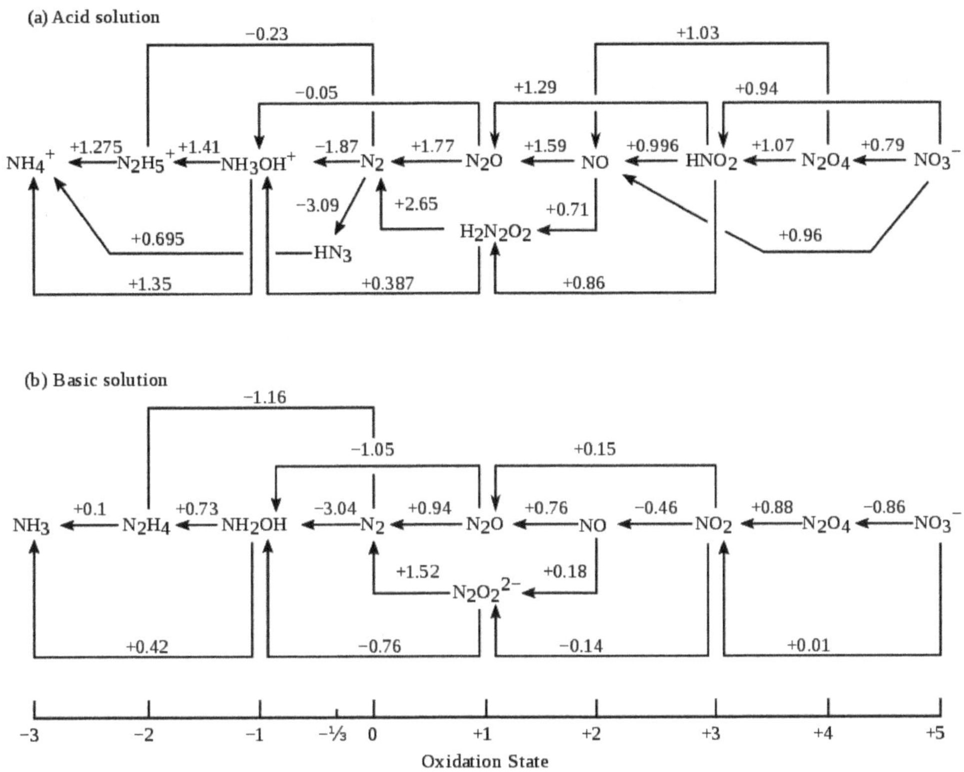

Standard reduction potentials for nitrogen-containing species. Top diagram shows potentials at pH 0; bottom diagram shows potentials at pH 14.

Industrially, ammonia (NH_3) is the most important compound of nitrogen and is prepared in larger amounts than any other compound, because it contributes significantly to the nutritional needs of terrestrial organisms by serving as a precursor to food and fertilisers. It is a colourless alkaline gas with a characteristic pungent smell. The presence of hydrogen bonding has very significant effects on ammonia, conferring on it its high melting (−78 °C) and boiling (−33 °C) points. As a liquid, it is a very good solvent with a high heat of vaporisation (enabling it to be used in vacuum flasks), that also has a low viscosity and electrical conductivity and high dielectric constant, and is less dense than water. However, the hydrogen bonding in NH_3 is weaker than that in H_2O due to the lower electronegativity of nitrogen compared to oxygen and the presence of only one lone pair in NH_3 rather than two in H_2O. It is a weak base in aqueous solution (pK_b 4.74); its conjugate acid is ammonium, NH^{+4}. It can also act as an extremely weak acid, losing a proton to produce the amide anion, NH^{-2}. It thus undergoes self-dissociation, similar to water, to produce ammonium and amide. Ammonia burns in air or oxygen, though not readily, to produce nitrogen gas; it burns in fluorine with a greenish-yellow flame to give nitrogen trifluoride. Reactions with the other nonmetals are very complex an

REFERENCES

1. Conventional Atomic Weights 2013. Commission on Isotopic Abundances and Atomic Weights
2. Standard Atomic Weights 2013. Commission on Isotopic Abundances and Atomic Weights
3. Greenwood and Earnshaw, pp. 406–7
4. Rutherford, Daniel (1772) "Dissertatio Inauguralis de aere fixo, aut mephitico" (Inaugural dissertation on the air [called] fixed or mephitic), M.D. dissertation, University of Edinburgh, Scotland. English translation: Dobbin, Leonard (1935). "Daniel Rutherford's inaugural dissertation". Journal of Chemical Education. 12 (8): 370–375. doi:10.1021/ed012p370.
5. Weeks, Mary Elvira (1932). "The discovery of the elements. IV. Three important gases". Journal of Chemical Education. 9 (2): 215. Bibcode:1932JChEd...9..215W. doi:10.1021/ed009p215.
6. Aaron J. Ihde, The Development of Modern Chemistry, New York 1964.
7. Carl Wilhelm Scheele, Chemische Abhandlung von der Luft und dem Feuer [Chemical treatise on air and fire] (Upsala, Sweden: Magnus Swederus, 1777 ; and Leipzig, (Germany): Siegfried Lebrecht Crusius, 1777).
8. Priestley, Joseph (1772). "Observations on different kinds of air". Philosophical Transactions of the Royal Society of London. 62: 147–256. doi:10.1098/rstl.1772.0021. ; see p. 225.
9. Priestley, Joseph (1772). "Observations on different kinds of air". Philosophical Transactions of the Royal Society of London. 62: 147–256. doi:10.1098/rstl.1772.0021. ; see: "VII. Of air infected with the fumes of burning charcoal." pp. 225–228.
10. Elements of Chemistry, trans. Robert Kerr (Edinburgh, 1790; New York: Dover, 1965), p. 52.
11. Chaptal, J. A. and Nicholson, William trans. (1800) Elements of Chemistry, 3rd ed. London, England: C.C. and J. Robinson, vol. 1, pp. xxxv-xxxvi. From pp. xxxv-xxxvi: "In order to correct the Nomenclature on this head [i.e., in this regard], nothing more is necessary than to substitute to [i.e., for] this word a denomination which is derived from the general system made use of; and I have presumed to propose that of Nitrogene Gas. In the first place, it is deduced from the characteristic and exclusive property of this gas, which forms the radical of the nitric acid. By this means we shall preserve to the combinations [i.e., compounds] of this substance the received [i.e., prevailing] denominations, such as those of the Nitric Acid, Nitrates, Nitrites, &c."
12. nitrogen. Etymonline.com. Retrieved 2011-10-26.

13. Strutt, R. J. (1911) "Bakerian Lecture. A chemically active modification of nitrogen, produced by the electric discharge," Proceedings of the Royal Society A, 85 (577): 219–229.
14. Lord Rayleigh's Active Nitrogen. Lateralscience.co.uk. Retrieved 2011-10-26.
15. Erisman, Jan Willem; Sutton, Mark A.; Galloway, James; Klimont, Zbigniew; Winiwarter, Wilfried (2008). "How a century of ammonia synthesis changed the world". Nature Geoscience. 1 (10): 636. Bibcode:2008NatGe...1..636E. doi:10.1038/ngeo325.
16. GB 190200698, Ostwald, Wilhelm, "Improvements in the Manufacture of Nitric Acid and Nitrogen Oxides", published January 9, 1902, issued March 20, 1902
17. GB 190208300, Ostwald, Wilhelm, "Improvements in and relating to the Manufacture of Nitric Acid and Oxides of Nitrogen", published December 18, 1902, issued February 26, 1903
18. Greenwood and Earnshaw, pp. 411–2
19. Greenwood and Earnshaw, p. 550
20. Kaupp, Martin (1 December 2006). "The role of radial nodes of atomic orbitals for chemical bonding and the periodic table" (PDF). Journal of Computational Chemistry. 28 (1): 320–5. doi:10.1002/jcc.20522. Retrieved 14 October 2016.
21. Greenwood and Earnshaw, pp. 412–6
22. Bethe, H. A. (1939). "Energy Production in Stars". Physical Review. 55 (5): 434–56. Bibcode:1939PhRv...55..434B. doi:10.1103/PhysRev.55.434.
23. CIAAW (2003). "Atomic Weight of Nitrogen". ciaaw.org. CIAAW. Retrieved 13 October 2016.
24. Flanagan, Lawrence B.; Ehleringer, James R; Pataki, Diane E. (15 December 2004). Stable Isotopes and Biosphere - Atmosphere Interactions: Processes and Biological Controls. pp. 74–75. ISBN 978-0-08-052528-0.
25. Greenwood and Earnshaw, p. 408
26. «Evaluated Nuclear Data File (ENDF) Retrieval & Plotting". National Nuclear Data Center.
27. Arthur G Palmer (2007). Protein NMR Spectroscopy. Elsevier Academic Press. ISBN 0-12-164491-X.
28. Katzenberg, MA (2008). "Chapter 13: Stable Isotope Analysis: A Tool for Studying Past Diet, Demography, and Life History". Biological Anthropology of the Human Skeleton (2nd ed.). ISBN 978-0-471-79372-4.
29. Audi, G.; Wapstra, A. H.; Thibault, C.; Blachot, J. & Bersillon, O. (2003). "The NUBASE evaluation of nuclear and decay properties"(PDF). Nuclear Physics A. 729: 3–128. Bibcode:2003NuPhA.729....3A. doi:10.1016/j.nuclphysa.2003.11.001.
30. Carlson, Neil (January 22, 2012). Physiology of Behavior. Methods and Strategies of Research. 11th edition. Pearson. p. 151. ISBN 0-205-23939-0.
31. Neeb, Karl Heinz (1997). The Radiochemistry of Nuclear Power Plants with Light Water Reactors. Berlin-New York: Walter de Gruyter. p. 227. ISBN 3-11-

013242-7.

32. Lewars, Errol G. (2008). Modeling Marvels: Computational Anticipation of Novel molecules. Springer Science+Business Media. pp. 141–163. doi:10.1007/978-1-4020-6973. ISBN 978-1-4020-6972-7.
33. Greenwood and Earnshaw, p. 483
34. Polymeric nitrogen synthesized". physorg.com. 5 August 2004. Retrieved 2009-06-22.
35. Gray, Theodore (2009). The Elements: A Visual Exploration of Every Known Atom in the Universe. New York: Black Dog & Leventhal Publishers. ISBN 978-1-57912-814-2.
36. Schuch, A. F.; Mills, R. L. (1970). "Crystal Structures of the Three Modifications of Nitrogen 14 and Nitrogen 15 at High Pressure". The Journal of Chemical Physics. 52 (12): 6000–6008. Bibcode:1970JChPh..52.6000S. doi:10.1063/1.1672899.
37. Iancu, C. V.; Wright, E. R.; Heymann, J. B.; Jensen, G. J. (2006). "A comparison of liquid nitrogen and liquid helium as cryogens for electron cryotomography". Journal of Structural Biology. 153 (3): 231–240. doi:10.1016/j.jsb.2005.12.004. PMID 16427786.
38. Flowing nitrogen ice glaciers seen on surface of Pluto after New Horizons flyby". ABC. 25 July 2015. Retrieved 6 October 2015.
39. McKinnon, William B.; Kirk, Randolph L. (2014). "Triton". In Spohn, Tilman; Breuer, Doris; Johnson, Torrence. Encyclopedia of the Solar System (3rd ed.). Amsterdam; Boston: Elsevier. pp. 861–882. ISBN 978-0-12-416034-7.
40. Neptune: Moons: Triton". NASA. Archived from the original on October 5, 2011. Retrieved September 21, 2007.
41. Fryzuk, M. D. & Johnson, S. A. (2000). "The continuing story of dinitrogen activation". Coordination Chemistry Reviews. 200–202: 379. doi:10.1016/S0010-8545(00)00264-2.
42. Schrock, R. R. (2005). "Catalytic Reduction of Dinitrogen to Ammonia at a Single Molybdenum Center". Acc. Chem. Res. 38 (12): 955–962. doi:10.1021/ar0501121. PMC 2551323. PMID 16359167.
43. Greenwood and Earnshaw, pp. 417–20
44. Greenwood and Earnshaw, pp. 434–8
45. Greenwood and Earnshaw, pp. 420–6
46. Greenwood and Earnshaw, pp. 426–33
47. Vieira, R.; C. Pham-Huu; N. Keller; M. J. Ledoux (2002). "New carbon nanofiber/graphite felt composite for use as a catalyst support for hydrazine catalytic decomposition". Chemical Communications (9): 954–955. doi:10.1039/b202032g.
48. Greenwood and Earnshaw, pp. 438–42
49. Bowden, F. P. (1958). "Initiation of Explosion by Neutrons, α-Particles, and

Fission Products". Proceedings of the Royal Society of London A. 246 (1245): 216–219. doi:10.1098/rspa.1958.0123.

50. Ford, L. A.; Grundmeier, E. W. (1993). Chemical Magic. Dover. p. 76. ISBN 0-486-67628-5.
51. Greenwood and Earnshaw, pp. 443–58
52. Rahm, Martin; Dvinskikh, Sergey V.; Furó, István; Brinck, Tore (23 December 2010). "Experimental Detection of Trinitramide, $N(NO_2)_3$". Angewandte Chemie International Edition. 50 (5): 1145–8. doi:10.1002/anie.201007047.
53. Hou, YC; Janczuk, A; Wang, PG (1999). "Current trends in the development of nitric oxide donors". Current pharmaceutical design. 5 (6): 417–41. PMID 10390607.
54. Talawar, M. B.; et al. (2005). "Establishment of Process Technology for the Manufacture of Dinitrogen Pentoxide and its Utility for the Synthesis of Most Powerful Explosive of Today—CL-20". Journal of Hazardous Materials. 124: 153–64. doi:10.1016/j.jhazmat.2005.04.021.
55. Greenwood and Earnshaw, pp. 459–72
56. March, Jerry (1985), Advanced Organic Chemistry: Reactions, Mechanisms, and Structure (3rd ed.), New York: Wiley, ISBN 0-471-85472-7
57. Kjeldahl Method". Encyclopedia of Genetics, Genomics, Proteomics and Informatics. 2008. p. 1063. doi:10.1007/978-1-4020-6754-9_9066. ISBN 978-1-4020-6753-2.
58. Greenwood and Earnshaw, pp. 407–9
59. Nielsen, M. K.; Jørgensen, B. M. (Jun 2004). "Quantitative relationship between trimethylamine oxide aldolase activity and formaldehyde accumulation in white muscle from gadiform fish during frozen storage". Journal of Agricultural and Food Chemistry. 52 (12): 3814–3822. doi:10.1021/jf035169l. PMID 15186102.
60. Knox, G. A. (2007). Biology of the Southern Ocean. CRC Press. p. 392. ISBN 0-8493-3394-6.
61. Vickerstaff Joneja; Janice M. (2004). Digestion, diet, and disease: irritable bowel syndrome and gastrointestinal function. Rutgers University Press. p. 121. ISBN 0-8135-3387-2.
62. Froehlich, Peter (May 2013). "A Sustainable Approach to the Supply of Nitrogen". www.parker.com. Parker Hannifin Corporation. Retrieved 24 November 2016.
63. Reich, Murray; Kapenekas, Harry (1957). "Nitrogen Purfication. Pilot Plant Removal of Oxygen". Industrial & Engineering Chemistry. 49 (5): 869–873. doi:10.1021/ie50569a032.
64. Greenwood and Earnshaw, pp. 409–11
65. Bartlett, J. K. (1967). "Analysis for nitrite by evolution of nitrogen: A general chemistry laboratory experiment". Journal of Chemical Education. 44 (8): 475. Bibcode:1967JChEd..44..475B. doi:10.1021/ed044p475.
66. Eremets, M. I.; Popov, M. Y.; Trojan, I. A.; Denisov, V. N.; Boehler, R.;

Hemley, R. J. (2004). "Polymerization of nitrogen in sodium azide". The Journal of Chemical Physics. 120 (22): 10618–10623. Bibcode:2004JChPh.12010618E. doi:10.1063/1.1718250. PMID 15268087.

67. Ministers, Nordic Council of (2002). "Food Additives in Europe 2000": 591. ISBN 978-92-893-0829-8.

68. Harding, Charlie, ed. (2002). Elements of the p Block. Cambridge: Royal Society of Chemistry. ISBN 978-0-85404-690-4.

69. Gavriliuk, V. G.; Berns, Hans (1999). High nitrogen steels: structure, properties, manufacture, applications. Springer. ISBN 3-540-66411-4.

70. Meka, S.R.; Chauhan, A.; Steiner, T.; Bischoff, E.; Ghosh, P.K.; Mittemeijer, E.J. (2015). "Generating duplex microstructures by nitriding; nitriding of iron based Fe–Mn alloy". Materials Science and Technology: 1743284715Y.000. doi:10.1179/1743284715Y.0000000098.

71. Centre Fuel Tank Inerting". B737.org.uk. Retrieved 2013-08-21.

72. Why don›t they use normal air in race car tires?". Howstuffworks. Retrieved 2006-07-22.

73. Kemmochi, Y; Tsutsumi, K; Arikawa, A; Nakazawa, H (2002). "Centrifugal concentrator for the substitution of nitrogen blow-down micro-concentration in dioxin/polychlorinated biphenyl sample preparation". Journal of Chromatography A. 943 (2): 295–297. doi:10.1016/S0021-9673(01)01466-2. PMID 11833649.

74. Baxter, E. Denise; Hughes, Paul S. (2001). Beer: Quality, Safety and Nutritional Aspects. Royal Society of Chemistry. p. 22. ISBN 978-0-85404-588-4.

75. «How does the widget in a beer can work?". Howstuffworks.

76. Denny, Mark (1 November 2009). Froth!: The Science of Beer. p. 131. ISBN 978-0-8018-9569-2.

77. Kennett, Andrew J. (2008). "Design of a pneumatically assisted shifting system for Formula SAE® racing applications". Dept. of Mechanical Engineering, Massachusetts Institute of Technology. hdl:1721.1/45820.

78. Sanburn, Josh (2015-04-10). "The Dawn of a New Form of Capital Punishment". Time. Retrieved 2015-04-11.

79. Sexton, Mike (18 December 2012). "Euthanasia campaigner under scrutiny". ABC. Retrieved 6 May 2013.

80. Betterton, E. A. (2003). "Environmental Fate of Sodium Azide Derived from Automobile Airbags". Critical Reviews in Environmental Science and Technology. 33 (4): 423–458. doi:10.1080/10643380390245002.

81. Kaganer, M. G.; Kozheurov, V. & Levina, Zh. L. (1967). "Vessels for the storage and transport of liquid oxygen and nitrogen". Chemical and Petroleum Engineering. 3 (12): 918–922. doi:10.1007/BF01136404.

82. Ahmed I; Agarwal S; Ilchyshyn A; Charles-Holmes S; Berth-Jones J (May 2001). "Liquid nitrogen cryotherapy of common warts: cryo-spray vs. cotton wool bud". Br. J. Dermatol. 144 (5): 1006–9. doi:10.1046/j.1365-2133.2001.04190.x. PMID 11359389.

83. Kent, Allen; Williams, James G. (1994). Encyclopedia of Computer Science and Technology. 30. CRC Press. p. 318. ISBN 0-8247-2283-3.

84. Biology Safety – Cryogenic materials. The risks posed by them.". University of Bath. Archived from the original on February 6, 2007. Retrieved 2007-01-03.

85. Space Shuttle Columbia Fast Facts". CNN. September 30, 2013.

86. Fowler, B.; Ackles, K.N.; Porlier, G. (1985). "Effects of inert gas narcosis on behavior—a critical review". Undersea Biomed. Res. 12 (4): 369–402. PMID 4082343. Retrieved 2008-09-21.

87. Rogers, W. H.; Moeller, G. (1989). "Effect of brief, repeated hyperbaric exposures on susceptibility to nitrogen narcosis". Undersea Biomed. Res. 16 (3): 227–32. OCLC 2068005. PMID 2741255. Retrieved 2008-09-21.

88. Acott, C. (1999). "A brief history of diving and decompression illness". South Pacific Underwater Medicine Society Journal. 29 (2). OCLC 16986801. Retrieved 2008-09-21.

89. Kindwall, E. P.; Baz, A.; Lightfoot, E. N.; Lanphier, E. H.; Seireg, A. (1975). "Nitrogen elimination in man during decompression". Undersea Biomed. Res. 2 (4): 285–97. OCLC 2068005. PMID 1226586. Retrieved 2008-09-21.

90. US Navy Diving Manual, 6th revision. United States: US Naval Sea Systems Command. 2006. Retrieved 2008-04-24.

91. Walker, Jearl. "Boiling and the Leidenfrost Effect" (PDF). Fundamentals of Physics: 1–4. Retrieved 11 October 2014.

92. Liquid nitrogen cocktail leaves teen in hospital, BBC News, October 8, 2012.

93. Mattox, Brent S. "Investigative Report on Chemistry 301A Cylinder Explosion" (reprint). Texas A&M University.

94. British Compressed Gases Association (2000) BCGA Code of Practice CP30. The Safe Use of Liquid nitrogen Dewars up to 50 litres. ISSN 0260-4809.

95. Confined Space Entry - Worker and Would-be Rescuer Asphyxiated, Valero Refinery Asphyxiation Incident Case Study.

96. Inquiry after man dies in chemical leak, BBC News, October 25, 1999.

97. Liquid Nitrogen – Code of practice for handling. United Kingdom: Birkbeck, University of London. 2007. Retrieved 2012-02-08.

98. Levey, Christopher G. "Liquid Nitrogen Safety". Thayer School of Engineering at Dartmouth.

CHAPTER 6

Carbohydrates Biochemical Aspects

Chapter Contents

- **Introduction**
- **Structure**
- **Monosaccharides**
- **Nutrition**

Carbohydrates, or saccharides, are sugars and starches, which provide energy for humans and animals, and cellulose which make up many plant structures. "Carbs," as they are now commonly referred to, have become both a blessing and a curse, as the process of modern food production has changed the way they are consumed. There are two types of carbohydrates, simple, or monosaccharides and complex, or polysaccharides.

INTRODUCTION

A carbohydrate is a biological molecule consisting of carbon (C), hydrogen (H) and oxygen (O) atoms, usually with a hydrogen–oxygen atom ratio of 2:1 (as in water); in other words, with the empirical formula$C_m(H_2O)_n$ (where *m* could be different from *n*). Some exceptions exist; for example, deoxyribose, a sugar component of DNA, has the empirical formula $C_5H_{10}O_4$. Carbohydrates are technically hydrates of carbon; structurally it is more accurate to view them as polyhydroxy aldehydes and ketones.

The term is most common in biochemistry, where it is a synonym of **saccharide**, a group that includes sugars, starch, and cellulose. The saccharides are divided into four chemical groups: monosaccharides, disaccharides, oligosaccharides, and polysaccharides. In general, the monosaccharides and disaccharides, which are smaller (lower molecular weight) carbohydrates, are commonly referred to as sugars. The word *saccharide* comes from the Greek word *σάκχαρον* (*sákkharon*), meaning "sugar". While the scientific nomenclature of carbohydrates is complex, the names of the monosaccharides and disaccharides very often end in the suffix -ose. For example, grape sugar is the monosaccharide glucose, cane sugar is the disaccharide sucrose, and milk sugar is the disaccharide lactose (see illustration).

Carbohydrates perform numerous roles in living organisms. Polysaccharides serve for the storage of energy (e.g. starch and glycogen) and as structural components (e.g. cellulose in plants and chitin in arthropods). The 5-carbon monosaccharide ribose is an important component of coenzymes (e.g. ATP, FAD and NAD) and the backbone of the genetic molecule known as RNA. The related deoxyribose is a component of DNA. Saccharides and their derivatives include many other important biomolecules that play key roles in the immune system, fertilization, preventing pathogenesis, blood clotting, and development.

In food science and in many informal contexts, the term carbohydrate often means any food that is particularly rich in the complex carbohydrate starch (such as cereals, bread and pasta) or simple carbohydrates, such as sugar (found in candy, jams, and desserts).

Often in lists of nutritional information, such as the USDA National Nutrient Database, the term "carbohydrate" (or "carbohydrate by difference") is used for everything other than water, protein, fat, ash, and ethanol. This will include chemical compounds such as acetic or lactic acid, which are not normally considered carbohydrates. It also includes "dietary fiber" which is a carbohydrate but which does not contribute much in the way of food energy (calories), even though it is often included in the calculation of total food energy just as though it were a sugar.

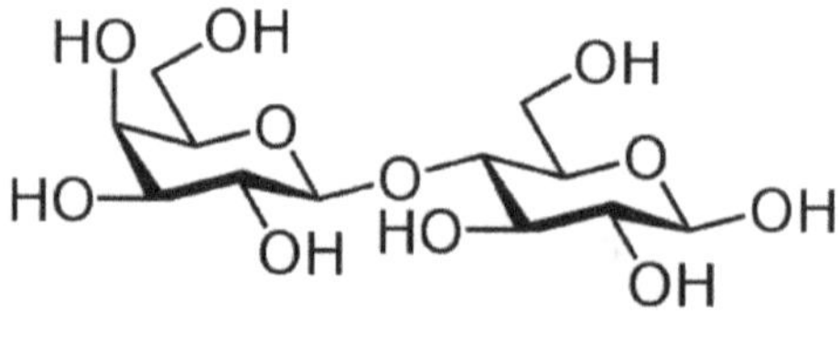

Lactose is a disaccharide found in milk. It consists of a molecule of D-galactose and a molecule of D-glucose bonded by beta-1-4 glycosidic linkage. It has a formula of $C_{12}H_{22}O_{11}$.

STRUCTURE

Formerly the name "carbohydrate" was used in chemistry for any compound with the formula $C_m (H_2O)_n$. Following this definition, some chemists considered formaldehyde (CH_2O) to be the simplest carbohydrate, while others claimed that title for glycolaldehyde. Today, the term is generally understood in the biochemistry sense, which excludes compounds with only one or two carbons and includes many biological carbohydrates which deviate from this formula. For example, while the above representative formulas would seem to capture the commonly known carbohydrates, ubiquitous and abundant carbohydrates often deviate from this. For example, carbohydrates often display chemical groups such as: *N*-acetyl (e.g. chitin), sulphate (e.g. glycosaminoglycans), carboxylic acid (e.g. sialic acid) and deoxy modifications (e.g. fucose and sialic acid).

Natural saccharides are generally built of simple carbohydrates called monosaccharides with general formula $(CH_2O)_n$ where n is three or more. A typical monosaccharide has the structure $H–(CHOH)_x(C{=}O)–(CHOH)_y–H$, that is, an aldehyde or ketone with many hydroxyl groups added, usually one on each carbon atom that is not part of the aldehyde or ketone functional group. Examples of monosaccharides are glucose, fructose, and glyceraldehydes. However, some biological substances commonly called "monosaccharides" do not conform to this formula (e.g. uronic acids and deoxy-sugars such as fucose) and there are many chemicals that do conform to this formula but are not considered to be monosaccharides (e.g. formaldehyde CH_2O and inositol $(CH_2O)_6$).

The open-chain form of a monosaccharide often coexists with a closed ring form where the aldehyde/ketone carbonyl group carbon (C=O) and hydroxyl group (–OH) react forming a hemiacetal with a new C–O–C bridge.

Monosaccharides can be linked together into what are called polysaccharides (or oligosaccharides) in a large variety of ways. Many carbohydrates contain one or more modified monosaccharide units that have had one or more groups replaced or removed. For example, deoxyribose, a component of DNA, is a modified version of ribose; chitin is composed of repeating units of N-acetyl glucosamine, a nitrogen-containing form of glucose.

Division

Carbohydrates are polyhydroxy aldehydes, ketones, alcohols, acids, their simple derivatives and their polymers having linkages of the acetal type. They may be classified according to their degree of polymerization and may be divided initially into three principal groups, namely sugars, oligosaccharides and polysaccharides

The major dietary carbohydrates

Class (DP*)	Subgroup	Components
Sugars (1–2)	Monosaccharides	Glucose, galactose, fructose, xylose
	Disaccharides	Sucrose, lactose, maltose, trehalose
	Polyols	Sorbitol, mannitol
Oligosaccharides (3–9)	Malto-oligosaccharides	Maltodextrins
	Other oligosaccharides	Raffinose, stachyose, fructo-oligosaccharides
Polysaccharides (>9)	Starch	Amylose, amylopectin, modified starches
	Non-starch polysaccharides	Cellulose, hemicellulose, pectins, hydrocolloids

DP * = Degree of polymerization

MONOSACCHARIDES

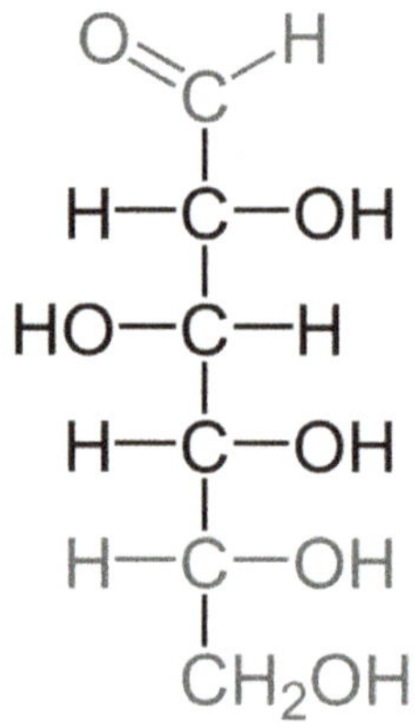

D-glucose is an aldohexose with the formula $(C{\cdot}H_2O)_6$. The red atoms highlight the aldehyde group and the blue atoms highlight the asymmetric center furthest from the aldehyde; because this -OH is on the right of the Fischer projection, this is a D sugar.

Monosaccharides are the simplest carbohydrates in that they cannot be hydrolyzed to smaller carbohydrates. They are aldehydes or ketones with two or more hydroxyl groups. The general chemical formula of an unmodified monosaccharide is $(C{\bullet}H_2O)_n$, literally a "carbon hydrate". Monosaccharides are important fuel molecules as well as building blocks for nucleic acids. The smallest monosaccharides, for which n=3, are dihydroxyacetone and D- and L-glyceraldehydes.

Classification of Monosaccharides

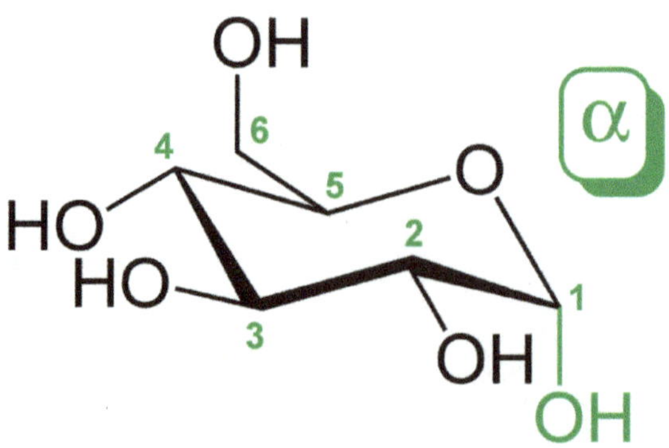

Monosaccharides are classified according to three different characteristics: the placement of its carbonyl group, the number of carbon atoms it contains, and its chiral handedness. If the carbonyl group is an aldehyde, the monosaccharide is an aldose; if the carbonyl group is a ketone, the monosaccharide is a ketose. Monosaccharides with three carbon atoms are called trioses, those with four are called tetroses, five are called pentoses, six are hexoses, and so on. These two systems of classification are often combined. For example, glucose is an aldohexose (a six-carbon aldehyde), ribose is an aldopentose (a five-carbon aldehyde), and fructose is a ketohexose (a six-carbon ketone).

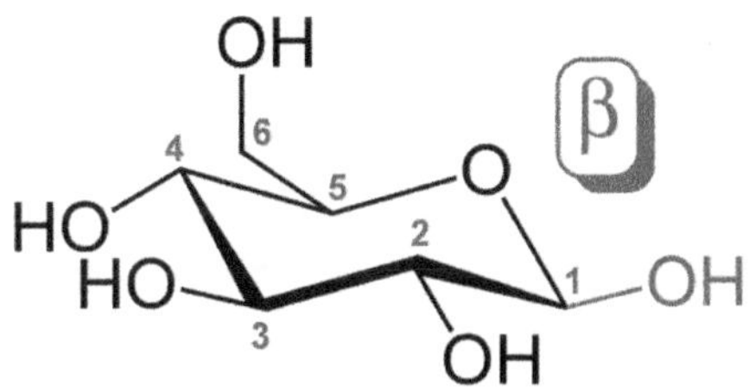

The α and β anomers of glucose. Note the position of the hydroxyl group (red or green) on the anomeric carbon relative to the CH_2OH group bound to carbon 5: they either have identical absolute configurations (R,R or S,S) (α), or opposite absolute configurations (R,S or S,R) (β).

Each carbon atom bearing a hydroxyl group (-OH), with the exception of the first and last carbons, are asymmetric, making them stereo centers with two possible configurations each (R or S). Because of this asymmetry, a number of isomers may exist for any given monosaccharide formula. Using Le Bel-van›t Hoff rule, the aldohexose D-glucose, for example, has the formula $(C·H_2O)_6$, of which four of its six carbons atoms are stereogenic, making D-glucose one of 2^4=16 possible stereoisomers. In the case of glyceraldehydes, an aldotriose, there is one pair of possible stereoisomers, which are enantiomers and epimers. 1, 3-dihydroxyacetone, the ketose corresponding to the aldose glyceraldehydes, is a symmetric molecule with no stereo centers. The assignment of D or L is made according to the orientation of the asymmetric carbon furthest from the carbonyl group: in a standard Fischer projection if the hydroxyl group is on the right the molecule is a D sugar, otherwise it is an L sugar. The "D-" and "L-" prefixes should not be confused with "d-" or "l-", which indicate the direction that the sugar rotates plane polarized light. This usage of "d-" and "l-" is no longer followed in carbohydrate chemistry.

Structure and Nomenclature

With few exceptions (e.g., deoxyribose), monosaccharides have this chemical formula: $C_x(H_2O)_y$, where conventionally $x \geq 3$. Monosaccharides can be classified by the number x of carbon atoms they contain: triose (3) tetrose (4), pentose (5), hexose (6), heptose (7), and so on.

The most important monosaccharide, glucose, is a hexose. Examples of heptoses include the ketoses mannoheptulose and sedoheptulose. Monosaccharides with eight or more carbons are rarely observed as they are quite unstable. In aqueous solutions monosaccharides exist as rings.

Linear-Chain Monosaccharides

Simple monosaccharides have a linear and unbranched carbon skeleton with one carbonyl (C=O) functional group, and one hydroxyl (OH) group on each of the remaining carbon atoms. Therefore, the molecular structure of a simple monosaccharide can be written as $H(CHOH)_n(C{=}O)(CHOH)_mH$, where $n + 1 + m = x$; so that its elemental formula is $C_xH_{2x}O_x$.

By convention, the carbon atoms are numbered from 1 to x along the backbone, starting from the end that is closest to the C=O group.

If the carbonyl is at position 1 (that is, n or m is zero), the molecule begins with a formyl group H(C=O)− and is technically an aldehyde. In that case, the compound is termed an aldose. Otherwise, the molecule has a keto group, a carbonyl −(C=O)− between two carbons; then it is formally a ketone, and is termed a ketose. Ketoses of biological interest usually have the carbonyl at position 2.

The various classifications above can be combined, resulting in names such as aldohexose and ketotriose.

A more general nomenclature for open-chain monosaccharides combines a Greek prefix to indicate the number of carbons (tri-, tetr-, pent-, hex-, etc.) with the suffixes -ose for aldoses and -ulose for ketoses. In the latter case, if the carbonyl is not at position 2, its position is then indicated by a numeric infix. So, for example, $H(C{=}O)(CHOH)_4H$ is pentose, $H(CHOH)(C{=}O)(CHOH)_3H$ is pentulose, and $H(CHOH)_2(C{=}O)(CHOH)_2H$ is pent-3-ulose.

Open-Chain Stereoisomers

Two monosaccharides with equivalent molecular graphs (same chain length and same carbonyl position) may still be distinct stereoisomers, whose molecules differ in the three-dimensional arrangement of the bonds of certain atoms. This happens only if the molecule contains a stereogenic center, specifically a carbon atom that is chiral (connected to four distinct molecular sub-structures). Those four bonds can have any of two configurations in space distinguished by their handedness. In a simple open-chain monosaccharide, every carbon is chiral except the first and the last atoms of the chain, and (in ketoses) the carbon with the keto group.

For example, the triketose H(CHOH)(C=O)(CHOH)H (glycerone, dihydroxyacetone) has no stereogenic center, and therefore exists as a single stereoisomer. The other triose, the aldose $H(C{=}O)(CHOH)_2H$ (glyceraldehyde), has one chiral carbon — the central one, number 2 — which is bonded to groups −H, −OH, $-C(OH)H_2$, and −(C=O)H. Therefore, it exists as two stereoisomers whose molecules are mirror images of each other (like a left and a right glove). Monosaccharides with four or more carbons may contain multiple chiral carbons, so they typically have more than two stereoisomers. The number of distinct stereoisomers with the same diagram is bounded by 2^c, where c is the total number of chiral carbons.

The Fischer projection is a systematic way of drawing the skeletal formula of an acyclic monosaccharide so that the handedness of each chiral carbon is well specified. Each stereoisomer of a simple open-chain monosaccharide can be identified by the positions (right or left) in the Fischer diagram of the chiral hydroxyls (the hydroxyls attached to the chiral carbons).

Most stereoisomers are themselves chiral (distinct from their mirror images). In the Fischer projection, two mirror-image isomers differ by having the positions of all chiral hydroxyls reversed right-to-left. Mirror-image isomers are chemically identical in non-chiral environments, but usually have very different biochemical properties and occurrences in nature.

While most stereoisomers can be arranged in pairs of mirror-image forms, there are some non-chiral stereoisomers that are identical to their mirror images, in spite of having chiral centers. This happens whenever the molecular graph is symmetrical, as in the 3-ketopentoses $H(CHOH)_2(CO)(CHOH)_2H$, and the two halves are mirror images of each other. In that case, mirroring is equivalent to a half-turn rotation. For this reason, there are only three distinct 3-ketopentose stereoisomers, even though the molecule has two chiral carbons.

Distinct stereoisomers that are not mirror-images of each other usually have different chemical properties, even in non-chiral environments. Therefore, each mirror pair and each non-chiral stereoisomer may be given a specific monosaccharide name. For example, there are 16 distinct aldohexose stereoisomers, but the name "glucose" means a specific pair of mirror-image aldohexoses. In the Fischer projection, one of the two glucose isomers has the hydroxyl at left on C3, and at right on C4 and C5; while the other isomer has the reversed pattern. These specific monosaccharide names have conventional three-letter abbreviations, like "Glu" for glucose and "Thr" for threose.

Generally, a monosaccharide with n asymmetrical carbons has 2^n stereoisomers. The number of open chain stereoisomers for an aldose monosaccharide is larger by one than that of a ketose monosaccharide of the same length. Every ketose will have $2^{(n-3)}$ stereoisomers where $n > 2$ is the number of carbons. Every aldose will have $2^{(n-2)}$ stereoisomers where $n > 2$ is the number of carbons. These are also referred to as epimers which have the different arrangement of −OH and −H groups at the asymmetric or chiral carbon atoms (this does not apply to those carbons having the carbonyl functional group).

Configuration of Monosaccharides

Like many chiral molecules, the two stereoisomers of glyceraldehyde will gradually rotate the polarization direction of linearly polarized light as it passes through it, even in solution. The two stereoisomers are identified with the prefixes D- and L-, according to the sense of rotation: D-glyceraldehyde is dextrorotatory (rotates the polarization axis clockwise), while L-glyceraldehyde is levorotatory (rotates it counterclockwise).

The D- and L- prefixes are also used with other monosaccharides, to distinguish two particular stereoisomers that are mirror-images of each other. For this purpose, one considers the chiral carbon that is furthest removed from the C=O group. Its four bonds must connect to −H, −OH, −C(OH)H, and the rest of the molecule. If the molecule can be rotated in space so that the directions of those four groups match those of the analog groups in D-glyceraldehyde's C2, then the isomer receives the D- prefix. Otherwise, it receives the L- prefix.

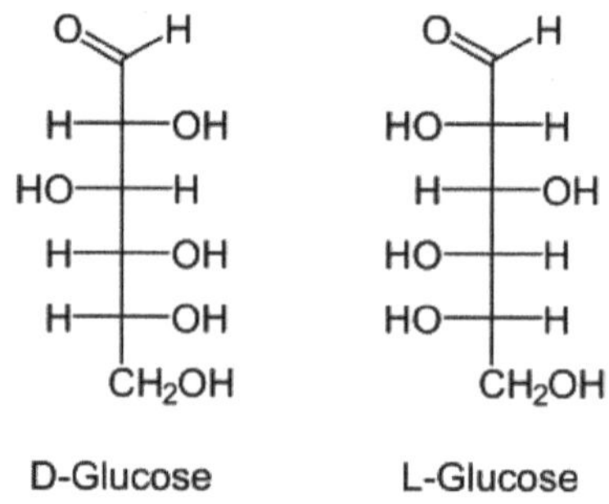

D- and L-glucose.

In the Fischer projection, the D- and L- prefixes specifies the configuration at the carbon atom that is second from bottom: D- if the hydroxyl is on the right side, and L- if it is on the left side.

Note that the D- and L- prefixes do not indicate the direction of rotation of polarized light, which is a combined effect of the arrangement at all chiral centers. However, the two enantiomers will always rotate the light in opposite directions, by the same amount.

Cyclisation of Monosaccharides

A monosaccharide often switches from the acyclic (open-chain) form to a cyclic form, through a nucleophilic addition reaction between the carbonyl group and one of the hydroxyls of the same molecule. The reaction creates a ring of carbon atoms closed by one bridging oxygen atom. The resulting molecule has an hemiacetal or hemiketal group, depending on whether the linear form was an aldose or a ketose. The reaction is easily reversed, yielding the original open-chain form.

In these cyclic forms, the ring usually has **5** or **6** atoms. These forms are called furanoses and pyranoses, respectively — by analogy with furan and pyran, the simplest compounds with the same carbon-oxygen ring (although they lack the double bonds of these two molecules). For example, the aldohexose glucose may form a hemiacetal linkage between the hydroxyl on carbon 1 and the oxygen on carbon 4, yielding a molecule with a 5-membered ring, called glucofuranose. The same reaction can take place between carbons 1 and 5 to form a molecule with a 6-membered ring, called glucopyranose. Cyclic forms with a 7-atom ring (the same of oxepane), rarely encountered, are called heptoses.

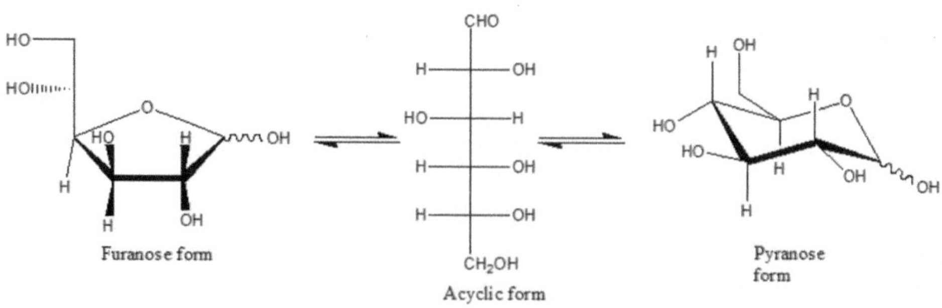

Conversion between the furanose, acyclic, and pyranose forms of D-glucose.

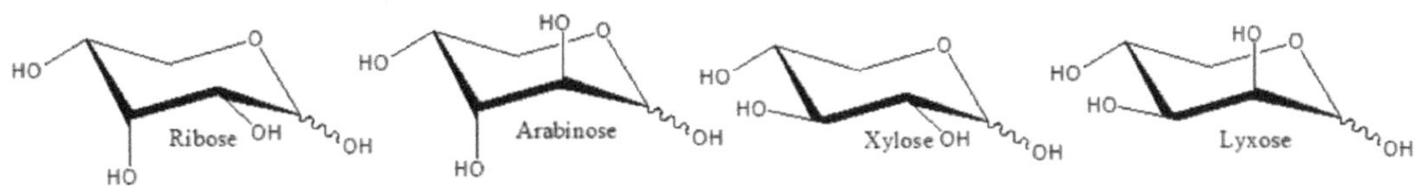

Pyranose forms of some pentose sugars.

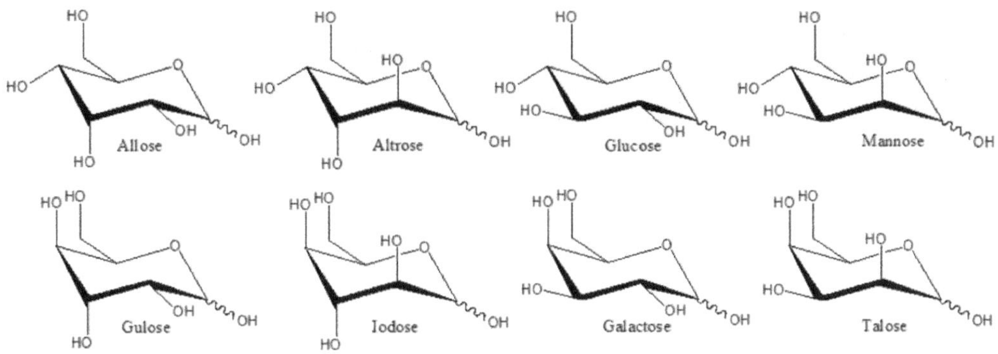

Pyranose forms of some hexose sugars.

For many monosaccharides (including glucose), the cyclic forms predominate, in the solid state and in solutions, and therefore the same name commonly is used for the open- and closed-chain isomers. Thus, for example, the term "glucose" may signify glucofuranose, glucopyranose, the open-chain form, or a mixture of the three.

Cyclization creates a new stereogenic center at the carbonyl-bearing carbon. The −OH group that replaces the carbonyl's oxygen may end up in two distinct positions relative to the ring's midplane. Thus each open-chain monosaccharide yields two cyclic isomers (anomers), denoted by the prefixes α- and β-. The molecule can change between these two forms by a process called mutarotation, that consists in a reversal of the ring-forming reaction followed by another ring formation.

Haworth Projection

The stereochemical structure of a cyclic monosaccharide can be represented in a Haworth projection. In this diagram, the α-isomer has the -OH of the anomeric carbon below the plane of the carbon atoms, and the β-isomer has the -OH of the anomeric carbon above the plane. Pyranoses typically adopt a chair conformation, similar to that of cyclohexane. In this conformation, the α-isomer has the -OH of the anomeric carbon in an axial position, whereas the β-isomer has the OH- of the anomeric carbon in equatorial position.

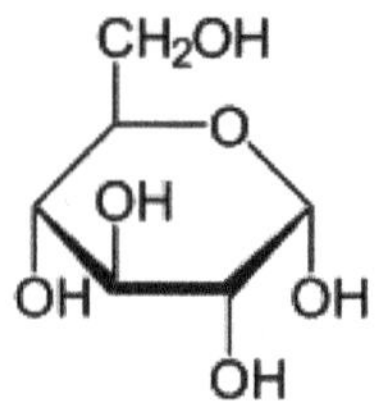

α-D-Glucopyranose

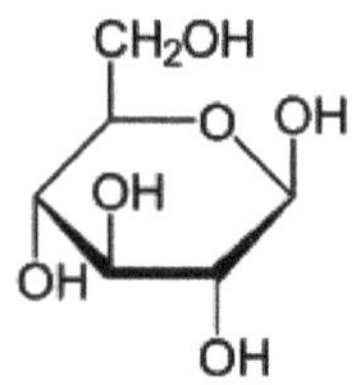

β-D-Glucopyranose

Derivatives

A large number of biologically important modified monosaccharides exist:

- Amino sugars such as:
  - Galactosamine
  - Glucosamine
  - Sialic acid
  - *N*-Acetylglucosamine
- Sulfosugars such as:
  - Sulfoquinovose

Ring-Straight Chain Isomerism

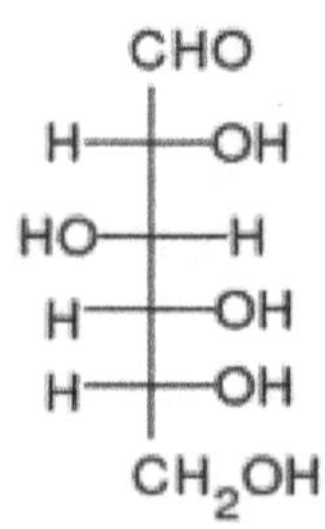

Glucose can exist in both a straight-chain and ring form.

The aldehyde or ketone group of a straight-chain monosaccharide will react reversibly with a hydroxyl group on a different carbon atom to form a hemiacetal or hemiketal, forming a heterocyclic ring with an oxygen bridge between two carbon atoms. Rings with five and six atoms are called furanose and pyranose forms, respectively, and exist in equilibrium with the straight-chain form.

During the conversion from straight-chain form to the cyclic form, the carbon atom containing the carbonyl oxygen, called the anomeric carbon, becomes a stereogenic center with two possible configurations: The oxygen atom may take a position either above or below the plane of the ring. The resulting possible pair of stereoisomers is called anomers. In the *α anomer*, the -OH substituent on the anomeric carbon rests on the opposite side (trans) of the ring from the CH_2OH side branch. The alternative form, in which the CH_2OH substituent and the anomeric hydroxyl are on the same side (cis) of the plane of the ring, is called the *β anomer*.

Use in Living Organisms

Monosaccharides are the major source of fuel for metabolism, being used both as an energy source (glucose being the most important in nature) and in biosynthesis. When monosaccharides are not immediately needed by many cells they are often converted to more space-efficient forms, often polysaccharides. In many animals, including humans, this storage form is glycogen, especially in liver and muscle cells. In plants, starch is used for the same purpose. The most abundant carbohydrate, cellulose, is a structural component of the cell wall of plants and many forms of algae. Ribose is a component of RNA. Deoxyribose is a component of DNA. Lyxose is a component of lyxoflavin found in the human heart. Ribulose and xylulose occur in the pentose phosphate pathway. Galactose, a component of milk sugar lactose, is found in galactolipids in plant cell membranes and in glycoproteins in many tissues. Mannose occurs in human metabolism, especially in the glycosylation of certain proteins. Fructose, or fruit sugar, is found in many plants and in humans, it is metabolized in the liver, absorbed directly into the intestines during digestion, and found in semen. Trehalose, a major sugar of insects, is rapidly hydrolyzed into two glucose molecules to support continuous flight.

Disaccharides

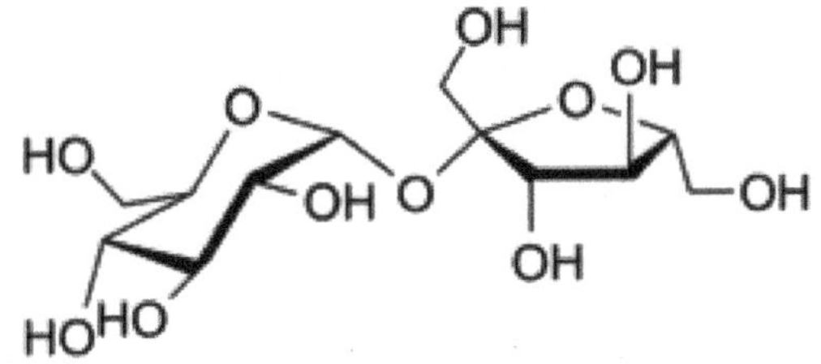

Sucrose, also known as table sugar, is a common disaccharide. It is composed of two monosaccharides: D-glucose (left) and D-fructose (right).

Two joined monosaccharides are called a disaccharide and these are the simplest polysaccharides. Examples include sucrose and lactose. They are composed of two monosaccharide units bound together by a covalent bond known as a glycosidic linkage formed via a dehydration reaction, resulting in the loss of a hydrogen atom from one monosaccharide and a hydroxyl group from the other. The formula of unmodified disaccharides is $C_{12}H_{22}O_{11}$. Although there are numerous kinds of disaccharides, a handful of disaccharides are particularly notable.

Sucrose, pictured to the right, is the most abundant disaccharide, and the main form in which carbohydrates are transported in plants. It is composed of one D-glucose molecule and one D-fructose molecule. The systematic name for sucrose, *O*-α-D-glucopyranosyl-(1→2)-D-fructofuranoside, indicates four things:

- Its monosaccharides: glucose and fructose
- Their ring types: glucose is a pyranose and fructose is a furanose
- How they are linked together: the oxygen on carbon number 1 (C1) of α-D-glucose is linked to the C2 of D-fructose.
- The *-oside* suffix indicates that the anomeric carbon of both monosaccharides participates in the glycosidic bond.

Lactose, a disaccharide composed of one D-galactose molecule and one D-glucose molecule, occurs naturally in mammalian milk. The systematic name for lactose is *O*-β-D-galactopyranosyl-(1→4)-D-glucopyranose. Other notable disaccharides include maltose (two D-glucoses linked α-1,4) and cellulobiose (two D-glucoses linked β-1,4). Disaccharides can be classified into two types: reducing and non-reducing disaccharides. If the functional group is present in bonding with another sugar unit, it is called a reducing disaccharide or biose.

NUTRITION

Carbohydrate consumed in food yields 3.87 calories of energy per gram for simple sugars, and 3.57 to 4.12 calories per gram for complex carbohydrate in most other foods. Relatively high levels of carbohydrate are associated with processed foods or refined foods made from plants, including sweets, cookies and candy, table sugar, honey, soft drinks, breads and crackers, jams and fruit products, pastas and breakfast cereals. Lower amounts of carbohydrate are usually associated with unrefined foods, including beans, tubers, rice, and unrefined fruit. Animal-based foods generally have the lowest carbohydrate levels, although milk does contain a high proportion of lactose.

Carbohydrates are a common source of energy in living organisms; however, no carbohydrate is an essential nutrient in humans. Humans are able to obtain all of their energy requirement from protein and fats, though the potential for some negative health effects of extreme carbohydrate restriction remains, as the issue has not been studied extensively yet. However, in the case of dietary fiber – indigestible carbohydrates which are not a source of energy – inadequate intake can lead to significant increases in mortality.

Grain products: rich sources of carbohydrates

Following a diet consisting of very low amounts of daily carbohydrate for several days will usually result in higher levels of blood ketone bodies than an isocaloric diet with similar protein content. This relatively high level of ketone bodies is commonly known as ketosis and is very often confused with the potentially fatal condition often seen in type 1 diabetics known as diabetic ketoacidosis. Somebody suffering ketoacidosis will have much higher levels of blood ketone bodies along with high blood sugar, dehydration and electrolyte imbalance.

Long-chain fatty acids cannot cross the blood–brain barrier, but the liver can break these down to produce ketones. However, the medium-chain fatty acids octanoic and heptanoic acids can cross the barrier and be used by the brain, which normally relies upon glucose for its energy. Gluconeogenesis allows humans to synthesize some glucose from specific amino acids: from the glycerol backbone in triglycerides and in some cases from fatty acids.

Organisms typically cannot metabolize all types of carbohydrate to yield energy. Glucose is a nearly universal and accessible source of energy. Many organisms also have the ability to metabolize other monosaccharides and disaccharides but glucose is often metabolized first. In *Escherichia coli*, for example, the lac operon will express enzymes for the digestion of lactose when it is present, but if both lactose and glucose are present the *lac* operon is repressed, resulting in the glucose being used first (see: Diauxie). Polysaccharides are also common sources of energy. Many organisms can easily break down starches into glucose; most organisms, however, cannot metabolize cellulose or other polysaccharides like chitin and arabinoxylans. These carbohydrate types can be metabolized by some bacteria and protists. Ruminants and termites, for example, use microorganisms to process cellulose. Even though these complex carbohydrates are not very digestible, they represent an important dietary element for humans, called dietary fiber. Fiber enhances digestion, among other benefits.

Based on the effects on risk of heart disease and obesity, the Institute of Medicine recommends that American and Canadian adults get between 45–65% of dietary energy from carbohydrates. The Food and Agriculture Organization and World Health Organization jointly recommend that national dietary guidelines set a goal of 55–75% of total energy from carbohydrates, but only 10% directly from sugars (their term for simple carbohydrates).

Classification

Nutritionists often refer to carbohydrates as either simple or complex. However, the exact distinction between these groups can be ambiguous. The term *complex carbohydrate* was first used in the U.S. Senate Select Committee on Nutrition and Human Needs publication *Dietary Goals for the United States* (1977) where it was intended to distinguish sugars from other carbohydrates (which were perceived to be nutritionally superior). However, the report put "fruit, vegetables and whole-grains" in the complex carbohydrate column, despite the fact that these may contain sugars as well as polysaccharides. This confusion persists as today some nutritionists use the term complex carbohydrate to refer to any sort of digestible saccharide present in a whole food, where fiber, vitamins and minerals are also found (as opposed to processed carbohydrates, which provide energy but few other nutrients). The standard usage, however, is to classify carbohydrates chemically: simple if they are sugars (monosaccharides and disaccharides) and complex if they are polysaccharides (or oligosaccharides).

In any case, the simple vs. complex chemical distinction has little value for determining the nutritional quality of carbohydrates. Some simple carbohydrates (e.g. fructose) raise blood glucose slowly, while some complex carbohydrates (starches), especially if processed, raise blood sugar rapidly. The speed of digestion is determined by a variety of factors including which other nutrients are consumed with the carbohydrate, how the food is prepared, individual differences in metabolism, and the chemistry of the carbohydrate.

The USDA›s *Dietary Guidelines for Americans 2010* call for moderate- to high-carbohydrate consumption from a balanced diet that includes six one-ounce servings of grain foods each day, at least half from whole grain sources and the rest from enriched.

The glycemic index (GI) and glycemic load concepts have been developed to characterize food behavior during human digestion. They rank carbohydrate-rich foods based on the rapidity and magnitude of their effect on blood glucose levels. Glycemic index is a measure of how quickly food glucose is absorbed, while glycemic load is a measure of the total absorbable glucose in foods. The insulin index is a similar, more recent classification method that ranks foods based on their effects on blood insulin levels, which are caused by glucose (or starch) and some amino acids in food.

Metabolism

Carbohydrate metabolism denotes the various biochemical processes responsible for the formation, breakdown and interconversion of carbohydrates in living organisms.

The most important carbohydrate is glucose, a simple sugar (monosaccharide) that is metabolized by nearly all known organisms. Glucose and other carbohydrates are part of a wide variety of metabolic pathways across species: plants synthesize carbohydrates from carbon dioxide and water by photosynthesis storing the absorbed energy internally, often in the form of starch or lipids. Plant components are consumed by animals and fungi, and used as fuel for cellular respiration. Oxidation of one gram of carbohydrate yields approximately 4 kcal of energy, while the oxidation of one gram of lipids yields about 9 kcal. Energy obtained from metabolism (e.g., oxidation of glucose) is usually stored temporarily within cells in the form of ATP. Organisms capable of aerobic respiration metabolize glucose and oxygen to release energy with carbon dioxide and water as byproducts.

Catabolism

Catabolism is the metabolic reaction which cells undergo to extract energy. There are two major metabolic pathways of monosaccharide catabolism: glycolysis and the citric acid cycle.

In glycolysis, oligo/polysaccharides are cleaved first to smaller monosaccharides by enzymes called glycoside hydrolases. The monosaccharide units can then enter into monosaccharide catabolism. In some cases, as with humans, not all carbohydrate types are usable as the digestive and metabolic enzymes necessary are not present.

Carbohydrate Chemistry

Carbohydrate chemistry is a large and economically important branch of organic chemistry. Some of the main organic reactions that involve carbohydrates are:

- Carbohydrate acetalisation
- Cyanohydrin reaction
- Lobry-de Bruyn-van Ekenstein transformation
- Amadori rearrangement
- Nef reaction
- Wohl degradation
- Koenigs–Knorr reaction
- Carbohydrate digestion

REFERENCES

1. Western Kentucky University (May 29, 2013). "WKU BIO 113 Carbohydrates". wku.edu.
2. Eldra Pearl Solomon; Linda R. Berg; Diana W. Martin (2004). Biology. Cengage Learning. p. 52. ISBN 978-0534278281 – via google.books.com.
3. National Institute of Standards and Technology (2011). "Material Measurement Library D-erythro-Pentose, 2-deoxy-". nist.gov.
4. Long Island University (May 29, 2013). "The Chemistry of Carbohydrates" (PDF). brooklyn.liu.edu.

5. Purdue University (May 29, 2013). "Carbohydrates: The Monosaccharides". purdue.edu.
6. Flitsch, Sabine L.; Ulijn, Rein V (2003). "Sugars tied to the spot". Nature. 421 (6920): 219–20. doi:10.1038/421219a. PMID 12529622.
7. Maton, Anthea; Jean Hopkins; Charles William McLaughlin; Susan Johnson; Maryanna Quon Warner; David LaHart; Jill D. Wright (1993). Human Biology and Health. Englewood Cliffs, New Jersey, USA: Prentice Hall. pp. 52–59. ISBN 0-13-981176-1.
8. USDA National Nutrient Database, 2015, p. 13
9. USDA National Nutrient Database, 2015, p. 14
10. John Merle Coulter, Charler Reid Barnes, Henry Chandler Cowles (1930), A Textbook of Botany for Colleges and Universities"
11. Carl A. Burtis, Edward R. Ashwood, Norbert W. Tietz (2000), Tietz fundamentals of clinical chemistry
12. Matthews, C. E.; K. E. Van Holde; K. G. Ahern (1999) Biochemistry. 3rd edition. Benjamin Cummings. ISBN 0-8053-3066-6
13. "Carbohydrates in human nutrition - Chapter 1 - The role of carbohydrates in nutrition". Food and Agriculture Organization of the United Nations. FAO.
14. Carolyn R Bertozzi. "Structural Basis of Glycan Diversity - Essentials of Glycobiology - NCBI Bookshelf". nih.gov.
15. Campbell, Neil A.; Brad Williamson; Robin J. Heyden (2006). Biology: Exploring Life. Boston, Massachusetts: Pearson Prentice Hall. ISBN 0-13-250882-6.
16. Pigman, Ward; Horton, D. (1972). "Chapter 1: Stereochemistry of the Monosaccharides". In Pigman and Horton. The Carbohydrates: Chemistry and Biochemistry Vol 1A (2nd ed.). San Diego: Academic Press. pp. 1–67.
17. Pigman, Ward; Anet, E.F.L.J. (1972). "Chapter 4: Mutarotations and Actions of Acids and Bases". In Pigman and Horton. The Carbohydrates: Chemistry and Biochemistry Vol 1A (2nd ed.). San Diego: Academic Press. pp. 165–194.
18. "lyxoflavin". Merriam-Webster.
19. "Show Foods". usda.gov.
20. "CALCULATION OF THE ENERGY CONTENT OF FOODS - ENERGY CONVERSION FACTORS". fao.org.
21. "Carbohydrate reference list" (PDF). www.diabetes.org.uk. Retrieved October 30, 2016.
22. Westman, EC (2002). "Is dietary carbohydrate essential for human nutrition?". The American Journal of Clinical Nutrition. 75 (5): 951–3; author reply 953–4. PMID 11976176.
23. Park, Y; Subar, AF; Hollenbeck, A; Schatzkin, A (2011). "Dietary fiber intake and mortality in the NIH-AARP diet and health study". Archives of Internal Medicine. 171 (12): 1061–8. doi:10.1001/archinternmed.2011.18. PMC 3513325 . PMID 21321288.

24. "Ketogenic low-carbohydrate diets have no metabolic advantage over nonketogenic low-carbohydrate diets"(PDF). nutrition.org.
25. Douglas Ebert. "Energy Contribution of Octanoate to Intact Rat Brain Metabolism Measured by 13C Nuclear Magnetic Resonance Spectroscopy". jneurosci.org.
26. "Journal of Cerebral Blood Flow & Metabolism - Abstract of article: Heptanoate as a neural fuel: energetic and neurotransmitter precursors in normal and glucose transporter I-deficient (G1D) brain". nature.com.
27. "Integration of Metabolism". medbio.info.
28. Pichon, L; Huneau, JF; Fromentin, G; Tomé, D (2006). "A high-protein, high-fat, carbohydrate-free diet reduces energy intake, hepatic lipogenesis, and adiposity in rats". The Journal of Nutrition. 136 (5): 1256–60. PMID 16614413.
29. Tighe, P; Duthie, G; Vaughan, N; Brittenden, J; Simpson, WG; Duthie, S; Mutch, W; Wahle, K; Horgan, G; Thies, F. (2010). "Effect of increased consumption of whole-grain foods on blood pressure and other cardiovascular risk markers in healthy middle-aged persons: a randomized, controlled trial". The American Journal of Clinical Nutrition. 92 (4): 733–40. doi:10.3945/ajcn.2010.29417. PMID 20685951.
30. Food and Nutrition Board (2002/2005). Dietary Reference Intakes for Energy, Carbohydrate, Fiber, Fat, Fatty Acids, Cholesterol, Protein and Amino Acids. Washington, D.C.: The National Academies Press. Page 769. ISBN 0-309-08537-3.
31. Joint WHO/FAO expert consultation (2003). (PDF). Geneva: World Health Organization. pp. 55–56. ISBN 92-4-120916-X.
32. Joint WHO/FAO expert consultation (1998), Carbohydrates in human nutrition, chapter 1. ISBN 92-5-104114-8.
33. "Carbohydrates". The Nutrition Source. Harvard School of Public Health. Retrieved April 3, 2013.
34. Jenkins, David; Alexandra L. Jenkins; Thomas M.S. Woleve; Lilian H. Thompson; A. Venkat Rao (February 1986). "Simple and Complex Carbohydrates". Nutrition Reviews. 44 (2): 44–49. doi:10.1111/j.1753-4887.1986.tb07585.x.
35. DHHS and USDA, Dietary Guidelines for Americans 2010.
36. Energetics of Cellular Respiration (Glucose Metabolism)

CHAPTER 7

Amino Acid Biochemical Description of Amino Acids

Chapter Contents

Amino acids are the building blocks of proteins. They band together in chains to form the stuff from which life is born. This is a two-step process: first, they get together and form peptides or polypeptides, and it is from these groupings that proteins are made.

INTRODUCTION

Amino acids are biologically important organic compounds containing amine (-NH_2) and carboxyl (-COOH) functional groups, along with a side-chain (R group) specific to each amino acid. The key elements of an amino acid are carbon, hydrogen, oxygen, and nitrogen, though other elements are found in the side-chains of certain amino acids. About 500 amino acids are known (though only 20 appear in the genetic code) and can be classified in many ways. They can be classified according to the core structural functional groups' locations as alpha- (α-), beta- (β-), gamma- (γ-) or delta- (δ-) amino acids; other categories relate to polarity, pH level, and side-chain group type (aliphatic, acyclic, aromatic, containing hydroxyl or sulfur, etc.). In the form of proteins, amino acids comprise the second-largest component (water is the largest) of human muscles, cells and other tissues. Outside proteins, amino acids perform critical roles in processes such as neurotransmitter transport and biosynthesis.

In biochemistry, amino acids having both the amine and the carboxylic acid groups attached to the first (alpha-) carbon atom have particular importance. They are known as 2-, alpha-, or **α-amino acids** (generic formula $H_2NCHRCOOH$ in most cases, where R is an organicsubstituent known as a "side-chain"); often the term "amino acid" is used to refer specifically to these. They include the 23 proteinogenic ("protein-building") amino acids, which combine into peptide chains ("polypeptides") to form the building-blocks of a vast array of proteins. These are all L-stereoisomers ("left-handed" isomers), although a few D-amino acids ("right-handed") occur in bacterial envelopes, as a neuromodulator (D-serine), and in some antibiotics.Twenty of the proteinogenic amino acids are encoded directly by triplet codons in the genetic code and are known as standard amino acids. The other three (non-standard or non-canonical) are selenocysteine (present in many noneukaryotes as well as most eukaryotes, but not coded directly by DNA), pyrrolysine (found only in some archea and one bacterium) and *N*-formylmethionine (which is often the initial amino acid of proteins in bacteria, mitochondria, and chloroplasts). Pyrrolysine and selenocysteine are encoded via variant codons; for example, selenocysteine is encoded by stop codon and SECIS element.Codon–tRNA combinations not found in nature can also be used to "expand" the genetic code and create novel proteins known as alloproteins incorporating non-proteinogenic amino acids.

Many important proteinogenic and non-proteinogenic amino acids also play critical non-protein roles within the body. For example, in the human brain, glutamate (standard glutamic acid) and gamma-amino-butyric acid (GABA, non-standard gamma-amino acid) are, respectively, the main excitatory and inhibitory neurotransmitters; hydroxyproline (a major component of the connective tissue collagen) is synthesised from proline; the standard amino acid glycine is used to synthesise porphyrins used in red blood cells; and the non-standard carnitine is used in lipid transport.

Nine proteinogenic amino acids are called essential for humans because they cannot be created from other compounds by the human body and so must be taken in as food. Others may be conditionally essential for certain ages or medical conditions. Essential amino acids may also differ between species.

Because of their biological significance, amino acids are important in nutrition and are commonly used in nutritional supplements, fertilizers, and food technology. Industrial uses include the production of drugs, biodegradable plastics, and chiral catalysts.

The first few amino acids were discovered in the early 19th century. In 1806, French chemists Louis-Nicolas Vauquelin and Pierre Jean Robiquet isolated a compound in asparagus that was subsequently named asparagine, the first amino acid to be discovered. Cystine was discovered in 1810, although its monomer, cysteine, remained undiscovered until 1884. Glycine and leucine were discovered in 1820. The last of the 20 common amino acids to be discovered was threonine in 1935 by William Cumming Rose, who also determined the essential amino acids and established the minimum daily requirements of all amino acids for optimal growth.

Usage of the term *amino acid* in the English language is from 1898. Proteins were found to yield amino acids after enzymatic digestion or acid hydrolysis. In 1902, Emil

Fischer and Franz Hofmeister proposed that proteins are the result of the formation of bonds between the amino group of one amino acid with the carboxyl group of another, in a linear structure that Fischer termed peptide.

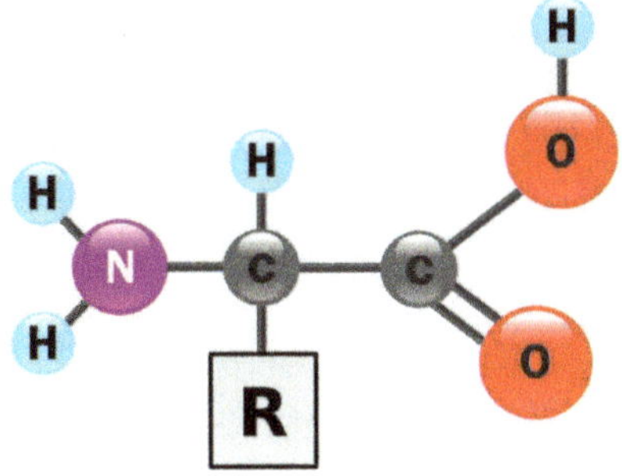

The structure of an alpha amino acid in its un-ionized form

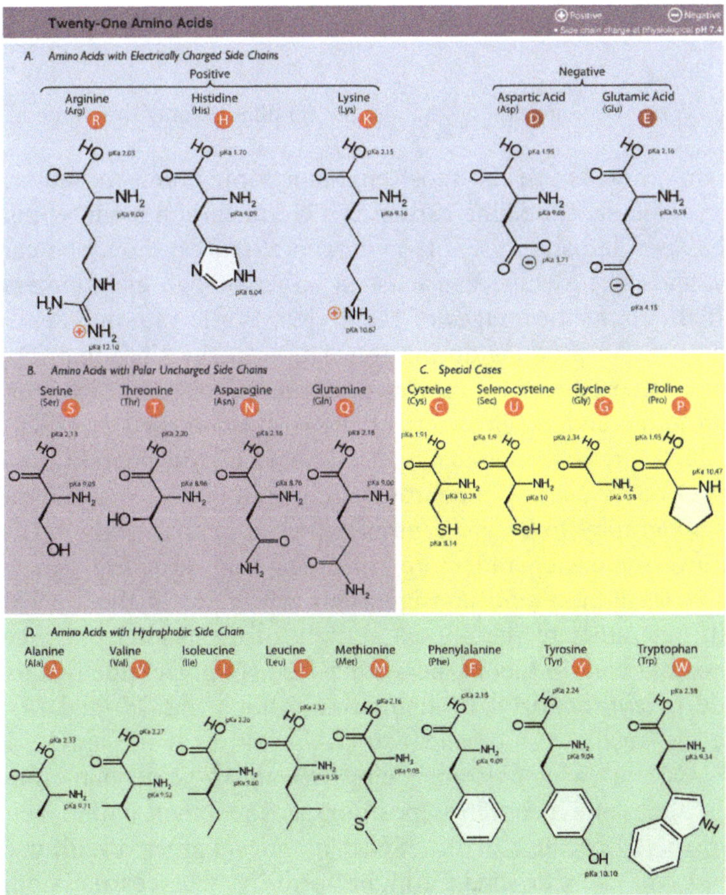

The 21 proteinogenic α-amino acids found in eukaryotes, grouped according to their side-chains' pK_a values and charges carried at physiological pH 7.4

GENERAL STRUCTURE

In the structure shown at the top of the page, **R** represents a side-chain specific to each amino acid. The carbon atom next to the carboxyl group (which is therefore numbered 2 in the carbon chain starting from that functional group) is called the α–carbon. Amino acids containing an amino group bonded directly to the alpha carbon are referred to as *alpha amino acids*. These include amino acids such as proline which contain secondary amines, which used to be often referred to as "imino acids".

Isomerism

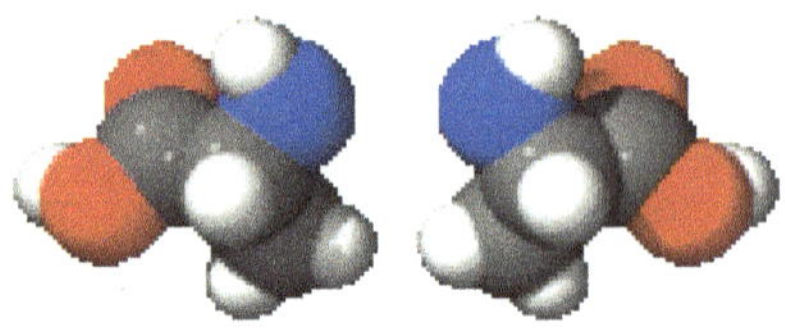

The two enantiomers of alanine, D-alanine and L-alanine

The alpha amino acids are the most common form found in nature, but only when occurring in the L-isomer. The alpha carbon is a chiral carbon atom, with the exception of glycine which has two indistinguishable hydrogen atoms on the alpha carbon. Therefore, all alpha amino acids but glycine can exist in either of two enantiomers, called L or D amino acids, which are mirror images of each other. While L-amino acids represent all of the amino acids found in proteins during translation in the ribosome, D-amino acids are found in some proteins produced by enzyme posttranslational modifications after translation and translocation to the endoplasmic reticulum, as in exotic sea-dwelling organisms such as cone snails. They are also abundant components of the peptidoglycan cell walls of bacteria, and D-serine may act as a neurotransmitter in the brain. D-amino acids are used in racemic crystallography to create centrosymmetric crystals, which (depending on the protein) may allow for easier and more robust protein structure determination. The L and D convention for amino acid configuration refers not to the optical activity of the amino acid itself but rather to the optical activity of the isomer of glyceraldehyde from which that amino acid can, in theory, be synthesized (D-glyceraldehyde is dextrorotatory; L-glyceraldehyde is levorotatory). In alternative fashion, the *(S)* and *(R)* designators are used to indicate the absolute stereochemistry. Almost all of the amino acids in proteins are *(S)* at the α carbon, with cysteine being *(R)* and glycine non-chiral. Cysteine has its side-chain in the same geometric position as the other amino acids, but the *R/S* terminology is reversed because of the higher atomic number of sulfur compared to the carboxyl oxygen gives the side-chain a higher priority, whereas the atoms in most other side-chains give them lower priority.

Side Chains

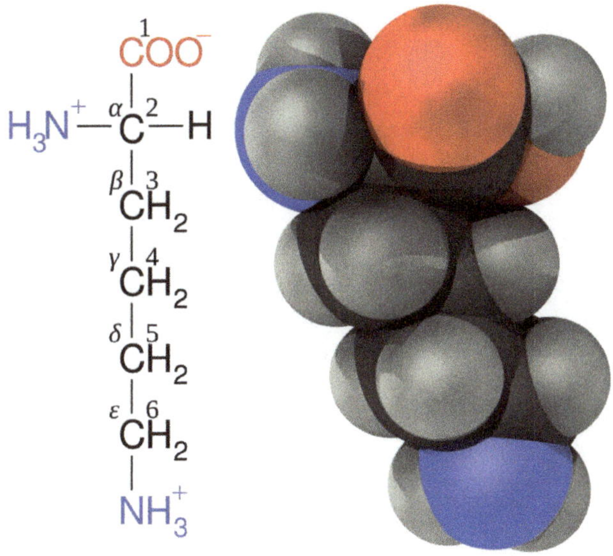

Lysine with carbon atoms labeled

In amino acids that have a carbon chain attached to the α–carbon (such as lysine, shown to the right) the carbons are labeled in order as α, β, γ, δ, and so on. In some amino acids, the amine group is attached to the β or γ-carbon, and these are therefore referred to as *beta* or *gamma amino acids*.

Amino acids are usually classified by the properties of their side-chain into four groups. The side-chain can make an amino acid a weak acid or a weak base, and a hydrophile if the side-chain is polar or a hydrophobe if it is nonpolar. The chemical structures of the 22 standard amino acids, along with their chemical properties, are described more fully in the article on these proteinogenic amino acids.

The phrase branched-chain amino acids or BCAA refers to the amino acids having aliphatic side-chains that are non-linear; these are leucine, isoleucine, and valine. Proline is the only proteinogenic amino acid whose side-group links to the α-amino group and, thus, is also the only proteinogenic amino acid containing a secondary amine at this position. In chemical terms, proline is, therefore, an imino acid, since it lacks a primary amino group, although it is still classed as an amino acid in the current biochemical nomenclature, and may also be called an N-alkylated alpha-amino acid.

Zwitterions

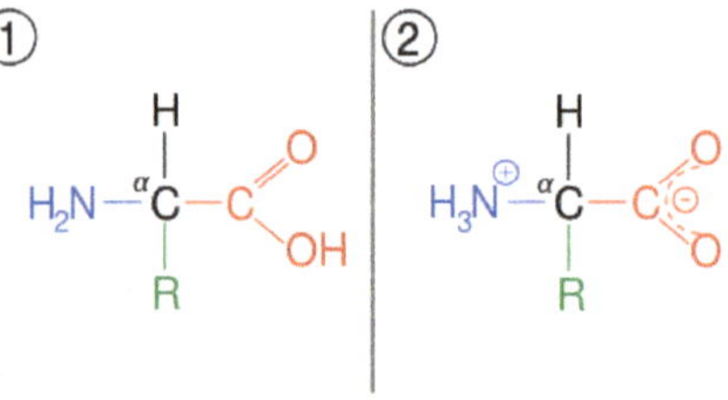

An amino acid in its (1) un-ionized and (2) zwitterionic forms

The α-carboxylic acid group of amino acids is a weak acid, meaning that it releases a hydron (such as a proton) at moderate pH values. In other words, carboxylic acid groups ($-CO_2H$) can be deprotonated to become negative carboxylates ($-CO_2^-$). The negatively charged carboxylate ion predominates at pH values greater than the pKa of the carboxylic acid group (mean for the 20 common amino acids is about 2.2, see the table of amino acid structures above). In a complementary fashion, the α-amine of amino acids is a weak base, meaning that it accepts a proton at moderate pH values. In other words, α-amino groups (NH_2-) can be protonated to become positive α-ammonium groups ($^+NH_3-$). The positively charged α-ammonium group predominates at pH values less than the pKa of the α-ammonium group (mean for the 20 common α-amino acids is about 9.4).

Because all amino acids contain amine and carboxylic acid functional groups, they share amphiprotic properties. Below pH 2.2, the predominant form will have a neutral carboxylic acid group and a positive α-ammonium ion (net charge +1), and above pH 9.4, a negative carboxylate and neutral α-amino group (net charge −1). But at pH between 2.2 and 9.4, an amino acid usually contains both a negative carboxylate and a positive α-ammonium group, as shown in structure (2) on the right, so has net zero charge. This molecular state is known as a zwitterion, from the German **Zwitter** meaning *hermaphrodite* or *hybrid*. The fully neutral form (structure (1) on the right) is a very minor species in aqueous solution throughout the pH range (less than 1 part in 10^7). Amino acids exist as zwitterions also in the solid phase, and crystallize with salt-like properties unlike typical organic acids or amines.

Isoelectric Point

The variation in titration curves when the amino acids are grouped by category can be seen here. With the exception of tyrosine, using titration to differentiate between hydrophobic amino acids is problematic.

At pH values between the two pKa values, the zwitterion predominates, but coexists in dynamic equilibrium with small amounts of net negative and net positive ions. At the exact midpoint between the two pKa values, the trace amount of net negative and trace of net positive ions exactly balance, so that average net charge of all forms present is zero.

This pH is known as the isoelectric point pI, so pI = ½(pKa_1 + pKa_2). The individual amino acids all have slightly different pKa values, so have different isoelectric points. For amino acids with charged side-chains, the pKa of the side-chain is involved. Thus for Asp, Glu with negative side-chains, pI = ½(pKa_1 + pKa_R), where pKa_R is the side-chain pKa. Cysteine also has potentially negative side-chain with pKa_R = 8.14, so pI should be calculated as for Asp and Glu, even though the side-chain is not significantly charged at neutral pH. For His, Lys, and Arg with positive side-chains, pI = ½(pKa_R + pKa_2). Amino acids have zero mobility in electrophoresis at their isoelectric point, although this behaviour is more usually exploited for peptides and proteins than single amino acids. Zwitterions have minimum solubility at their isoelectric point and some amino acids (in particular, with non-polar side-chains) can be isolated by precipitation from water by adjusting the pH to the required isoelectric point.

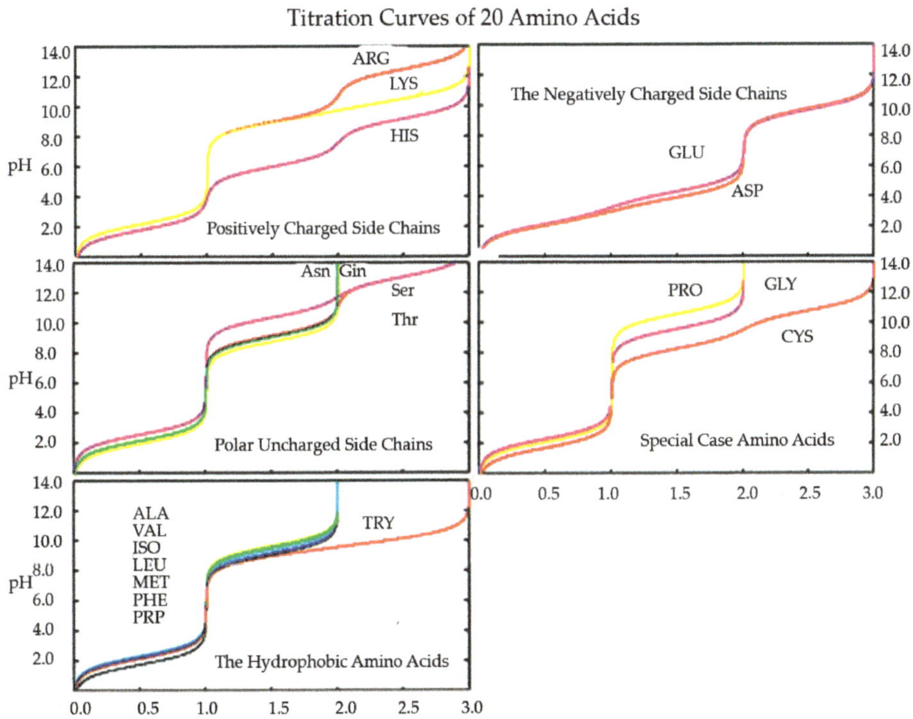

Composite of titration curves of twenty proteinogenic amino acids grouped by side chain category.

OCCURRENCE AND FUNCTIONS IN BIOCHEMISTRY

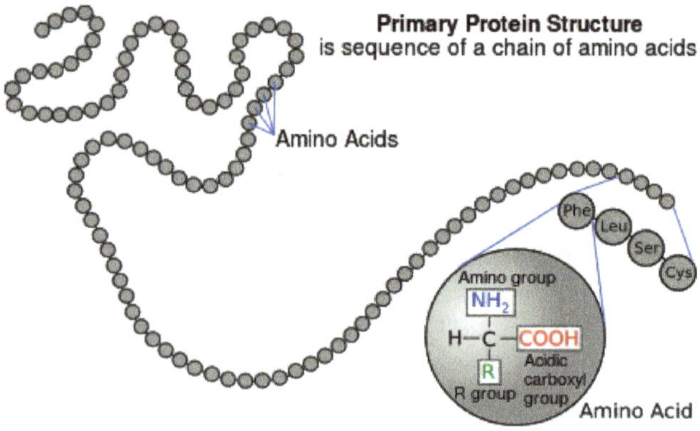

A polypeptide is an unbranched chain of amino acids.

Proteinogenic Amino Acids

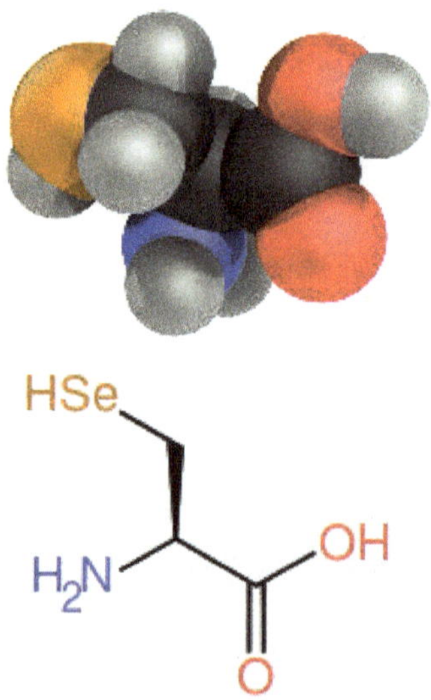

The amino acid selenocysteine.

Amino acids are the structural units (monomers) that make up proteins. They join together to form short polymer chains called peptides or longer chains called either polypeptides or proteins. These polymers are linear and unbranched, with each amino acid within the chain attached to two neighboring amino acids. The process of making proteins is called *translation* and involves the step-by-step addition of amino acids to a growing protein chain by a ribozyme that is called a ribosome. The order in which the amino acids are added is read through the genetic code from an mRNA template, which is a RNA copy of one of the organism's genes.

Twenty-two amino acids are naturally incorporated into polypeptides and are called proteinogenic or natural amino acids. Of these, 20 are encoded by the universal genetic code. The remaining 2, selenocysteine and pyrrolysine, are incorporated into proteins by unique synthetic mechanisms. Selenocysteine is incorporated when the mRNA being translated includes a SECIS element, which causes the UGA codon to encode selenocysteine instead of a stop codon. Pyrrolysine is used by some methanogenic archaea in enzymes that they use to produce methane. It is coded for with the codon UAG, which is normally a stop codon in other organisms. This UAG codon is followed by a PYLIS downstream sequence.

Non-Proteinogenic Amino Acids

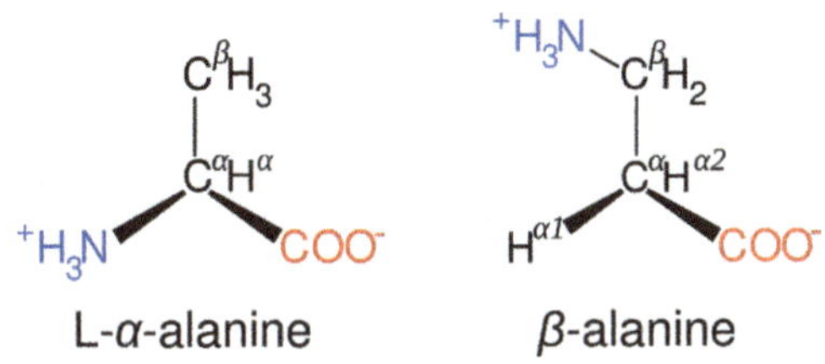

β-alanine and its α-alanine isomer

Aside from the 22 proteinogenic amino acids, there are many other amino acids that are called *non-proteinogenic*. Those either are not found in proteins (for example carnitine, GABA) or are not produced directly and in isolation by standard cellular machinery (for example, hydroxyproline and selenomethionine).

Non-proteinogenic amino acids that are found in proteins are formed by post-translational modification, which is modification after translation during protein synthesis. These modifications are often essential for the function or regulation of a protein; for example, the carboxylation of glutamate allows for better binding of calcium cations, and the hydroxylation of proline is critical for maintaining connective tissues.Another example is the formation of hypusine in the translation initiation factor EIF5A, through modification of a lysine residue. Such modifications can also determine the localization of the protein, e.g., the addition of long hydrophobic groups can cause a protein to bind to a phospholipid membrane.

Some non-proteinogenic amino acids are not found in proteins. Examples include 2-aminoisobutyric acid and the neurotransmitter gamma-aminobutyric acid. Non-proteinogenic amino acids often occur as intermediates in the metabolic pathways for standard amino acids – for example, ornithine and citrulline occur in the urea cycle, part of amino acid catabolism (see below). A rare exception to the dominance of α-amino acids in biology is the β-amino acid beta alanine (3-aminopropanoic acid), which is used in plants and microorganisms in the synthesis of pantothenic acid (vitamin B_5), a component of coenzyme A.

D-Amino Acid Natural Abundance

d-isomers are uncommon in live organisms. For instance, gramicidin is a polypeptide made up from mixture of d- and l-amino acids. Other compounds containing D-amino acid are tyrocidine and valinomycin. These compounds disrupt bacterial cell walls, particularly in Gram-positive bacteria. Only 837 d-amino acids were found in Swiss-Prot database (187 million amino acids analysed).

Non-Standard Amino Acids

The 20 amino acids that are encoded directly by the codons of the universal genetic code are called *standard* or *canonical* amino acids. The others are called *non-standard* or *non-canonical*. Most of the non-standard amino acids are also non-proteinogenic (i.e. they cannot be used to build proteins), but three of them are proteinogenic, as they can be used to build proteins by exploiting information not encoded in the universal genetic code.

The three non-standard proteinogenic amino acids are selenocysteine (present in many non-eukaryotes as well as most eukaryotes, but not coded directly by DNA), pyrrolysine (found only in some archaea and one bacterium), and *N*-formylmethionine (which is often the initial amino acid of proteins in bacteria, mitochondria, and chloroplasts). For example, 25 human proteins include selenocysteine (Sec) in their primary structure,and the structurally characterized enzymes (selenoenzymes) employ Sec as the catalytic moiety in their active sites. Pyrrolysine and selenocysteine are encoded via variant codons. For example, selenocysteine is encoded by stop codon and SECIS element.

In Human Nutrition

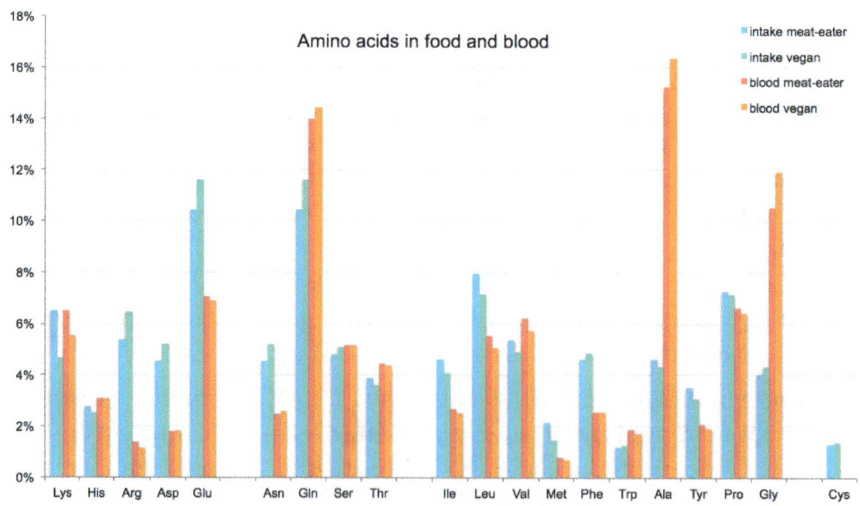

Share of amino acid in different human diets and the resulting mix of amino acids in human blood serum. Glutamate and glutamine are the most frequent in food at over 10%, while alanine, glutamine, and glycine are the most common in blood.

When taken up into the human body from the diet, the 20 standard amino acids either are used to synthesize proteins and other biomolecules or are oxidized to urea and carbon dioxide as a source of energy. The oxidation pathway starts with the removal of the amino group by a transaminase; the amino group is then fed into the urea cycle. The other product of transamidation is a keto acid that enters the citric acid cycle. Glucogenic amino acids can also be converted into glucose, through gluconeogenesis. Of the 20 standard amino acids, nine (His, Ile, Leu, Lys, Met, Phe, Thr, Trp and Val), are called essential amino acids because the human body cannot synthesize them from other compounds at the level needed for normal growth, so they must be obtained from food. In addition, cysteine, taurine, tyrosine, and arginine are considered semiessential amino-acids in children (though taurine is not technically an amino acid), because the metabolic pathways that synthesize these amino acids are not fully developed. The amounts required also depend on the age and health of the individual, so it is hard to make general statements about the dietary requirement for some amino acids. Dietary exposure to the non-standard amino acid BMAA has been linked to human neurodegenerative diseases, including ALS.

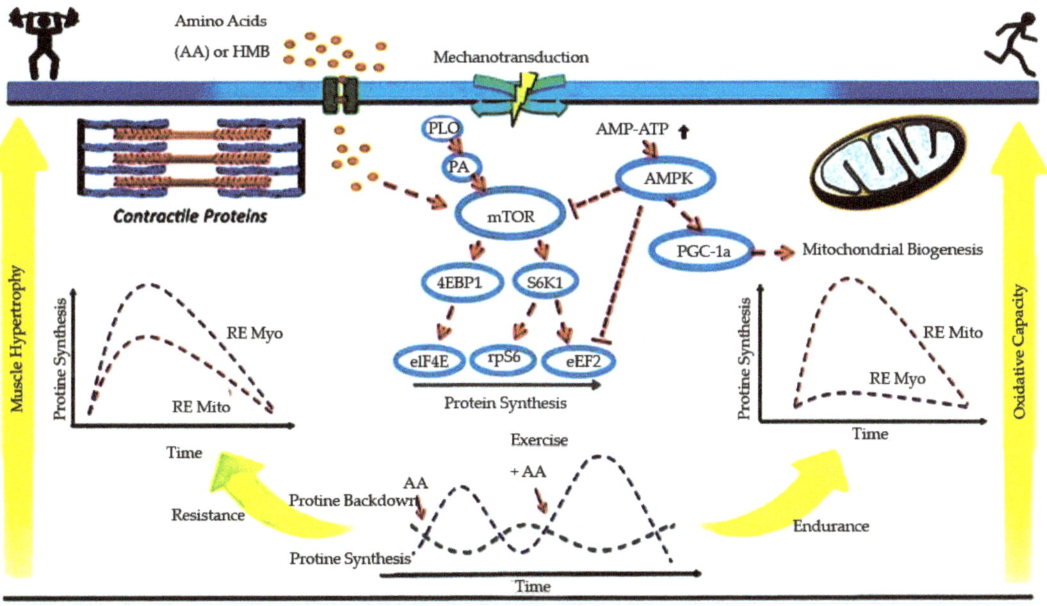

Diagram of the molecular signaling cascades that are involved in myofibrillar muscle protein synthesis and mitochondrial biogenesis in response to physical exercise and specific amino acids or their derivatives (primarily l-leucine and HMB). Many amino acids derived from food protein promote the activation of mTORC1 and increase protein synthesis by signaling through Rag GTPases.

Abbreviations and Representations

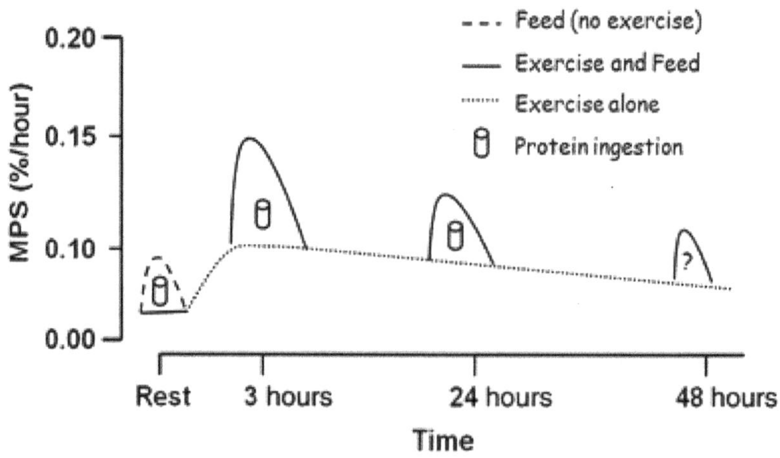

Resistance training stimulates muscle protein synthesis (MPS) for a period of up to 48 hours following exercise (shown by dotted line). Ingestion of a protein-rich meal at any point during this period will augment the exercise-induced increase in muscle protein synthesis (shown by solid lines).

Non-Protein Functions

In humans, non-protein amino acids also have important roles as metabolic intermediates, such as in the biosynthesis of the neurotransmitter gamma-amino-butyric acid (GABA). Many amino acids are used to synthesize other molecules, for example:

- Tryptophan is a precursor of the neurotransmitter serotonin.
- Tyrosine (and its precursor phenylalanine) are precursors of the catecholamineneurotransmitters dopamine, epinephrine and norepinephrine and various trace amines.
- Phenylalanine is a precursor of phenethylamine and tyrosine in humans. In plants, it is a precursor of various phenylpropanoids, which are important in plant metabolism.
- Glycine is a precursor of porphyrins such as heme.
- Arginine is a precursor of nitric oxide.
- Ornithine and S-adenosylmethionine are precursors of polyamines.
- Aspartate, glycine, and glutamine are precursors of nucleotides.

However, not all of the functions of other abundant non-standard amino acids are known.

Some non-standard amino acids are used as defenses against herbivores in plants. For example, canavanine is an analogue of arginine that is found in many legumes, and in particularly large amounts in *Canavalia gladiata* (sword bean). This amino acid protects the plants from predators such as insects and can cause illness in people if some types of legumes are eaten without processing. The non-protein amino acid mimosine is found in other species of legume, in particular *Leucaena leucocephala*. This compound is an analogue of tyrosine and can poison animals that graze on these plants.

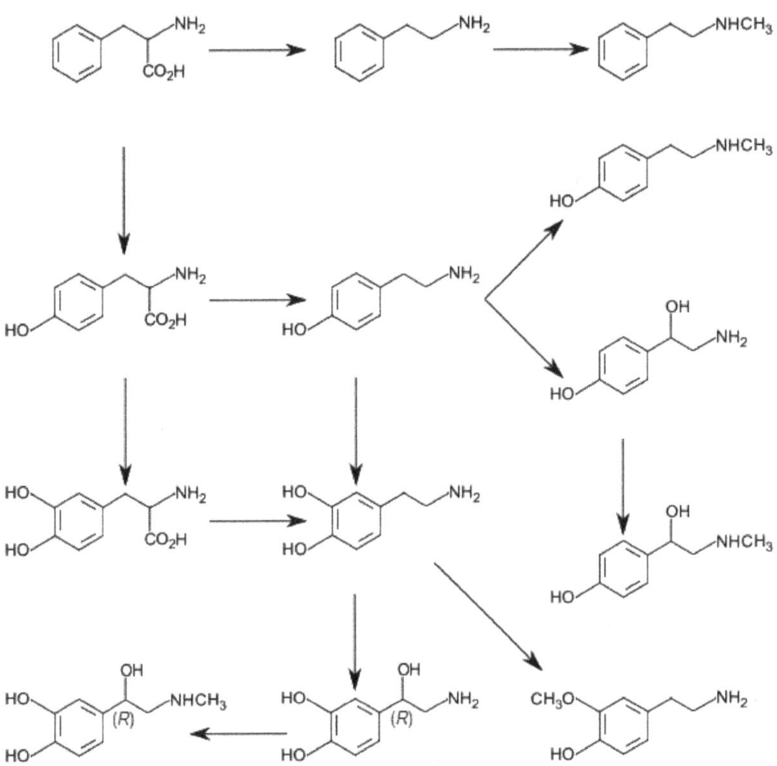

Catecholamines and trace amines are synthesized from phenylalanine and tyrosine in humans.

USES IN INDUSTRY

Amino acids are used for a variety of applications in industry, but their main use is as additives to animal feed. This is necessary, since many of the bulk components of these feeds, such as soybeans, either have low levels or lack some of the essential amino acids: lysine, methionine, threonine, and tryptophan are most important in the production of these feeds. In this industry, amino acids are also used to chelate metal cations in order to improve the absorption of minerals from supplements, which may be required to improve the health or production of these animals.

The food industry is also a major consumer of amino acids, in particular, glutamic acid, which is used as a flavor enhancer, and aspartame (aspartyl-phenylalanine-1-methyl ester) as a low-calorie artificial sweetener. Similar technology to that used for animal nutrition is employed in the human nutrition industry to alleviate symptoms of mineral deficiencies, such as anemia, by improving mineral absorption and reducing negative side effects from inorganic mineral supplementation.

The chelating ability of amino acids has been used in fertilizers for agriculture to facilitate the delivery of minerals to plants in order to correct mineral deficiencies, such as iron chlorosis. These fertilizers are also used to prevent deficiencies from occurring and improving the overall health of the plants. The remaining production of amino acids is used in the synthesis of drugs and cosmetics.

Similarly, some amino acids derivatives are used in pharmaceutical industry. They include 5-HTP (5-hydroxytryptophan) used for experimental treatment of depression, L-DOPA (L-dihydroxyphenylalanine) for Parkinson's treatment, and eflornithine drug that inhibits ornithine decarboxylase and used in the treatment of sleeping sickness.

Expanded Genetic Code

Since 2001, 40 non-natural amino acids have been added into protein by creating a unique codon (recoding) and a corresponding transfer-RNA:aminoacyl – tRNA-synthetase pair to encode it with diverse physicochemical and biological properties in order to be used as a tool to exploring protein structure and function or to create novel or enhanced proteins.

Nullomers

Nullomers are codons that in theory code for an amino acid, however in nature there is a selective bias against using this codon in favor of another, for example bacteria prefer to use CGA instead of AGA to code for arginine. This creates some sequences that do not appear in the genome. This characteristic can be taken advantage of and used to create new selective cancer-fighting drugs and to prevent cross-contamination of DNA samples from crime-scene investigations.

Chemical Building Blocks

Amino acids are important as low-cost feedstocks. These compounds are used in chiral pool synthesis as enantiomerically pure building-blocks.

Amino acids have been investigated as precursors chiral catalysts, e.g., for asymmetric hydrogenation reactions, although no commercial applications exist.

Biodegradable Plastics

Amino acids are under development as components of a range of biodegradable polymers. These materials have applications as environmentally friendly packaging and in medicine in drug delivery and the construction of prosthetic implants. These polymers include polypeptides, polyamides, polyesters, polysulfides, and polyurethanes with amino acids either forming part of their main chains or bonded as side-chains. These modifications alter the physical properties and reactivities of the polymers. An interesting example of such materials is polyaspartate, a water-soluble biodegradable polymer that may have applications in disposable diapers and agriculture. Due to its solubility and ability to chelate metal ions, polyaspartate is also being used as a biodegradeable anti-scaling agent and a corrosion inhibitor. In addition, the aromatic amino acid tyrosine is being developed as a possible replacement for toxic phenols such as bisphenol A in the manufacture of polycarbonates.

Reactions

As amino acids have both a primary amine group and a primary carboxyl group, these chemicals can undergo most of the reactions associated with these functional groups. These include nucleophilic addition, amide bond formation, and imine formation for the amine group, and esterification, amide bond formation, and decarboxylation for the carboxylic acid group. The combination of these functional groups allow amino acids to be effective polydentate ligands for metal-amino acid chelates. The multiple side-chains of amino acids can also undergo chemical reactions. The types of these reactions are determined by the groups on these side-chains and are, therefore, different between the various types of amino acid.

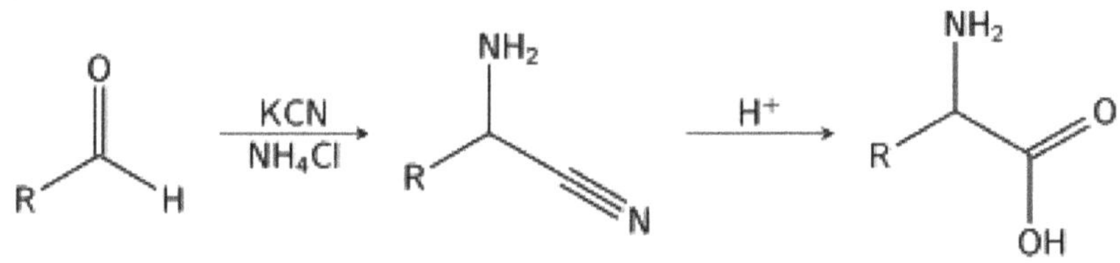

The Strecker amino acid synthesis

Chemical Synthesis

Several methods exist to synthesize amino acids. One of the oldest methods begins with the bromination at the α-carbon of a carboxylic acid. Nucleophilic substitution with ammonia then converts the alkyl bromide to the amino acid. In alternative fashion, the Strecker amino acid synthesis involves the treatment of an aldehyde with potassium cyanide and ammonia, this produces an α-amino nitrile as an intermediate. Hydrolysis of the nitrile in acid then yields a α-amino acid. Using ammonia or ammonium salts in this reaction gives unsubstituted amino acids, whereas substituting primary and secondary amines will yield substituted amino acids. Likewise, using ketones, instead of aldehydes, gives α,α-disubstituted amino acids. The classical synthesis gives racemic mixtures of α-amino acids as products, but several alternative procedures using asymmetric auxiliaries or asymmetric catalysts have been developed.

At the current time, the most-adopted method is an automated synthesis on a solid support (e.g., polystyrene beads), using protecting groups (e.g., Fmoc and t-Boc) and activating groups (e.g., DCC and DIC).

Peptide Bond Formation

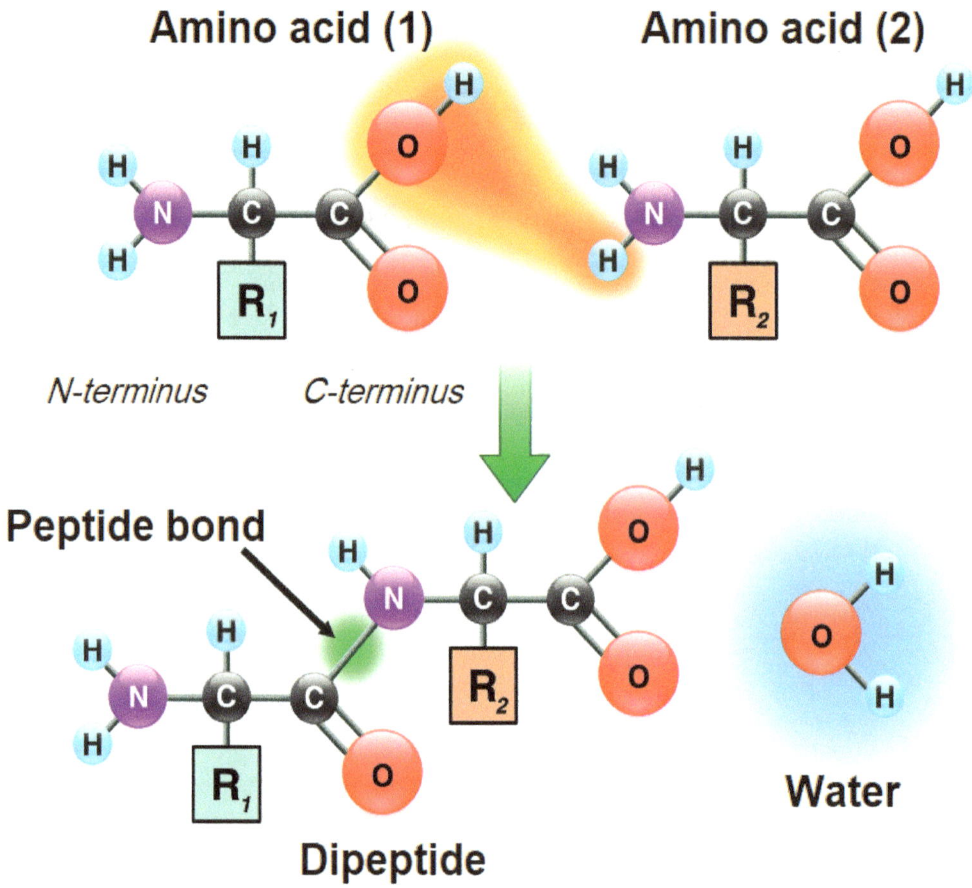

The condensation of two amino acids to form a *dipeptide* through a *peptide bond.*

As both the amine and carboxylic acid groups of amino acids can react to form amide bonds, one amino acid molecule can react with another and become joined through an amide linkage. This polymerization of amino acids is what creates proteins. This condensation reaction yields the newly formed peptide bond and a molecule of water. In cells, this reaction does not occur directly; instead, the amino acid is first activated by attachment to a transfer RNA molecule through an ester bond. This aminoacyl-tRNA is produced in an ATP-dependent reaction carried out by an aminoacyl tRNA synthetase. This aminoacyl-tRNA is then a substrate for the ribosome, which catalyzes the attack of the amino group of the elongating protein chain on the ester bond. As a result of this mechanism, all proteins made by ribosomes are synthesized starting at their N-terminus and moving toward their C-terminus.

However, not all peptide bonds are formed in this way. In a few cases, peptides are synthesized by specific enzymes. For example, the tripeptide glutathione is an essential part of the defenses of cells against oxidative stress. This peptide is synthesized in two steps from free amino acids. In the first step, gamma-glutamylcysteine synthetase condenses cysteine and glutamic acid through a peptide bond formed between the side-chain carboxyl of the glutamate (the gamma carbon of this side-chain) and the amino group of the cysteine. This dipeptide is then condensed with glycine by glutathione synthetase to form glutathione.

In chemistry, peptides are synthesized by a variety of reactions. One of the most-used in solid-phase peptide synthesis uses the aromatic oxime derivatives of amino acids as activated units. These are added in sequence onto the growing peptide chain, which is attached to a solid resin support. The ability to easily synthesize vast numbers of different peptides by varying the types and order of amino acids (using combinatorial chemistry) has made peptide synthesis particularly important in creating libraries of peptides for use in drug discovery through high-throughput screening.

Biosynthesis

In plants, nitrogen is first assimilated into organic compounds in the form of glutamate, formed from alpha-ketoglutarate and ammonia in the mitochondrion. In order to form other amino acids, the plant uses transaminases to move the amino group to another alpha-keto carboxylic acid. For example, aspartate aminotransferase converts glutamate and oxaloacetate to alpha-ketoglutarate and aspartate. Other organisms use transaminases for amino acid synthesis, too.

Nonstandard amino acids are usually formed through modifications to standard amino acids. For example, homocysteine is formed through the transsulfuration pathway or by the demethylation of methionine via the intermediate metabolite S-adenosyl methionine, while hydroxyproline is made by a posttranslational modification of proline.

Microorganisms and plants can synthesize many uncommon amino acids. For example, some microbes make 2-aminoisobutyric acid and lanthionine, which is a sulfide-bridged derivative of alanine. Both of these amino acids are found in peptidic lantibiotics such as alamethicin.However, in plants, 1-aminocyclopropane-1-carboxylic acid is a small disubstituted cyclic amino acid that is a key intermediate in the production of the plant hormone ethylene.

Catabolism

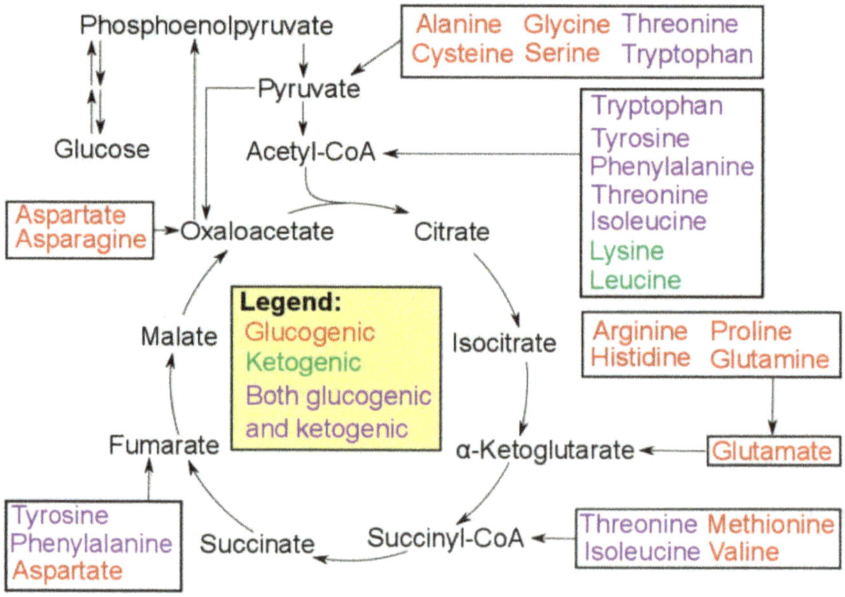

Catabolism of proteinogenic amino acids. Amino acids can be classified according to the properties of their main products as either of the following: *Glucogenic*, with the products having the ability to form glucose by gluconeogenesis.

Ketogenic, with the products not having the ability to form glucose. These products may still be used for ketogenesis or lipid synthesis. Amino acids catabolized into both glucogenic and ketogenic products.

Amino acids must first pass out of organelles and cells into blood circulation via amino acid transporters, since the amine and carboxylic acid groups are typically ionized. Degradation of an amino acid, occurring in the liver and kidneys, often involves deamination by moving its amino group to alpha-ketoglutarate, forming glutamate. This process involves transaminases, often the same as those used in amination during synthesis. In many vertebrates, the amino group is then removed through the urea cycle and is excreted in the form of urea. However, amino acid degradation can produce uric acid or ammonia instead. For example, serine dehydratase converts serine to pyruvate and ammonia. After removal of one or more amino groups, the remainder of the molecule can sometimes be used to synthesize new amino acids, or it can be used for energy by entering glycolysis or the citric acid cycle, as detailed in image at right.

Physicochemical Properties of Amino Acids

The 20 amino acids encoded directly by the genetic code can be divided into several groups based on their properties. Important factors are charge, hydrophilicity or hydrophobicity, size, and functional groups. These properties are important for protein structure and protein–protein interactions. The water-soluble proteins tend to have their hydrophobic residues (Leu, Ile, Val, Phe, and Trp) buried in the middle of the protein, whereas hydrophilic side-chains are exposed to the aqueous solvent. (Note that in biochemistry, a residue refers to a specific monomer within the polymeric chain of a polysaccharide, protein or nucleic acid.) The integral membrane proteins tend to have outer rings of exposed hydrophobic amino acids that anchor them into the lipid bilayer. In the case part-way between these two extremes, some peripheral membrane proteins have a patch of hydrophobic amino acids on their surface that locks onto the membrane. In similar fashion, proteins that have to bind to positively charged molecules have surfaces rich with negatively charged amino acids like glutamate and aspartate, while proteins binding to negatively charged molecules have surfaces rich with positively charged chains like lysine and arginine. There are different hydrophobicity scales of amino acid residues.

Some amino acids have special properties such as cysteine, that can form covalent disulfide bonds to other cysteine residues, proline that forms a cycle to the polypeptide backbone, and glycine that is more flexible than other amino acids.

Many proteins undergo a range of posttranslational modifications, when additional chemical groups are attached to the amino acids in proteins. Some modifications can produce hydrophobic lipoproteins, or hydrophilic glycoproteins. These type of modification allow the reversible targeting of a protein to a membrane. For example, the addition and removal of the fatty acid palmitic acid to cysteine residues in some signaling proteins causes the proteins to attach and then detach from cell membranes.

Able Of Standard Amino Acid Abbreviations and Properties

Amino Acid	**3-Letter**	**1-Letter**	**Side-chain class**	**Side-chain polarity**	**Side-chain charge (pH 7.4)**	**Hydropathy index**	**Absorbance λ_{max}(nm)**	**ε at λ_{max}(mM^{-1} cm^{-1})**	**MW (Weight)**	**Occurrence in proteins (%)**
Alanine	Ala	A	aliphatic	nonpolar	neutral	1.8			89.094	8.76
Arginine	Arg	R	basic	basic polar	positive	−4.5			174.203	5.78
Asparagine	Asn	N	amide	polar	neutral	−3.5			132.119	3.93
Aspartic acid	Asp	D	acid	acidic polar	negative	−3.5			133.104	5.49
Cysteine	Cys	C	sulfur-containing	nonpolar	neutral	2.5	250	0.3	121.154	1.38
Glutamic acid	Glu	E	acid	acidic polar	negative	−3.5			147.131	6.32
Glutamine	Gln	Q	amide	polar	neutral	−3.5			146.146	3.9
Glycine	Gly	G	aliphatic	nonpolar	neutral	−0.4			75.067	7.03
Histidine	His	H	basic aromatic	basic polar	positive(10%) neutral(90%)	−3.2	211	5.9	155.156	2.26
Isoleucine	Ile	I	aliphatic	nonpolar	neutral	4.5			131.175	5.49
Leucine	Leu	L	aliphatic	nonpolar	neutral	3.8			131.175	9.68
Lysine	Lys	K	basic	basic polar	positive	−3.9			146.189	5.19
Methionine	Met	M	sulfur-containing	nonpolar	neutral	1.9			149.208	2.32
Phenylalanine	Phe	F	aromatic	nonpolar	neutral	2.8	257, 206, 188	0.2, 9.3, 60.0	165.192	3.87
Proline	Pro	P	cyclic	nonpolar	neutral	−1.6			115.132	5.02
Serine	Ser	S	hydroxyl-containing	polar	neutral	−0.8			105.093	7.14
Threonine	Thr	T	hydroxyl-containing	polar	neutral	−0.7			119.119	5.53
Tryptophan	Trp	W	aromatic	nonpolar	neutral	−0.9	280, 219	5.6, 47.0	204.228	1.25
Tyrosine	Tyr	Y	aromatic	polar	neutral	−1.3	274, 222, 193	1.4, 8.0, 48.0	181.191	2.91
Valine	Val	V	aliphatic	nonpolar	neutral	4.2			117.148	6.73

Two additional amino acids are in some species coded for by codons that are usually interpreted as stop codons:

21st and 22nd amino acids	3-Letter	1-Letter	MW(Weight)
Selenocysteine	Sec	U	168.064
Pyrrolysine	Pyl	O	255.313

In addition to the specific amino acid codes, placeholders are used in cases where chemical or crystallographic analysis of a peptide or protein cannot conclusively determine the identity of a residue. They are also used to summarise conserved protein sequence motifs. The use of single letters to indicate sets of similar residues is similar to the use of abbreviation codes for degenerate bases.

Ambiguous Amino Acids	3-Letter	1-Letter	
Any / unknown	Xaa	X	All
Asparagine or aspartic acid	Asx	B	D, N
Glutamine or glutamic acid	Glx	Z	E, Q
Leucine or Isoleucine	Xle	J	I, L
Hydrophobic		Φ	V, I, L, F, W, Y, M
Aromatic		Ω	F, W, Y, H
Aliphatic (Non-Aromatic)		Ψ	V, I, L, M
Small		π	P, G, A, S
Hydrophilic		ζ	S, T, H, N, Q, E, D, K, R
Positively charged		+	K, R, H
Negatively charged		−	D, E

Unk is sometimes used instead of **Xaa**, but is less standard.

In addition, many non-standard amino acids have a specific code. For example, several peptide drugs, such as Bortezomib and MG132, are artificially synthesized and retain their protecting groups, which have specific codes. Bortezomib is Pyz-Phe-boroLeu, and MG132 is Z-Leu-Leu-Leu-al. To aid in the analysis of protein structure, photo-reactive amino acid analogs are available. These include photoleucine (**pLeu**) and photomethionine (**pMet**).

REFERENCES

1. "Amino". Dictionary.com. 2015. Retrieved 3 July 2015.
2. "amino acid". Cambridge Dictionaries Online. Cambridge University Press. 2015. Retrieved 3 July 2015.
3. "amino". FreeDictionary.com. Farlex. 2015. Retrieved 3 July 2015.
4. Wagner I, Musso H (November 1983). "New Naturally Occurring Amino Acids". Angewandte Chemie International Edition in English. 22 (11): 816–28. doi:10.1002/anie.198308161.

5. Latham, Michael C. (1997). "Chapter 8. Body composition, the functions of food, metabolism and energy". Human nutrition in the developing world. Food and Nutrition Series – No. 29. Rome: Food and Agriculture Organization of the United Nations.
6. Proline is an exception to this general formula. It lacks the NH2 group because of the cyclization of the side-chain and is known as an imino acid; it falls under the category of special structured amino acids.
7. Clark, Jim (August 2007). "an introduction to amino acids". chemguide. Retrieved 4 July 2015.
8. Jakubke, Hans-Dieter; Sewald, Norbert (2008). "Amino acids". Peptides from A to Z: A Concise Encyclopedia. Germany: Wiley-VCH. p. 20. ISBN 9783527621170 – via Google Books.
9. Pollegioni, Loredano; Servi, Stefano, eds. (2012). Unnatural Amino Acids: Methods and Protocols. Methods in Molecular Biology – Volume 794. Humana Press. p. v. ISBN 978-1-61779-331-8. OCLC 756512314.
10. Hertweck C (2011). "Biosynthesis and Charging of Pyrrolysine, the 22nd Genetically Encoded Amino Acid". Angewandte Chemie International Edition. 50 (41): 9540–1. doi:10.1002/anie.201103769. PMID 21796749.
11. "Chapter 1: Proteins are the Body's Worker Molecules". The Structures of Life. National Institute of General Medical Sciences. 27 October 2011. Retrieved 20 May 2008.
12. Michal, Gerhard; Schomburg, Dietmar, eds. (2012). Biochemical Pathways: An Atlas of Biochemistry and Molecular Biology (2nd ed.). Wiley. p. 5.
13. Modeling Electrostatic Contributions to Protein Folding and Binding – Tjong, p.1 footnote
14. Frontiers in Drug Design and Discovery ed. Atta-Ur-Rahman & others, p.299
15. Elzanowski A, Ostell J (7 April 2008). "The Genetic Codes". National Center for Biotechnology Information (NCBI). Retrieved 10 March 2010.
16. Xie J, Schultz PG (December 2005). "Adding amino acids to the genetic repertoire". Current Opinion in Chemical Biology. 9 (6): 548–54. doi:10.1016/j.cbpa.2005.10.011. PMID 16260173.
17. Wang Q, Parrish AR, Wang L (March 2009). "Expanding the genetic code for biological studies". Chem. Biol. 16 (3): 323–36. doi:10.1016/j.chembiol.2009.03.001. PMC 2696486 . PMID 19318213.
18. Simon M (2005). Emergent computation: emphasizing bioinformatics. New York: AIP Press/Springer Science+Business Media. pp. 105–106. ISBN 0-387-22046-1.
19. Petroff OA (December 2002). "GABA and glutamate in the human brain". Neuroscientist. 8 (6): 562–573. doi:10.1177/1073858402238515. PMID 12467378.
20. For example, ruminants such as cows obtain a number of amino acids via microbes in the first two stomach chambers.
21. Vauquelin LN, Robiquet PJ (1806). "The discovery of a new plant principle in Asparagus sativus". Annales de Chimie. 57: 88–93.

22. Anfinsen CB, Edsall JT, Richards FM (1972). Advances in Protein Chemistry. New York: Academic Press. pp. 99, 103. ISBN 978-0-12-034226-6.
23. Wollaston WH (1810). "On cystic oxide, a new species of urinary calculus". Philosophical Transactions of the Royal Society. 100: 223–30. doi:10.1098/rstl.1810.0015.
24. Baumann E (1884). "Über cystin und cystein. Z Physiol Chem. 8 (4): 299–305. Archived from the original on 14 March 2011. Retrieved 28 March 2011.
25. Braconnot HM (1820). Sur la conversion des matières animales en nouvelles substances par le moyen de l›acide sulfurique. Annales de Chimie et de Physique. 2nd Series. 13: 113–25.
26. Robert D. Simoni; Robert L. Hill; Martha Vaughan (13 September 2002). The Discovery of the Amino Acid Threonine: the Work of William C. Rose. Journal of Biological Chemistry. 277 (37): 56–58. PMID 12218068.
27. McCoy, R. H.; Meyer, C. E. & Rose, W. C. (1935). Feeding Experiments with Mixtures of Highly Purified Amino Acids. VIII. Isolation and Identification of a New Essential Amino Acid. Journal of Biological Chemistry. 112: 283–302.
28. etymonline.com entry for amino. www.etymonline.com. Retrieved 19 July 2010.
29. Joseph S. Fruton (1990). Chapter 5- Emil Fischer and Franz Hofmeister. Contrasts in Scientific Style: Research Groups in the Chemical and Biochemical Sciences,. 191. American Philosophical Society. pp. 163–165. ISBN 0-87169-191-4.
30. Alpha amino acid – Medical definition. Merriam-Webster dictionary.
31. Proline at the US National Library of Medicine Medical Subject Headings (MeSH)
32. http://opbs.okstate.edu/5753/Amino%20Acids.html
33. IUPAC, Compendium of Chemical Terminology, 2nd ed. (the Gold Book) (1997). Online corrected version: (2006–) Imino acids.
34. Creighton, Thomas H. (1993). Chapter 1. Proteins: structures and molecular properties. San Francisco: W. H. Freeman. ISBN 978-0-7167-7030-5.
35. Pisarewicz K, Mora D, Pflueger FC, Fields GB, Marí F (May 2005). Polypeptide chains containing D-gamma-hydroxyvaline. Journal of the American Chemical Society. 127 (17): 6207–15. doi:10.1021/ja050088m. PMID 15853325.
36. van Heijenoort J (March 2001). Formation of the glycan chains in the synthesis of bacterial peptidoglycan. Glycobiology. 11 (3): 25R–36R. doi:10.1093/glycob/11.3.25R. PMID 11320055.
37. Wolosker H, Dumin E, Balan L, Foltyn VN (July 2008). D-amino acids in the brain: D-serine in neurotransmission and neurodegeneration. The FEBS Journal. 275 (14): 3514–26. doi:10.1111/j.1742-4658.2008.06515.x. PMID 18564180.
38. Matthews BW (2009). Racemic crystallography—easy crystals and easy structures: What›s not to like?. Protein Science. 18 (6): 1135–1138. doi:10.1002/pro.125. PMC 2774423 . PMID 19472321.
39. Hatem, Salama Mohamed Ali (2006). Gas chromatographic determination of Amino Acid Enantiomers in tobacco and bottled wines. University of Giessen.

Retrieved 17 November 2008.

40. Nomenclature and Symbolism for Amino Acids and Peptides. IUPAC-IUB Joint Commission on Biochemical Nomenclature. 1983. Archived from the original on 9 October 2008. Retrieved 17 November 2008.
41. Jodidi, S. L. (1 March 1926). The Formol Titration of Certain Amino Acids. Journal of the American Chemical Society. 48 (3): 751–753. doi:10.1021/ja01414a033.
42. Liebecq, Claude, ed. (1992). Biochemical Nomenclature and Related Documents (2nd ed.). Portland Press. pp. 39–69. ISBN 978-1-85578-005-7.
43. Smith, Anthony D. (1997). Oxford dictionary of biochemistry and molecular biology. Oxford: Oxford University Press. p. 535. ISBN 978-0-19-854768-6. OCLC 37616711.
44. Simmons, William J.; Gerhard Meisenberg (2006). Principles of medical biochemistry. Mosby Elsevier. p. 19. ISBN 0-323-02942-6.
45. Fennema OR. Food Chemistry 3rd Ed. CRC Press. pp. 327–8. ISBN 0-8247-9691-8.
46. Rodnina MV, Beringer M, Wintermeyer W (January 2007). How ribosomes make peptide bonds. Trends in Biochemical Sciences. 32 (1): 20–6. doi:10.1016/j.tibs.2006.11.007. PMID 17157507.
47. Driscoll DM, Copeland PR (2003). Mechanism and regulation of selenoprotein synthesis. Annual Review of Nutrition. 23 (1): 17–40. doi:10.1146/annurev.nutr.23.011702.073318. PMID 12524431.
48. Krzycki JA (December 2005). The direct genetic encoding of pyrrolysine. Current Opinion in Microbiology. 8 (6): 706–12. doi:10.1016/j.mib.2005.10.009. PMID 16256420.
49. Théobald-Dietrich A, Giegé R, Rudinger-Thirion J (2005). Evidence for the existence in mRNAs of a hairpin element responsible for ribosome dependent pyrrolysine insertion into proteins. Biochimie. 87 (9–10): 813–7. doi:10.1016/j.biochi.2005.03.006. PMID 16164991.
50. Vermeer C (March 1990). Gamma-carboxyglutamate-containing proteins and the vitamin K-dependent carboxylase. The Biochemical Journal. 266 (3): 625–36. doi:10.1042/bj2660625. PMC 1131186 . PMID 2183788.
51. Bhattacharjee A, Bansal M (March 2005). Collagen structure: the Madras triple helix and the current scenario. IUBMB Life. 57 (3): 161–72. doi:10.1080/15216540500090710. PMID 16036578.
52. Park MH (February 2006). The post-translational synthesis of a polyamine-derived amino acid, hypusine, in the eukaryotic translation initiation factor 5A (eIF5A). Journal of Biochemistry. 139 (2): 161–9. doi:10.1093/jb/mvj034. PMC 2494880 . PMID 16452303.
53. Blenis J, Resh MD (December 1993). Subcellular localization specified by protein acylation and phosphorylation. Current Opinion in Cell Biology. 5 (6): 984–9. doi:10.1016/0955-0674(93)90081-Z. PMID 8129952.

54. Curis E, Nicolis I, Moinard C, Osowska S, Zerrouk N, Bénazeth S, Cynober L (November 2005). Almost all about citrulline in mammals. Amino Acids. 29 (3): 177–205. doi:10.1007/s00726-005-0235-4. PMID 16082501.
55. Coxon KM, Chakauya E, Ottenhof HH, Whitney HM, Blundell TL, Abell C, Smith AG (August 2005). Pantothenate biosynthesis in higher plants. Biochemical Society Transactions. 33 (Pt 4): 743–6. doi:10.1042/BST0330743. PMID 16042590.
56. Ketchem, Randal R.; Hu, Weidong; Cross, Timothy A. (1993). High-Resolution Conformation of Gramicidin A in a Lipid Bilayer by Solid-State NMR. Science. 261 (5127): 1457–1460. doi:10.1126/science.7690158.
57. Khoury, George A.; Baliban, Richard C.; Floudas, Christodoulos A. (2011). Proteome-Wide Post-Translational Modification Statistics: Frequency Analysis and Curation of the Swiss-Prot Database. Scientific Reports. 1 (90). doi:10.1038/srep00090. PMC 3201773 . PMID 22034591.
58. Kryukov GV, Castellano S, Novoselov SV, Lobanov AV, Zehtab O, Guigó R, Gladyshev VN (2003). Characterization of mammalian selenoproteomes. Science. 300 (5624): 1439–1443. Bibcode:2003Sci...300.1439K. doi:10.1126/science.1083516. PMID 12775843.
59. Gromer, S., Urig, S., Becker, K. (2004) The Thioredoxin System – From Science to Clinic. Medicinal Research Reviews. 24(1):40–89.
60. Sakami W, Harrington H (1963). Amino acid metabolism. Annual Review of Biochemistry. 32 (1): 355–98. doi:10.1146/annurev.bi.32.070163.002035. PMID 14144484.
61. Brosnan JT (April 2000). Glutamate, at the interface between amino acid and carbohydrate metabolism. The Journal of Nutrition. 130 (4S Suppl): 988S–90S. PMID 10736367.
62. Young VR, Ajami AM (September 2001). Glutamine: the emperor or his clothes?. The Journal of Nutrition. 131 (9 Suppl): 2449S–59S; discussion 2486S–7S. PMID 11533293.
63. Young VR (August 1994). Adult amino acid requirements: the case for a major revision in current recommendations. The Journal of Nutrition. 124 (8 Suppl): 1517S–1523S. PMID 8064412.
64. Fürst P, Stehle P (June 2004). What are the essential elements needed for the determination of amino acid requirements in humans?. The Journal of Nutrition. 134 (6 Suppl): 1558S–1565S. PMID 15173430.
65. Reeds PJ (July 2000). Dispensable and indispensable amino acids for humans. The Journal of Nutrition. 130 (7): 1835S–40S. PMID 10867060.
66. Imura K, Okada A (January 1998). Amino acid metabolism in pediatric patients. Nutrition. 14 (1): 143–8. doi:10.1016/S0899-9007(97)00230-X. PMID 9437700.
67. Lourenço R, Camilo ME (2002). Taurine: a conditionally essential amino acid in humans? An overview in health and disease. Nutrición Hospitalaria. 17 (6): 262–70. PMID 12514918.

68. Holtcamp, W. (2012). The emerging science of BMAA: do cyanobacteria contribute to neurodegenerative disease?. Environmental Health Perspectives. 120 (3): a110–a116. doi:10.1289/ehp.120-a110. PMC 3295368 . PMID 22382274.

69. Cox, PA, Davis, DA, Mash, DC, Metcalf, JS, Banack, SA. (2015). Dietary exposure to an environmental toxin triggers neurofibrillary tangles and amyloid deposits in the brain. Proceedings of the Royal Society B. 283 (1823): 20152397. doi:10.1098/rspb.2015.2397. PMC 4795023 . PMID 26791617.

70. Brook MS, Wilkinson DJ, Phillips BE, Perez-Schindler J, Philp A, Smith K, Atherton PJ (January 2016). Skeletal muscle homeostasis and plasticity in youth and ageing: impact of nutrition and exercise. Acta Physiologica. 216 (1): 15–41. doi:10.1111/apha.12532. PMC 4843955 . PMID 26010896.

71. Lipton JO, Sahin M (October 2014). The neurology of mTOR. Neuron. 84 (2): 275–291. doi:10.1016/j.neuron.2014.09.034. PMC 4223653 . PMID 25374355.

72. Figure 2: The mTOR Signaling Pathway

73. Phillips SM (May 2014). A brief review of critical processes in exercise-induced muscular hypertrophy. Sports Med. 44 Suppl 1: S71–S77. doi:10.1007/s40279-014-0152-3. PMC 4008813 . PMID 24791918.

74. Broadley KJ (March 2010). The vascular effects of trace amines and amphetamines. Pharmacol. Ther. 125 (3): 363–375. doi:10.1016/j.pharmthera.2009.11.005. PMID 19948186.

75. Lindemann L, Hoener MC (May 2005). A renaissance in trace amines inspired by a novel GPCR family. Trends Pharmacol. Sci. 26 (5): 274–281. doi:10.1016/j.tips.2005.03.007. PMID 15860375.

76. Wang X, Li J, Dong G, Yue J (February 2014). The endogenous substrates of brain CYP2D. Eur. J. Pharmacol. 724: 211–218. doi:10.1016/j.ejphar.2013.12.025. PMID 24374199. The highest level of brain CYP2D activity was found in the substantia nigra ... The in vitro and in vivo studies have shown the contribution of the alternative CYP2D-mediated dopamine synthesis to the concentration of this neurotransmitter although the classic biosynthetic route to dopamine from tyrosine is active. ... Tyramine levels are especially high in the basal ganglia and limbic system, which are thought to be related to individual behavior and emotion (Yu et al., 2003c). ... Rat CYP2D isoforms (2D2/2D4/2D18) are less efficient than human CYP2D6 for the generation of dopamine from p-tyramine. The Km values of the CYP2D isoforms are as follows: CYP2D6 (87–121 μm) ≈ CYP2D2 ≈ CYP2D18 > CYP2D4 (256 μm) for m-tyramine and CYP2D4 (433 μm) > CYP2D2 ≈ CYP2D6 > CYP2D18 (688 μm) for p-tyramine

77. Savelieva KV, Zhao S, Pogorelov VM, Rajan I, Yang Q, Cullinan E, Lanthorn TH (2008). Bartolomucci A, ed. Genetic disruption of both tryptophan hydroxylase genes dramatically reduces serotonin and affects behavior in models sensitive to antidepressants. PLoS ONE. 3 (10): e3301. Bibcode:2008PLoSO...3.3301S. doi:10.1371/journal.pone.0003301. PMC 2565062 . PMID 18923670.

78. Shemin D, Rittenberg D (1 December 1946). The biological utilization of glycine for the synthesis of the protoporphyrin of hemoglobin. Journal of Biological

Chemistry. 166 (2): 621–5. PMID 20276176.

79. Tejero J, Biswas A, Wang ZQ, Page RC, Haque MM, Hemann C, Zweier JL, Misra S, Stuehr DJ (November 2008). Stabilization and characterization of a heme-oxy reaction intermediate in inducible nitric-oxide synthase. The Journal of Biological Chemistry. 283 (48): 33498–507. doi:10.1074/jbc.M806122200. PMC 2586280 . PMID 18815130.
80. Rodríguez-Caso C, Montañez R, Cascante M, Sánchez-Jiménez F, Medina MA (August 2006). Mathematical modeling of polyamine metabolism in mammals. The Journal of Biological Chemistry. 281 (31): 21799–812. doi:10.1074/jbc.M602756200. PMID 16709566.
81. Stryer, Lubert; Berg, Jeremy Mark; Tymoczko, John L. (2002). Biochemistry. San Francisco: W.H. Freeman. pp. 693–8. ISBN 0-7167-4684-0.
82. Hylin, John W. (1969). Toxic peptides and amino acids in foods and feeds. Journal of Agricultural and Food Chemistry. 17 (3): 492–6. doi:10.1021/jf60163a003.
83. Turner BL, Harborne JB (1967). Distribution of canavanine in the plant kingdom. Phytochemistry. 6 (6): 863–66. doi:10.1016/S0031-9422(00)86033-1.
84. Ekanayake S, Skog K, Asp NG (May 2007). Canavanine content in sword beans (Canavalia gladiata): analysis and effect of processing. Food and Chemical Toxicology. 45 (5): 797–803. doi:10.1016/j.fct.2006.10.030. PMID 17187914.
85. Rosenthal GA (2001). L-Canavanine: a higher plant insecticidal allelochemical. Amino Acids. 21 (3): 319–30. doi:10.1007/s007260170017. PMID 11764412.
86. Hammond AC (May 1995). Leucaena toxicosis and its control in ruminants. Journal of Animal Science. 73 (5): 1487–92. PMID 7665380.
87. Leuchtenberger W, Huthmacher K, Drauz K (November 2005). Biotechnological production of amino acids and derivatives: current status and prospects. Applied Microbiology and Biotechnology. 69 (1): 1–8. doi:10.1007/s00253-005-0155-y. PMID 16195792.
88. Ashmead, H. DeWayne (1993). The Role of Amino Acid Chelates in Animal Nutrition. Westwood: Noyes Publications.
89. Garattini S (April 2000). Glutamic acid, twenty years later. The Journal of Nutrition. 130 (4S Suppl): 901S–9S. PMID 10736350.
90. Stegink LD (July 1987). The aspartame story: a model for the clinical testing of a food additive. The American Journal of Clinical Nutrition. 46 (1 Suppl): 204–15. PMID 3300262.
91. Albion Laboratories, Inc. Albion Ferrochel Website. Retrieved 12 July 2011.
92. Ashmead, H. DeWayne (1986). Foliar Feeding of Plants with Amino Acid Chelates. Park Ridge: Noyes Publications.
93. Turner EH, Loftis JM, Blackwell AD (March 2006). Serotonin a la carte: supplementation with the serotonin precursor 5-hydroxytryptophan. Pharmacology & Therapeutics. 109 (3): 325–38. doi:10.1016/j.pharmthera.2005.06.004. PMID 16023217.

94. Kostrzewa RM, Nowak P, Kostrzewa JP, Kostrzewa RA, Brus R (March 2005). Peculiarities of L: -DOPA treatment of Parkinson›s disease. Amino Acids. 28 (2): 157–64. doi:10.1007/s00726-005-0162-4. PMID 15750845.

95. Heby O, Persson L, Rentala M (August 2007). Targeting the polyamine biosynthetic enzymes: a promising approach to therapy of African sleeping sickness, Chagas› disease, and leishmaniasis. Amino Acids. 33 (2): 359–66. doi:10.1007/s00726-007-0537-9. PMID 17610127.

96. Cruz-Vera LR, Magos-Castro MA, Zamora-Romo E, Guarneros G (2004). Ribosome stalling and peptidyl-tRNA drop-off during translational delay at AGA codons. Nucleic Acids Research. 32. 18 (15): 4462–8. doi:10.1093/nar/gkh784. PMC 516057 . PMID 15317870.

97. Coghlan, Andy (225 October 2012). Molecules `too dangerous for nature› kill cancer cells. New Scientist. Check date values in: |date=(help)

98. Lethal DNA tags could keep innocent people out of jail. New Scientist. 2 May 2013.

99. Hanessian S (1993). Reflections on the total synthesis of natural products: Art, craft, logic, and the chiron approach. Pure and Applied Chemistry. 65 (6): 1189–204. doi:10.1351/pac199365061189.

100. Blaser, Hans Ulrich (1992). The chiral pool as a source of enantioselective catalysts and auxiliaries. Chemical Reviews. 92 (5): 935–52. doi:10.1021/cr00013a009.

101. Sanda F, Endo T (1999). Syntheses and functions of polymers based on amino acids. Macromolecular Chemistry and Physics. 200 (12): 2651–61. doi:10.1002/(SICI)1521-3935(19991201)200:12<2651::AID-MACP2651>3.0.CO;2-P.

102. Gross RA, Kalra B (2002). Biodegradable Polymers for the Environment. Science. 297 (5582): 803–807. Bibcode:2002Sci...297..803G. doi:10.1126/science.297.5582.803. PMID 12161646.

103. Low, K. C.; Wheeler, A. P.; Koskan, L. P. (1996). Commercial poly(aspartic acid) and Its Uses. Advances in Chemistry Series. 248. Washington, D.C.: American Chemical Society.

104. Thombre SM, Sarwade BD (2005). Synthesis and Biodegradability of Polyaspartic Acid: A Critical Review. Journal of Macromolecular Science, Part A. 42 (9): 1299–1315. doi:10.1080/10601320500189604.

105. Bourke SL, Kohn J (2003). Polymers derived from the amino acid l-tyrosine: polycarbonates, polyarylates and copolymers with poly(ethylene glycol). Advanced Drug Delivery Reviews. 55 (4): 447–466. doi:10.1016/S0169-409X(03)00038-3. PMID 12706045.

106. Elmore, Donald Trevor; Barrett, G. C. (1998). Amino acids and peptides. Cambridge, UK: Cambridge University Press. pp. 48–60. ISBN 0-521-46827-2.

107. Konara S, Gagnona K, Clearfield A, Thompson C, Hartle J, Ericson C, Nelson C (2010). Structural determination and characterization of copper and zinc bis-glycinates with X-ray crystallography and mass spectrometry. Journal of

Coordination Chemistry. 63 (19): 3335–47. doi:10.1080/00958972.2010.514336.

108. Gutteridge A, Thornton JM (November 2005). Understanding nature›s catalytic toolkit. Trends in Biochemical Sciences. 30 (11): 622–9. doi:10.1016/j.tibs.2005.09.006. PMID 16214343.

109. McMurry, John (1996). Organic chemistry. Pacific Grove, CA, USA: Brooks/Cole. p. 1064. ISBN 0-534-23832-7.

110. Strecker, Adolph (1850). Ueber die künstliche Bildung der Milchsäure und einen neuen, dem Glycocoll homologen Körper. Justus Liebigs Annalen der Chemie. 75 (1): 27–45. doi:10.1002/jlac.18500750103.

111. Strecker, Adolph (1854). Ueber einen neuen aus Aldehyd – Ammoniak und Blausäure entstehenden Körper. Justus Liebigs Annalen der Chemie. 91 (3): 349–51. doi:10.1002/jlac.18540910309.

112. Masumoto S, Usuda H, Suzuki M, Kanai M, Shibasaki M (May 2003). Catalytic enantioselective Strecker reaction of ketoimines. Journal of the American Chemical Society. 125 (19): 5634–5. doi:10.1021/ja034980. PMID 12733893.

113. Davis FA, Reddy RE, Portonovo PS (1994). Asymmetric strecker synthesis using enantiopure sulfinimines: A convenient synthesis of α-amino acids. Tetrahedron Letters. 35 (50): 9351–9354. doi:10.1016/S0040-4039(00)78540-6.

114. Ishitani H, Komiyama S, Hasegawa Y, Kobayashi S (2000). Catalytic Asymmetric Strecker Synthesis. Preparation of Enantiomerically Pure α-Amino Acid Derivatives from Aldimines and Tributyltin Cyanide or Achiral Aldehydes, Amines, and Hydrogen Cyanide Using a Chiral Zirconium Catalyst. Journal of the American Chemical Society. 122 (5): 762–6. doi:10.1021/ja9935207.

115. Huang J, Corey EJ (2004). A New Chiral Catalyst for the Enantioselective Strecker Synthesis of α-Amino Acids. Orgic Letters. 62 (6): 5027–9. doi:10.1021/ol047698w. PMID 15606127.

116. Duthaler, Rudolf O. (1994). Recent developments in the stereoselective synthesis of α-aminoacids. Tetrahedron. 50 (6): 1539–1650. doi:10.1016/S0040-4020(01)80840-1.

117. Ibba M, Söll D (May 2001). The renaissance of aminoacyl-tRNA synthesis. EMBO Reports. 2 (5): 382–7. doi:10.1093/embo-reports/kve095. PMC 1083889 . PMID 11375928.

118. Lengyel P, Söll D (June 1969). Mechanism of protein biosynthesis. Bacteriological Reviews. 33 (2): 264–301. PMC 378322 . PMID 4896351.

119. Wu G, Fang YZ, Yang S, Lupton JR, Turner ND (March 2004). Glutathione metabolism and its implications for health. The Journal of Nutrition. 134 (3): 489–92. PMID 14988435.

120. Meister A (November 1988). Glutathione metabolism and its selective modification. The Journal of Biological Chemistry. 263 (33): 17205–8. PMID 3053703.

121. Carpino, Louis A. (1992). 1-Hydroxy-7-azabenzotriazole. An efficient peptide coupling additive. Journal of the American Chemical Society. 115 (10): 4397–8.

doi:10.1021/ja00063a082.

122. Marasco D, Perretta G, Sabatella M, Ruvo M (October 2008). Past and future perspectives of synthetic peptide libraries. Current Protein & Peptide Science. 9 (5): 447–67. doi:10.2174/138920308785915209. PMID 18855697.

123. Jones, Russell Celyn; Buchanan, Bob B.; Gruissem, Wilhelm (2000). Biochemistry & molecular biology of plants. Rockville, Md: American Society of Plant Physiologists. pp. 371–2. ISBN 0-943088-39-9.

124. Brosnan JT, Brosnan ME (June 2006). The sulfur-containing amino acids: an overview. The Journal of Nutrition. 136 (6 Suppl): 1636S–1640S. PMID 16702333.

125. Kivirikko KI, Pihlajaniemi T (1998). Collagen hydroxylases and the protein disulfide isomerase subunit of prolyl 4-hydroxylases. Advances in Enzymology and Related Areas of Molecular Biology. Advances in Enzymology – and Related Areas of Molecular Biology. 72: 325–98. doi:10.1002/9780470123188.ch9. ISBN 9780470123188. PMID 9559057.

126. Whitmore L, Wallace BA (May 2004). Analysis of peptaibol sequence composition: implications for in vivo synthesis and channel formation. European Biophysics Journal. 33 (3): 233–7. doi:10.1007/s00249-003-0348-1. PMID 14534753.

127. Alexander L, Grierson D (October 2002). Ethylene biosynthesis and action in tomato: a model for climacteric fruit ripening. Journal of Experimental Botany. 53 (377): 2039–55. doi:10.1093/jxb/erf072. PMID 12324528.

128. Stipanuk, M. H. (2006). Biochemical, physiological, & molecular aspects of human nutrition (2 ed.): Saunders Elsevier.

129. Stryer, Lubert; Berg, Jeremy Mark; Tymoczko, John L. (2002). Biochemistry. San Francisco: W.H. Freeman. pp. 639–49. ISBN 0-7167-4684-0.

130. Urry DW (2004). The change in Gibbs free energy for hydrophobic association: Derivation and evaluation by means of inverse temperature transitions. Chemical Physics Letters. 399 (1–3): 177–83. doi:10.1016/S0009-2614(04)01565-9.

131. Magee T, Seabra MC (April 2005). Fatty acylation and prenylation of proteins: what›s hot in fat. Current Opinion in Cell Biology. 17 (2): 190–6. doi:10.1016/j.ceb.2005.02.003. PMID 15780596.

132. Pilobello KT, Mahal LK (June 2007). Deciphering the glycocode: the complexity and analytical challenge of glycomics. Current Opinion in Chemical Biology. 11 (3): 300–5. doi:10.1016/j.cbpa.2007.05.002. PMID 17500024.

133. Smotrys JE, Linder ME (2004). Palmitoylation of intracellular signaling proteins: regulation and function. Annual Review of Biochemistry. 73 (1): 559–87. doi:10.1146/annurev.biochem.73.011303.073954. PMID 15189153.

134. Hausman, Robert E.; Cooper, Geoffrey M. (2004). The cell: a molecular approach. Washington, D.C: ASM Press. p. 51. ISBN 0-87893-214-3.

135. Kyte J, Doolittle RF (May 1982). A simple method for displaying the hydropathic character of a protein. Journal of Molecular Biology. 157 (1): 105–32. doi:10.1016/0022-2836(82)90515-0. PMID 7108955.

136. Freifelder, D. (1983). Physical Biochemistry (2nd ed.). W. H. Freeman and Company. ISBN 0-7167-1315-2.[page needed]

137. Kozlowski, Lukasz P. (2016-10-26). Proteome-pI: proteome isoelectric point database. Nucleic Acids Research: gkw978. doi:10.1093/nar/gkw978. PMID 27789699.

138. Aasland, Rein; Abrams, Charles; Ampe, Christophe; Ball, Linda J.; Bedford, Mark T.; Cesareni, Gianni; Gimona, Mario; Hurley, James H.; Jarchau, Thomas (20 February 2002). Normalization of nomenclature for peptide motifs as ligands of modular protein domains. FEBS Letters. 513 (1): 141–144. doi:10.1016/S0014-5793(01)03295-1.

139. A One-Letter Notation for Amino Acid Sequences. European Journal of Biochemistry. 5 (2): 151–153. 1 July 1968. doi:10.1111/j.1432-1033.1968.tb00350.x.

140. Suchanek M, Radzikowska A, Thiele C (April 2005). Photo-leucine and photo-methionine allow identification of protein-protein interactions in living cells. Nature Methods. 2 (4): 261–7. doi:10.1038/nmeth752. PMID 15782218.

CHAPTER 8

Enzyme Biochemical Description of Enzymes

Chapter Contents

- **Introduction**
- **Etymology and History**
- **Structure**
- **Mechanism**
- **Thermodynamics**
- **Kinetics**
- **Biological Function**

Enzymes are proteins which enable chemical reactions in the body to occur many times more quickly than they otherwise would. In some cases, a reaction will not take placc at all unless a specific enzyme is present. Along with pH, temperature and substrate concentration, enzyme concentration has an important effect on the reaction process. Assuming the amount of the substance, or substrate, taking part in the chemical reaction stays the same, a low concentration can lead to a chemical reaction occurring slowly. On the other hand, a high concentration can speed up a reaction, until a maximum possible rate is reached.

INTRODUCTION

Enzymes are macromolecular biological catalysts. Enzymes accelerate, or catalyze, chemical reactions. The molecules at the beginning of the process upon which enzymes may act are called substrates and the enzyme converts these into different molecules, called products. Almost all metabolic processes in the cell need enzymes in order to occur at rates fast enough to sustain life. The set of enzymes made in a cell determines which metabolic pathways occur in that cell. The study of enzymes is called *enzymology*.

Enzymes are known to catalyze more than 5,000 biochemical reaction types. Most enzymes are proteins, although a few are catalytic RNA molecules. Enzymes' specificity comes from their unique three-dimensional structures.

Like all catalysts, enzymes increase the rate of a reaction by lowering its activation energy. Some enzymes can make their conversion of substrate to product occur many millions of times faster. An extreme example is orotidine 5'-phosphate decarboxylase, which allows a reaction that would otherwise take millions of years to occur in milliseconds. Chemically, enzymes are like any catalyst and are not consumed in chemical reactions, nor do they alter the equilibrium of a reaction. Enzymes differ from most other catalysts by being much more specific. Enzyme activity can be affected by other molecules: inhibitors are molecules that decrease enzyme activity, and activators are molecules that increase activity. Many drugs and poisons are enzyme inhibitors. An enzyme's activity decreases markedly outside its optimal temperature and pH.

Some enzymes are used commercially, for example, in the synthesis of antibiotics. Some household products use enzymes to speed up chemical reactions: enzymes in biological washing powders break down protein, starch or fat stains on clothes, and enzymes in meat tenderizer break down proteins into smaller molecules, making the meat easier to chew.

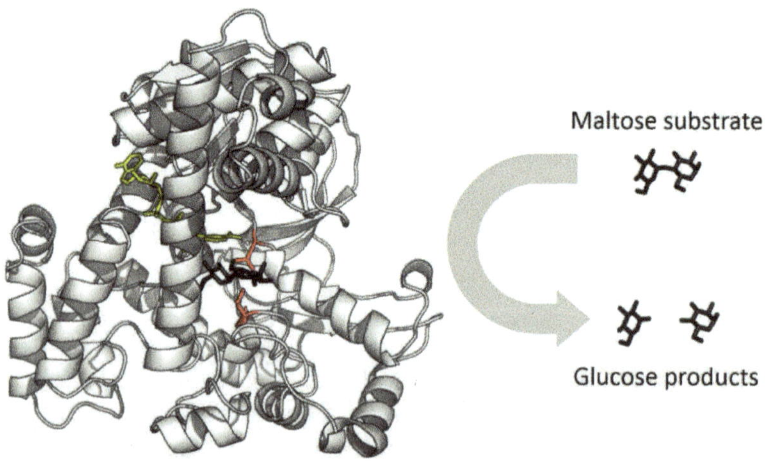

The enzyme glucosidase converts sugar maltose to two glucose sugars. Active site residues in red, maltose substrate in black, and NAD cofactor in yellow. (PDB: 1OBB)

ETYMOLOGY AND HISTORY

By the late 17th and early 18th centuries, the digestion of meat by stomach secretions and the conversion of starch to sugars by plant extracts and saliva were known but the mechanisms by which these occurred had not been identified.

French chemist Anselme Payen was the first to discover an enzyme, diastase, in 1833. A few decades later, when studying the fermentation of sugar to alcohol by yeast, Louis Pasteur concluded that this fermentation was caused by a vital force contained within the yeast cells called "ferments", which were thought to function only within living organisms.

He wrote that "alcoholic fermentation is an act correlated with the life and organization of the yeast cells, not with the death or putrefaction of the cells."

Eduard Buchner

In 1877, German physiologist Wilhelm Kühne (1837–1900) first used the term *enzyme*, which comes from Greek νζυμον, leavened, to describe this process. The word *enzyme* was used later to refer to nonliving substances such as pepsin, and the word *ferment* was used to refer to chemical activity produced by living organisms.

Eduard Buchner submitted his first paper on the study of yeast extracts in 1897. In a series of experiments at the University of Berlin, he found that sugar was fermented by yeast extracts even when there were no living yeast cells in the mixture. He named the enzyme that brought about the fermentation of sucrose zymase. In 1907, he received the Nobel Prize in Chemistry for "his discovery of cell-free fermentation". Following Buchner's example, enzymes are usually named according to the reaction they carry out: the suffix *-ase* is combined with the name of the substrate (e.g., lactase is the enzyme that cleaves lactose) or to the type of reaction (e.g., DNA polymerase forms DNA polymers).

The biochemical identity of enzymes was still unknown in the early 1900s. Many scientists observed that enzymatic activity was associated with proteins, but others (such as Nobel laureate Richard Willstätter) argued that proteins were merely carriers for the true enzymes and that proteins *per se* were incapable of catalysis. In 1926, James B. Sumner showed that the enzyme urease was a pure protein and crystallized it; he did likewise for the enzyme catalase in 1937. The conclusion that pure proteins can be enzymes was definitively demonstrated by John Howard Northrop and Wendell Meredith Stanley, who worked on the digestive enzymes pepsin (1930), trypsin and chymotrypsin. These three scientists were awarded the 1946 Nobel Prize in Chemistry.

The discovery that enzymes could be crystallized eventually allowed their structures to be solved by x-ray crystallography. This was first done for lysozyme, an enzyme found in tears, saliva and egg whites that digests the coating of some bacteria; the structure was solved by a group led by David Chilton Phillips and published in 1965. This high-resolution structure of lysozyme marked the beginning of the field of structural biology and the effort to understand how enzymes work at an atomic level of detail.

Naming Conventions

An enzyme's name is often derived from its substrate or the chemical reaction it catalyzes, with the word ending in *-ase*.Examples are lactase, alcohol dehydrogenase and DNA polymerase. Different enzymes that catalyze the same chemical reaction are called isozymes.

The International Union of Biochemistry and Molecular Biology have developed a nomenclature for enzymes, the EC numbers; each enzyme is described by a sequence of four numbers preceded by EC. The first number broadly classifies the enzyme based on its mechanism.

The top-level classification is:

- EC 1, Oxidoreductases: catalyze oxidation/reduction reactions
- EC 2, Transferases: transfer a functional group (*e.g.* a methyl or phosphate group)
- EC 3, Hydrolases: catalyze the hydrolysis of various bonds
- EC 4, Lyases: cleave various bonds by means other than hydrolysis and oxidation
- EC 5, Isomerases: catalyze isomerization changes within a single molecule
- EC 6, Ligases: join two molecules with covalent bonds.

These sections are subdivided by other features such as the substrate, products, and chemical mechanism. An enzyme is fully specified by four numerical designations. For example, hexokinase (EC 2.7.1.1) is a transferase (EC 2) that adds a phosphate group (EC 2.7) to a hexose sugar, a molecule containing an alcohol group (EC 2.7.1).

STRUCTURE

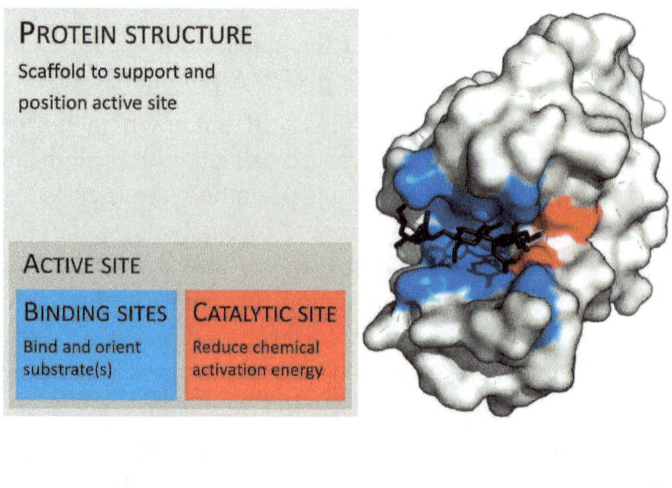

Organisation of enzyme structure and lysozyme example. Binding sites in blue, catalytic site in red and peptidoglycan substrate in black. (PDB: 9LYZ)

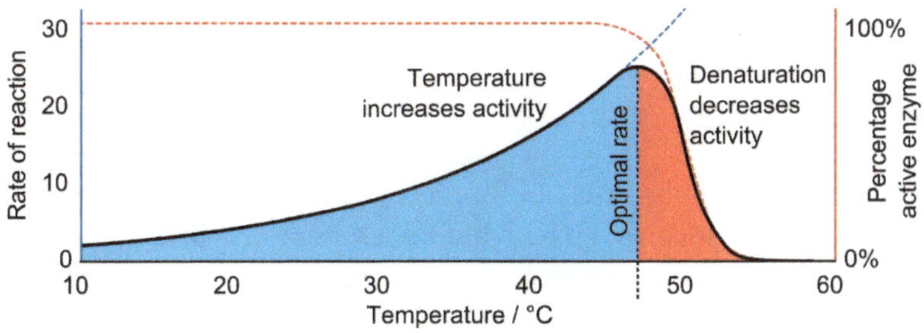

Enzyme activity initially increases with temperature (Q10 coefficient) until the enzyme's structure unfolds (denaturation), leading to an optimal rate of reaction at an intermediate temperature.

Enzymes are generally globular proteins, acting alone or in larger complexes. Like all proteins, enzymes are linear chains of amino acids that fold to produce a three-dimensional structure. The sequence of the amino acids specifies the structure which in turn determines the catalytic activity of the enzyme. Although structure determines function, a novel enzyme's activity cannot yet be predicted from its structure alone. Enzyme structures unfold (denature) when heated or exposed to chemical denaturants and this disruption to the structure typically causes a loss of activity. Enzyme denaturation is normally linked to temperatures above a species' normal level; as a result, enzymes from bacteria living in volcanic environments such as hot springs are prized by industrial users for their ability to function at high temperatures, allowing enzyme-catalysed reactions to be operated at a very high rate.

Enzymes are usually much larger than their substrates. Sizes range from just 62 amino acid residues, for the monomer of 4-oxalocrotonate tautomerase,to over 2,500 residues in the animal fatty acid synthase. Only a small portion of their structure (around 2–4 amino acids) is directly involved in catalysis: the catalytic site. This catalytic site is located next to one or more binding sites where residues orient the substrates. The catalytic site and binding site together comprise the enzyme's active site. The remaining majority of the enzyme structure serves to maintain the precise orientation and dynamics of the active site.

In some enzymes, no amino acids are directly involved in catalysis; instead, the enzyme contains sites to bind and orient catalytic cofactors. Enzyme structures may also contain allosteric sites where the binding of a small molecule causes a conformational change that increases or decreases activity.

A small number of RNA-based biological catalysts called ribozymes exist, which again can act alone or in complex with proteins. The most common of these is the ribosome which is a complex of protein and catalytic RNA components.

MECHANISM

Substrate Binding

Enzymes must bind their substrates before they can catalyse any chemical reaction. Enzymes are usually very specific as to what substrates they bind and then the chemical reaction catalysed. Specificity is achieved by binding pockets with complementary shape, charge and hydrophilic/hydrophobic characteristics to the substrates. Enzymes can therefore distinguish between very similar substrate molecules to be chemoselective, regioselective and stereospecific. Some of the enzymes showing the highest specificity and accuracy are involved in the copying and expression of the genome. Some of these enzymes have proof-reading mechanisms. Here, an enzyme such as DNA polymerase catalyzes a reaction in a first step and then checks that the product is correct in a second step. This two-step process results in average error rates of less than 1 error in 100 million reactions in high-fidelity mammalian polymerases. Similar proofreading mechanisms are also found in RNA polymerase, aminoacyl tRNA synthetasesand ribosomes.

Conversely, some enzymes display enzyme promiscuity, having broad specificity and acting on a range of different physiologically relevant substrates. Many enzymes possess small side activities which arose fortuitously (i.e. neutrally), which may be the starting point for the evolutionary selection of a new function.

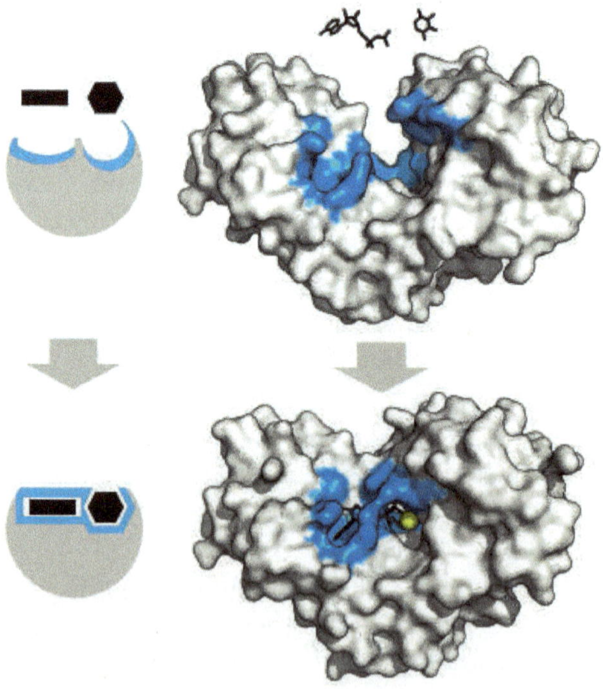

Enzyme changes shape by induced fit upon substrate binding to form enzyme-substrate complex. Hexokinase has a large induced fit motion that closes over the substrates adenosine triphosphate and xylose. Binding sites in blue, substrates in black and Mg^{2+} cofactor in yellow. (PDB: 2E2N, 2E2Q)

"Lock and Key" Model

To explain the observed specificity of enzymes, in 1894 Emil Fischer proposed that both the enzyme and the substrate possess specific complementary geometric shapes that fit exactly into one another. This is often referred to as "the lock and key" model.[:8.3.2] This early model explains enzyme specificity, but fails to explain the stabilization of the transition state that enzymes achieve.

Induced fit model

In 1958, Daniel Koshland suggested a modification to the lock and key model: since enzymes are rather flexible structures, the active site is continuously reshaped by interactions with the substrate as the substrate interacts with the enzyme. As a result, the substrate does not simply bind to a rigid active site; the amino acid side-chains that make up the active site are molded into the precise positions that enable the enzyme to perform its catalytic function. In some cases, such as glycosidases, the substrate molecule also changes shape slightly as it enters the active site. The active site continues to change until the substrate is completely bound, at which point the final shape and charge distribution is determined. Induced fit may enhance the fidelity of molecular recognition in the presence of competition and noise via the conformational proofreading mechanism.

Catalysis

Enzymes can accelerate reactions in several ways, all of which lower the activation energy ($\Delta G^{\ddagger}$, Gibbs free energy)

- By stabilizing the transition state:
  - Creating an environment with a charge distribution complementary to that of the transition state to lower its energy.
- By providing an alternative reaction pathway:
  - Temporarily reacting with the substrate, forming a covalent intermediate to provide a lower energy transition state.
- By destabilising the substrate ground state:
  - Distorting bound substrate(s) into their transition state form to reduce the energy required to reach the transition state.
  - By orienting the substrates into a productive arrangement to reduce the reaction entropy change. The contribution of this mechanism to catalysis is relatively small.

Enzymes may use several of these mechanisms simultaneously. For example, proteases such as trypsin perform covalent catalysis using a catalytic triad, stabilise charge build-up on the transition states using an oxyanion hole, complete hydrolysis using an oriented water substrate.

Dynamics

Enzymes are not rigid, static structures; instead they have complex internal dynamic motions – that is, movements of parts of the enzyme's structure such as individual amino acid residues, groups of residues forming a protein loop or unit of secondary structure, or even an entire protein domain. These motions give rise to a conformational ensemble of slightly different structures that interconvert with one another at equilibrium. Different states within this ensemble may be associated with different aspects of an enzyme's function. For example, different conformations of the enzyme dihydrofolate reductase are associated with the substrate binding, catalysis, cofactor release, and product release steps of the catalytic cycle.

Allosteric Modulation

Allosteric sites are pockets on the enzyme, distinct from the active site, that bind to molecules in the cellular environment. These molecules then cause a change in the conformation or dynamics of the enzyme that is transduced to the active site and thus affects the reaction rate of the enzyme. In this way, allosteric interactions can either inhibit or activate enzymes. Allosteric interactions with metabolites upstream or downstream in an enzyme's metabolic pathway cause feedback regulation, altering the activity of the enzyme according to the flux through the rest of the pathway.

Cofactors

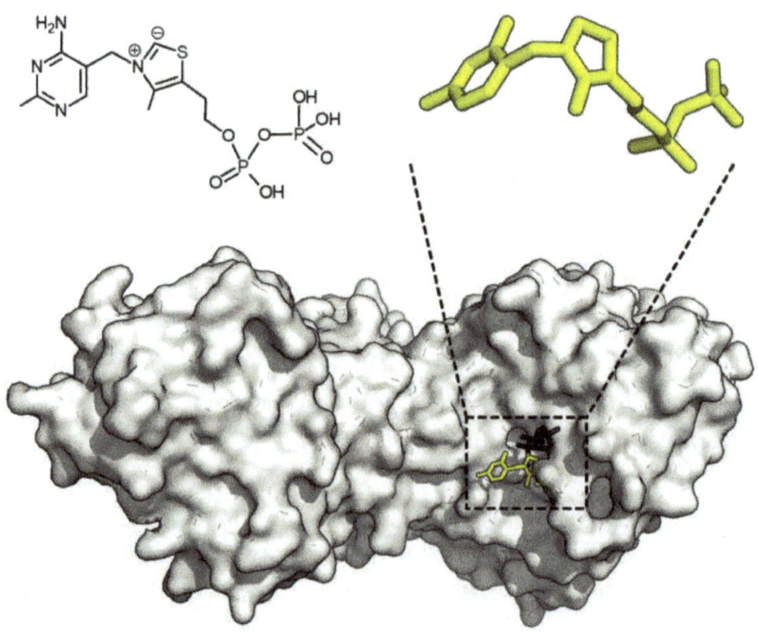

Chemical structure for thiamine pyrophosphate and protein structure of transketolase. Thiamine pyrophosphate cofactor in yellow and xylulose 5-phosphate substrate in black. (PDB: 4KXV)

Some enzymes do not need additional components to show full activity. Others require non-protein molecules called cofactors to be bound for activity. Cofactors can be either inorganic (e.g., metal ions and iron-sulfur clusters) or organic compounds (e.g., flavin and heme). Organic cofactors can be either coenzymes, which are released from the enzyme's active site during the reaction, or prosthetic groups, which are tightly bound to an enzyme. Organic prosthetic groups can be covalently bound (e.g., biotin in enzymes such as pyruvate carboxylase).

An example of an enzyme that contains a cofactor is carbonic anhydrase, which is shown in the ribbon diagram above with a zinc cofactor bound as part of its active site. These tightly bound ions or molecules are usually found in the active site and are involved in catalysis. For example, flavin and heme cofactors are often involved in redox reactions.

Enzymes that require a cofactor but do not have one bound are called *apoenzymes* or *apoproteins*. An enzyme together with the cofactor(s) required for activity is called a *holoenzyme* (or haloenzyme). The term *holoenzyme* can also be applied to enzymes that contain multiple protein subunits, such as the DNA polymerases; here the holoenzyme is the complete complex containing all the subunits needed for activity.[:8.1.1]

Coenzymes

Coenzymes are small organic molecules that can be loosely or tightly bound to an enzyme. Coenzymes transport chemical groups from one enzyme to another. Examples include NADH, NADPH and adenosine triphosphate (ATP). Some coenzymes, such as riboflavin, thiamine and folic acid, are vitamins, or compounds that cannot be synthesized by the body and must be acquired from the diet. The chemical groups carried include the hydride ion (H^-) carried by NAD or $NADP^+$, the phosphate group carried by adenosine triphosphate, the acetyl group carried by coenzyme A, formyl, methenyl or methyl groups carried by folic acid and the methyl group carried by S-adenosylmethionine.

Since coenzymes are chemically changed as a consequence of enzyme action, it is useful to consider coenzymes to be a special class of substrates, or second substrates, which are common to many different enzymes. For example, about 1000 enzymes are known to use the coenzyme NADH.

Coenzymes are usually continuously regenerated and their concentrations maintained at a steady level inside the cell. For example, NADPH is regenerated through the pentose phosphate pathway and *S*-adenosylmethionine by methionine adenosyltransferase. This continuous regeneration means that small amounts of coenzymes can be used very intensively. For example, the human body turns over its own weight in ATP each day.

THERMODYNAMICS

As with all catalysts, enzymes do not alter the position of the chemical equilibrium of the reaction. In the presence of an enzyme, the reaction runs in the same direction as it would without the enzyme, just more quickly.[:8.2.3] For example, carbonic anhydrase catalyzes its reaction in either direction depending on the concentration of its reactants:

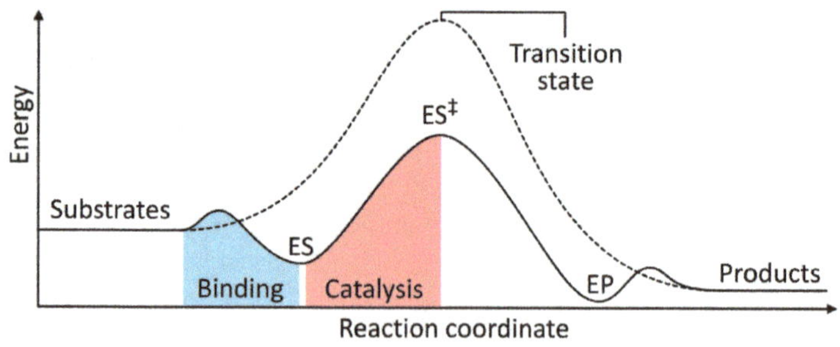

The energies of the stages of a chemical reaction. Uncatalysed (dashed line), substrates need a lot of activation energy to reach a transition state, which then decays into lower-energy products. When enzyme catalysed (solid line), the enzyme binds the substrates (ES), then stabilizes the transition state (ES‡) to reduce the activation energy required to produce products (EP) which are finally released.

$$CO_2 + H_2O \xrightarrow{\text{Carbonic anhydrase}} H_2CO_3 \quad \text{(in tissues; high } CO_2 \text{ concentration)} \tag{1}$$

$$H_2CO_3 \xrightarrow{\text{Carbonic anhydrase}} CO_2 + H_2O \quad \text{(in lungs; low } CO_2 \text{ concentration)} \tag{2}$$

The rate of a reaction is dependent on the activation energy needed to form the transition state which then decays into products. Enzymes increase reaction rates by lowering the energy of the transition state. First, binding forms a low energy enzyme-substrate complex (ES). Secondly the enzyme stabilises the transition state such that it requires less energy to achieve compared to the uncatalyzed reaction (ES‡). Finally the enzyme-product complex (EP) dissociates to release the products.[:8.3]

Enzymes can couple two or more reactions, so that a thermodynamically favorable reaction can be used to "drive" a thermodynamically unfavourable one so that the combined energy of the products is lower than the substrates. For example, the hydrolysis of ATP is often used to drive other chemical reactions.

KINETICS

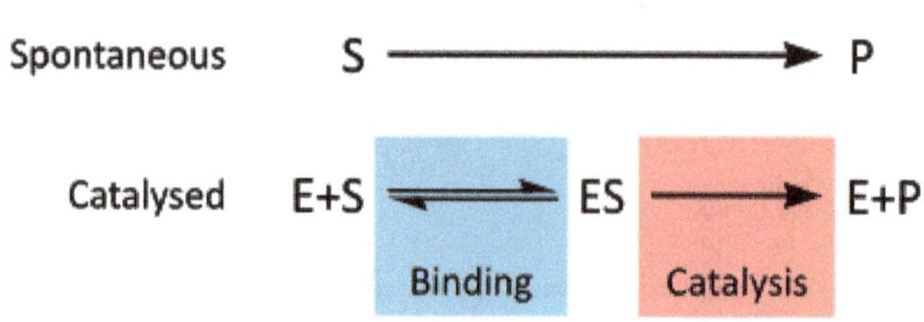

A chemical reaction mechanism with or without enzyme catalysis. The enzyme (E) binds substrate (S) to produce product (P).

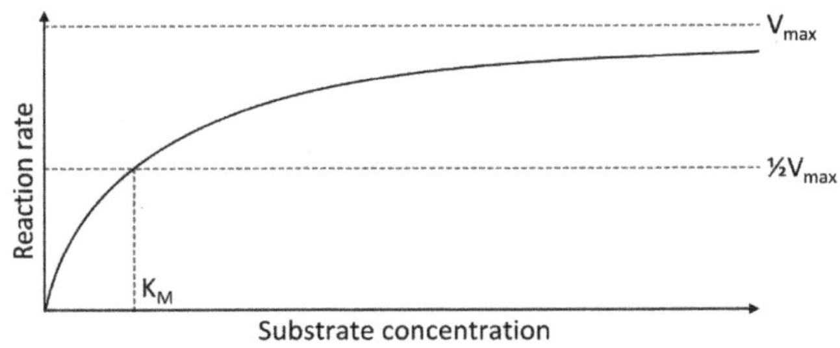

Saturation curve for an enzyme reaction showing the relation between the substrate concentration and reaction rate.

Enzyme kinetics is the investigation of how enzymes bind substrates and turn them into products. The rate data used in kinetic analyses are commonly obtained from enzyme assays. In 1913 Leonor Michaelis and Maud Leonora Menten proposed a quantitative theory of enzyme kinetics, which is referred to as Michaelis–Menten kinetics. The major contribution of Michaelis and Menten was to think of enzyme reactions in two stages. In the first, the substrate binds reversibly to the enzyme, forming the enzyme-substrate complex. This is sometimes called the Michaelis-Menten complex in their honor. The enzyme then catalyzes the chemical step in the reaction and releases the product. This work was further developed by G. E. Briggs and J. B. S. Haldane, who derived kinetic equations that are still widely used today.

Enzyme rates depend on solution conditions and substrate concentration. To find the maximum speed of an enzymatic reaction, the substrate concentration is increased until a constant rate of product formation is seen. This is shown in the saturation curve on the right. Saturation happens because, as substrate concentration increases, more and more of the free enzyme is converted into the substrate-bound ES complex. At the maximum reaction rate (V_{max}) of the enzyme, all the enzyme active sites are bound to substrate, and the amount of ES complex is the same as the total amount of enzyme.

V_{max} is only one of several important kinetic parameters. The amount of substrate needed to achieve a given rate of reaction is also important. This is given by the Michaelis-Menten constant (K_m), which is the substrate concentration required for an enzyme to reach one-half its maximum reaction rate; generally, each enzyme has a characteristic K_m for a given substrate. Another useful constant is k_{cat}, also called the *turnover number*, which is the number of substrate molecules handled by one active site per second.

The efficiency of an enzyme can be expressed in terms of k_{cat}/K_m. This is also called the specificity constant and incorporates the rate constants for all steps in the reaction up to and including the first irreversible step. Because the specificity constant reflects both affinity and catalytic ability, it is useful for comparing different enzymes against each other, or the same enzyme with different substrates. The theoretical maximum for the specificity constant is called the diffusion limit and is about 10^8 to 10^9 (M^{-1} s^{-1}). At this point every collision of the enzyme with its substrate will result in catalysis, and the rate of product formation is not limited by the reaction rate but by the diffusion rate. Enzymes with this property are called *catalytically perfect* or *kinetically perfect*. Example of such enzymes are triose-phosphate isomerase, carbonic anhydrase, acetylcholinesterase, catalase, fumarase, β-lactamase, and superoxide dismutase. The turnover of such enzymes can reach several million reactions per second.

Michaelis–Menten kinetics relies on the law of mass action, which is derived from the assumptions of free diffusion and thermodynamically driven random collision. Many biochemical or cellular processes deviate significantly from these conditions, because of macromolecular crowding and constrained molecular movement. More recent, complex extensions of the model attempt to correct for these effects.

Inhibition

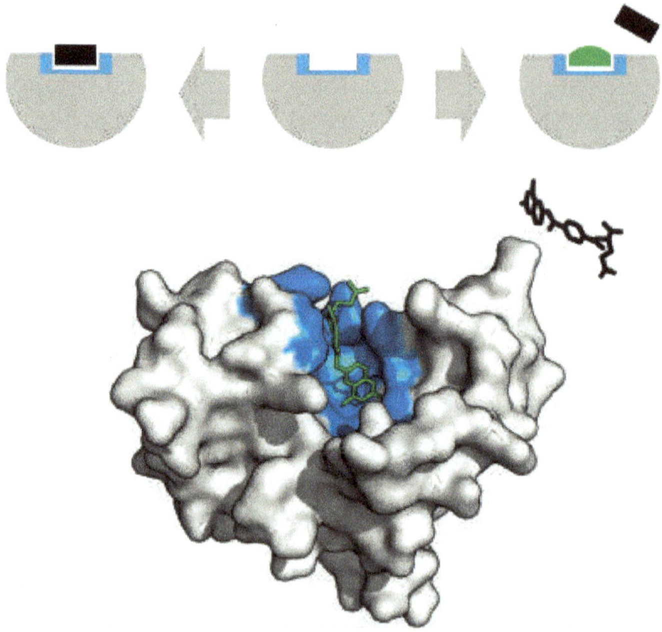

An enzyme binding site that would normally bind substrate can alternatively bind a competitive inhibitor, preventing substrate access. Dihydrofolate reductase is inhibited by methotrexate which prevents binding of its substrate, folic acid. Binding site in blue, inhibitor in green, and substrate in black. (PDB: 4QI9)

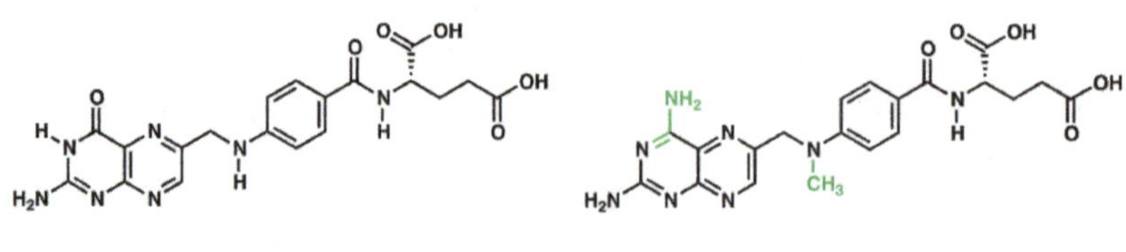

The coenzyme folic acid (left) and the anti-cancer drug methotrexate (right) are very similar in structure (differences show in green). As a result, methotrexate is a competitive inhibitor of many enzymes that use folates.

Enzyme reaction rates can be decreased by various types of enzyme inhibitors.[:73–74]

Types of Inhibition

Competitive

A competitive inhibitor and substrate cannot bind to the enzyme at the same time. Often competitive inhibitors strongly resemble the real substrate of the enzyme. For example, the drug methotrexate is a competitive inhibitor of the enzyme dihydrofolate reductase, which catalyzes the reduction of dihydrofolate to tetrahydrofolate. The similarity between the structures of dihydrofolate and this drug are shown in the accompanying figure. This type of inhibition can be overcome with high substrate concentration. In some cases, the inhibitor can bind to a site other than the binding-site of the usual substrate and exert an allosteric effect to change the shape of the usual binding-site.

Non-Competitive

A non-competitive inhibitor binds to a site other than where the substrate binds. The substrate still binds with its usual affinity and hence K_m remains the same. However the inhibitor reduces the catalytic efficiency of the enzyme so that V_{max} is reduced. In contrast to competitive inhibition, non-competitive inhibition cannot be overcome with high substrate concentration.

Uncompetitive

An uncompetitive inhibitor cannot bind to the free enzyme, only to the enzyme-substrate complex; hence, these types of inhibitors are most effective at high substrate concentration. In the presence of the inhibitor, the enzyme-substrate complex is inactive.[:78] This type of inhibition is rare.

Mixed

A mixed inhibitor binds to an allosteric site and the binding of the substrate and the inhibitor affect each other. The enzyme's function is reduced but not eliminated when bound to the inhibitor. This type of inhibitor does not follow the Michaelis-Menten equation.[:76–78]

Irreversible

An irreversible inhibitor permanently inactivates the enzyme, usually by forming a covalent bond to the protein. Penicillin and aspirin are common drugs that act in this manner.

Functions of Inhibitors

In many organisms, inhibitors may act as part of a feedback mechanism. If an enzyme produces too much of one substance in the organism, that substance may act as an inhibitor for the enzyme at the beginning of the pathway that produces it, causing production of the substance to slow down or stop when there is sufficient amount. This is a form of negative feedback. Major metabolic pathways such as the citric acid cycle make use of this mechanism.

Since inhibitors modulate the function of enzymes they are often used as drugs. Many such drugs are reversible competitive inhibitors that resemble the enzyme's native substrate, similar to methotrexate above; other well-known examples include statins used to treat high cholesterol, and protease inhibitors used to treat retroviral infections such as HIV. A common example of an irreversible inhibitor that is used as a drug is aspirin, which inhibits the COX-1 and COX-2 enzymes that produce the inflammation messenger prostaglandin. Other enzyme inhibitors are poisons. For example, the poison cyanide is an irreversible enzyme inhibitor that combines with the copper and iron in the active site of the enzyme cytochrome c oxidase and blocks cellular respiration.

BIOLOGICAL FUNCTION

Enzymes serve a wide variety of functions inside living organisms. They are indispensable for signal transduction and cell regulation, often via kinases and phosphatases. They also generate movement, with myosin hydrolyzing ATP to generate muscle contraction, and also transport cargo around the cell as part of the cytoskeleton. Other ATPases in the cell membrane are ion pumps involved in active transport. Enzymes are also involved in more exotic functions, such as luciferase generating light in fireflies. Viruses can also contain enzymes for infecting cells, such as the HIV integrase and reverse transcriptase, or for viral release from cells, like the influenza virus neuraminidase.

An important function of enzymes is in the digestive systems of animals. Enzymes such as amylases and proteases break down large molecules (starch or proteins, respectively) into smaller ones, so they can be absorbed by the intestines. Starch molecules, for example, are too large to be absorbed from the intestine, but enzymes hydrolyze the starch chains into smaller molecules such as maltose and eventually glucose, which can then be absorbed. Different enzymes digest different food substances. In ruminants, which have herbivorous diets, microorganisms in the gut produce another enzyme, cellulase, to break down the cellulose cell walls of plant fiber.

Metabolism

Several enzymes can work together in a specific order, creating metabolic pathways.[:30.1] In a metabolic pathway, one enzyme takes the product of another enzyme as a substrate. After the catalytic reaction, the product is then passed on to another enzyme. Sometimes more than one enzyme can catalyze the same reaction in parallel; this can allow more complex regulation: with, for example, a low constant activity provided by one enzyme but an inducible high activity from a second enzyme.

Enzymes determine what steps occur in these pathways. Without enzymes, metabolism would neither progress through the same steps and could not be regulated to serve the needs of the cell. Most central metabolic pathways are regulated at a few key steps, typically through enzymes whose activity involves the hydrolysis of ATP. Because this reaction releases so much energy, other reactions that are thermodynamically unfavorable can be coupled to ATP hydrolysis, driving the overall series of linked metabolic reactions.

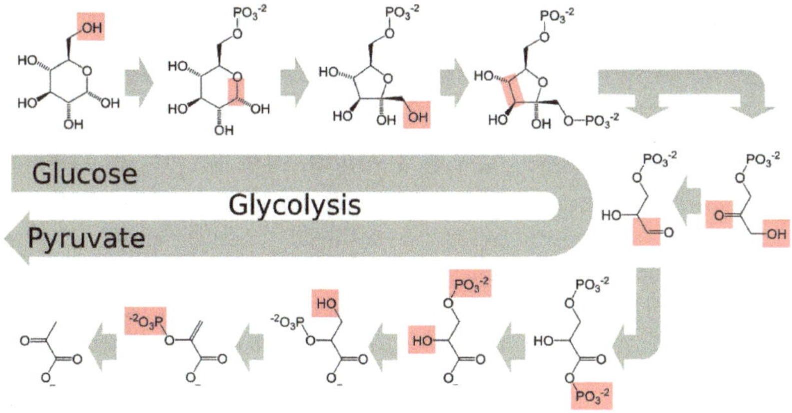

The metabolic pathway of glycolysis releases energy by converting glucose to pyruvate by via a series of intermediate metabolites. Each chemical modification (red box) is performed by a different enzyme.

Control of Activity

There are five main ways that enzyme activity is controlled in the cell.

Regulation

Enzymes can be either activated or inhibited by other molecules. For example, the end product(s) of a metabolic pathway are often inhibitors for one of the first enzymes of the pathway (usually the first irreversible step, called committed step), thus regulating the amount of end product made by the pathways. Such a regulatory mechanism is called a negative feedback mechanism, because the amount of the end product produced is regulated by its own concentration. Negative feedback mechanism can effectively adjust the rate of synthesis of intermediate metabolites according to the demands of the cells. This helps with effective allocations of materials and energy economy, and it prevents the excess manufacture of end products. Like other homeostatic devices, the control of enzymatic action helps to maintain a stable internal environment in living organisms.

Post-Translational Modification

Examples of post-translational modification include phosphorylation, myristoylation and glycosylation. For example, in the response to insulin, the phosphorylation of multiple enzymes, including glycogen synthase, helps control the synthesis or degradation of glycogen and allows the cell to respond to changes in blood sugar. Another example of post-translational modification is the cleavage of the polypeptide chain. Chymotrypsin, a digestive protease, is produced in inactive form as chymotrypsinogen in the pancreas and transported in this form to the stomach where it is activated. This stops the enzyme from digesting the pancreas or other tissues before it enters the gut. This type of inactive precursor to an enzyme is known as a zymogen or proenzyme.

Quantity

Enzyme production (transcription and translation of enzyme genes) can be enhanced or diminished by a cell in response to changes in the cell's environment. This form of gene regulation is called enzyme induction. For example, bacteria may become resistant to antibiotics such as penicillin because enzymes called beta-lactamases are induced that hydrolyse the crucial beta-lactam ring within the penicillin molecule.Another example comes from enzymes in the liver called cytochrome P450 oxidases, which are important in drug metabolism. Induction or inhibition of these enzymes can cause drug interactions. Enzyme levels can also be regulated by changing the rate of enzyme degradation.[:30.1.1]

Subcellular Distribution

Enzymes can be compartmentalized, with different metabolic pathways occurring in different cellular compartments. For example, fatty acids are synthesized by one set of enzymes in the cytosol, endoplasmic reticulum and Golgi and used by a different set of enzymes as a source of energy in the mitochondrion, through β-oxidation. In addition, trafficking of the enzyme to different compartments may change the degree of protonation (cytoplasm neutral and lysosome acidic) or oxidative state [e.g., oxidized (periplasm) or reduced (cytoplasm)] which in turn affects enzyme activity.

Organ Specialization

In multicellular eukaryotes, cells in different organs and tissues have different patterns of gene expression and therefore have different sets of enzymes (known as isozymes) available for metabolic reactions. This provides a mechanism for regulating the overall metabolism of the organism. For example, hexokinase, the first enzyme in the glycolysis pathway, has a specialized form called glucokinase expressed in the liver and pancreas that has a lower affinity for glucose yet is more sensitive to glucose concentration. This enzyme is involved in sensing blood sugar and regulating insulin production.

Involvement in Disease

Since the tight control of enzyme activity is essential for homeostasis, any malfunction (mutation, overproduction, underproduction or deletion) of a single critical enzyme can lead to a genetic disease. The malfunction of just one type of enzyme out of the thousands of types present in the human body can be fatal. An example of a fatal genetic disease due to enzyme insufficiency is Tay-Sachs disease, in which patients lack the enzyme hexosaminidase.

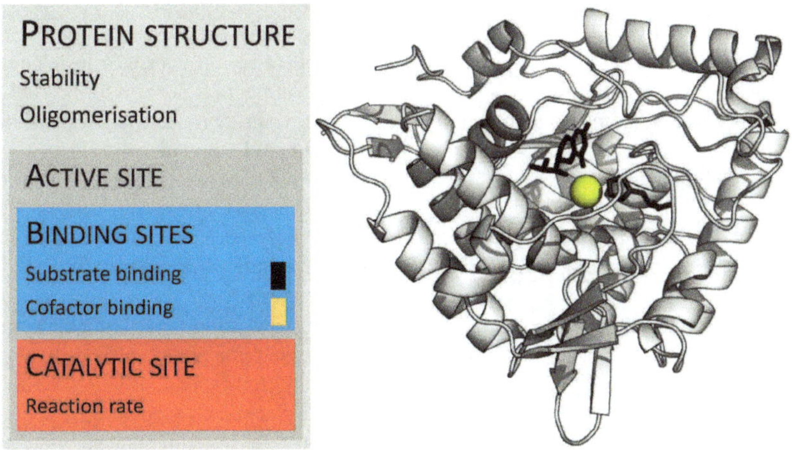

In phenylalanine hydroxylase over 300 different mutations throughout the structure cause phenylketonuria. Phenylalanine substrate and tetrahydrobiopterin coenzyme in black, and Fe^{2+}cofactor in yellow. (PDB: 1KW0)

One example of enzyme deficiency is the most common type of phenylketonuria. Many different single amino acid mutations in the enzyme phenylalanine hydroxylase, which catalyzes the first step in the degradation of phenylalanine, result in build-up of phenylalanine and related products. Some mutations are in the active site, directly disrupting binding and catalysis, but many are far from the active site and reduce activity by destabilising the protein structure, or affecting correct oligomerisation. This can lead to intellectual disability if the disease is untreated. Another example is pseudocholinesterase deficiency, in which the body's ability to break down choline ester drugs is impaired. Oral administration of enzymes can be used to treat some functional enzyme deficiencies, such as pancreatic insufficiency and lactose intolerance.

Another way enzyme malfunctions can cause disease comes from germline mutations in genes coding for DNA repair enzymes. Defects in these enzymes cause cancer because cells are less able to repair mutations in their genomes. This causes a slow accumulation of mutations and results in the development of cancers. An example of such a hereditary cancer syndrome is xeroderma pigmentosum, which causes the development of skin cancers in response to even minimal exposure to ultraviolet light.

Industrial Applications

Enzymes are used in the chemical industry and other industrial applications when extremely specific catalysts are required. Enzymes in general are limited in the number of reactions they have evolved to catalyze and also by their lack of stability in organic solvents and at high temperatures. As a consequence, protein engineering is an active area of research and involves attempts to create new enzymes with novel properties, either through rational design or *in vitro* evolution. These efforts have begun to be successful, and a few enzymes have now been designed "from scratch" to catalyze reactions that do not occur in nature.

Application	Enzymes used	Uses
Biofuel industry	Cellulases	Break down cellulose into sugars that can be fermented to produce cellulosic ethanol.
	Ligninases	Pretreatment of biomass for biofuel production.
Biological detergent	Proteases, amylases, lipases	Remove protein, starch, and fat or oil stains from laundry and dishware.
	Mannanases	Remove food stains from the common food additive guar gum.
Brewing industry	Amylase, glucanases, proteases	Split polysaccharides and proteins in the malt.
	Betaglucanases	Improve the wort and beer filtration characteristics.
	Amyloglucosidase and pullulanases	Make low-calorie beer and adjust fermentability.
	Acetolactate decarboxylase (ALDC)	Increase fermentation efficiency by reducing diacetyl formation.
Culinary uses	Papain	Tenderize meat for cooking.
Dairy industry	Rennin	Hydrolyze protein in the manufacture of cheese.
	Lipases	Produce Camembert cheese and blue cheeses such as Roquefort.
Food processing	Amylases	Produce sugars from starch, such as in making high-fructose corn syrup.
	Proteases	Lower the protein level of flour, as in biscuit-making.
	Trypsin	Manufacture hypoallergenic baby foods.
	Cellulases, pectinases	Clarify fruit juices.
Molecular biology	Nucleases, DNA ligase and polymerases	Use restriction digestion and the polymerase chain reaction to create recombinant DNA.
Paper industry	Xylanases, hemicellulases and lignin peroxidases	Remove lignin from kraft pulp.
Personal care	Proteases	Remove proteins on contact lenses to prevent infections.
Starch industry	Amylases	Convert starch into glucose and various syrups.

REFERENCES

1. Stryer L, Berg JM, Tymoczko JL (2002). Biochemistry (5th ed.). San Francisco: W.H. Freeman. ISBN 0-7167-4955-6.
2. Schomburg I, Chang A, Placzek S, Söhngen C, Rother M, Lang M, Munaretto C, Ulas S, Stelzer M, Grote A, Scheer M, Schomburg D (January 2013). "BRENDA in 2013: integrated reactions, kinetic data, enzyme function data, improved disease classification: new options and contents in BRENDA". Nucleic Acids Research. 41 (Database issue): D764–72. doi:10.1093/nar/gks1049. PMC 3531171 . PMID 23203881.
3. Radzicka A, Wolfenden R (January 1995). "A proficient enzyme". Science. 267 (5194): 90–931. Bibcode:1995Sci...267...90R. doi:10.1126/science.7809611. PMID 7809611.
4. Callahan BP, Miller BG (December 2007). "OMP decarboxylase—An enigma persists". Bioorganic Chemistry. 35 (6): 465–9. doi:10.1016/j.bioorg.2007.07.004. PMID 17889251.

5. de Réaumur RA (1752). "Observations sur la digestion des oiseaux". Histoire de l'academie royale des sciences. 1752: 266, 461.
6. Williams HS (1904). A History of Science: in Five Volumes. Volume IV: Modern Development of the Chemical and Biological Sciences. Harper and Brothers.
7. Payen A, Persoz JF (1833). "Mémoire sur la diastase, les principaux produits de ses réactions et leurs applications aux arts industriels" [Memoir on diastase, the principal products of its reactions, and their applications to the industrial arts]. Annales de chimie et de physique. 2nd (in French). 53: 73–92.
8. Manchester KL (December 1995). "Louis Pasteur (1822–1895)–chance and the prepared mind". Trends in Biotechnology. 13 (12): 511–5. doi:10.1016/S0167-7799(00)89014-9. PMID 8595136.
9. Kühne coined the word "enzyme" in: Kühne W (1877). "Über das Verhalten verschiedener organisirter und sog. ungeformter Fermente [On the behavior of various organized and so-called unformed ferments]
10. Holmes FL (2003). Enzymes. In Heilbron JL. The Oxford Companion to the History of Modern Science. Oxford: Oxford University Press. p. 270.
11. Eduard Buchner. Nobel Laureate Biography. Nobelprize.org. Retrieved 23 February 2015.
12. Eduard Buchner – Nobel Lecture: Cell-Free Fermentation. Nobelprize.org. 1907. Retrieved 23 February 2015.
13. The naming of enzymes by adding the suffix -ase to the substrate on which the enzyme acts, has been traced to French scientist Émile Duclaux (1840–1904), who intended to honor the discoverers of diastase – the first enzyme to be isolated – by introducing this practice in his book Duclaux E (1899). Traité de microbiologie: Diastases, toxines et venins [Microbiology Treatise: diastases , toxins and venoms] (in French). Paris, France: Masson and Co. See Chapter 1, especially page 9.
14. Willstätter R (1927). Faraday lecture. Problems and methods in enzyme research. Journal of the Chemical Society (Resumed): 1359. doi:10.1039/JR9270001359. quoted in Blow D (April 2000). So do we understand how enzymes work? (pdf). Structure (London, England : 1993). 8 (4): R77–R81. doi:10.1016/S0969-2126(00)00125-8. PMID 10801479.
15. Nobel Prizes and Laureates: The Nobel Prize in Chemistry 1946. Nobelprize.org. Retrieved 23 February 2015.
16. Blake CC, Koenig DF, Mair GA, North AC, Phillips DC, Sarma VR (May 1965). Structure of hen egg-white lysozyme. A three-dimensional Fourier synthesis at 2 Ångström resolution. Nature. 206 (4986): 757–61. Bibcode:1965Natur.206..757B. doi:10.1038/206757a0. PMID 5891407.
17. Johnson LN, Petsko GA (1999). David Phillips and the origin of structural enzymology. Trends Biochem. Sci. 24 (7): 287–9. doi:10.1016/S0968-0004(99)01423-1. PMID 10390620.

18. Nomenclature Committee. Classification and Nomenclature of Enzymes by the Reactions they Catalyse. International Union of Biochemistry and Molecular Biology (NC-IUBMB). School of Biological and Chemical Sciences, Queen Mary, University of London.
19. Nomenclature Committee. EC 2.7.1.1. International Union of Biochemistry and Molecular Biology (NC-IUBMB). School of Biological and Chemical Sciences, Queen Mary, University of London.
20. Anfinsen CB (July 1973). Principles that govern the folding of protein chains. Science. 181 (4096): 223–30. Bibcode:1973Sci...181..223A. doi:10.1126/science.181.4096.223. PMID 4124164.
21. Dunaway-Mariano D (November 2008). Enzyme function discovery. Structure (London, England : 1993). 16 (11): 1599–600. doi:10.1016/j.str.2008.10.001. PMID 19000810.
22. Petsko GA, Ringe D (2003). Chapter 1: From sequence to structure. Protein structure and function. London: New Science. p. 27. ISBN 978-1405119221.
23. Chen LH, Kenyon GL, Curtin F, Harayama S, Bembenek ME, Hajipour G, Whitman CP (September 1992). 4-Oxalocrotonate tautomerase, an enzyme composed of 62 amino acid residues per monomer. The Journal of Biological Chemistry. 267 (25): 17716–21. PMID 1339435.
24. Smith S (December 1994). The animal fatty acid synthase: one gene, one polypeptide, seven enzymes. FASEB Journal. 8 (15): 1248–59. PMID 8001737.
25. The Catalytic Site Atlas. The European Bioinformatics Institute. Retrieved 4 April 2007.
26. Suzuki H (2015). Chapter 7: Active Site Structure. How Enzymes Work: From Structure to Function. Boca Raton, FL: CRC Press. pp. 117–140. ISBN 978-981-4463-92-8.
27. Krauss G (2003). The Regulations of Enzyme Activity. Biochemistry of Signal Transduction and Regulation (3rd ed.). Weinheim: Wiley-VCH. pp. 89–114. ISBN 9783527605767.
28. Jaeger KE, Eggert T (August 2004). Enantioselective biocatalysis optimized by directed evolution. Current Opinion in Biotechnology. 15 (4): 305–13. doi:10.1016/j.copbio.2004.06.007. PMID 15358000.
29. Shevelev IV, Hübscher U (May 2002). The 3' 5' exonucleases. Nature Reviews Molecular Cell Biology. 3 (5): 364–76. doi:10.1038/nrm804. PMID 11988770.
30. Zenkin N, Yuzenkova Y, Severinov K (July 2006). Transcript-assisted transcriptional proofreading. Science. 313 (5786): 518–20. Bibcode:2006Sci...313..518Z. doi:10.1126/science.1127422. PMID 16873663.
31. Ibba M, Soll D (2000). Aminoacyl-tRNA synthesis. Annual Review of Biochemistry. 69: 617–50. doi:10.1146/annurev.biochem.69.1.617. PMID 10966471.
32. Rodnina MV, Wintermeyer W (2001). Fidelity of aminoacyl-tRNA selection on the ribosome: kinetic and structural mechanisms. Annual Review of Biochemistry.

70: 415–35. doi:10.1146/annurev.biochem.70.1.415. PMID 11395413.

33. Khersonsky O, Tawfik DS (2010). Enzyme promiscuity: a mechanistic and evolutionary perspective. Annual Review of Biochemistry. 79: 471–505. doi:10.1146/annurev-biochem-030409-143718. PMID 20235827.
34. O'Brien PJ, Herschlag D (April 1999). Catalytic promiscuity and the evolution of new enzymatic activities. Chemistry & Biology. 6 (4): R91–R105. doi:10.1016/S1074-5521(99)80033-7. PMID 10099128.
35. Fischer E (1894). Einfluss der Configuration auf die Wirkung der Enzyme [Influence of configuration on the action of enzymes]. Berichte der Deutschen chemischen Gesellschaft zu Berlin (in German). 27 (3): 2985–93. doi:10.1002/cber.18940270364. From page 2992: Um ein Bild zu gebrauchen, will ich sagen, dass Enzym und Glucosid wie Schloss und Schlüssel zu einander passen müssen, um eine chemische Wirkung auf einander ausüben zu können. (To use an image, I will say that an enzyme and a glucoside [i.e., glucose derivative] must fit like a lock and key, in order to be able to exert a chemical effect on each other.)
36. Cooper GM (2000). Chapter 2.2: The Central Role of Enzymes as Biological Catalysts. The Cell: a Molecular Approach (2nd ed.). Washington (DC): ASM Press. ISBN 0-87893-106-6.
37. Koshland DE (February 1958). Application of a Theory of Enzyme Specificity to Protein Synthesis. Proceedings of the National Academy of Sciences of the United States of America. 44 (2): 98–104. Bibcode:1958PNAS...44...98K. doi:10.1073/pnas.44.2.98. PMC 335371 . PMID 16590179.
38. Vasella A, Davies GJ, Böhm M (October 2002). Glycosidase mechanisms. Current Opinion in Chemical Biology. 6 (5): 619–29. doi:10.1016/S1367-5931(02)00380-0. PMID 12413546.
39. Boyer R (2002). Chapter 6: Enzymes I, Reactions, Kinetics, and Inhibition. Concepts in Biochemistry (2nd ed.). New York, Chichester, Weinheim, Brisbane, Singapore, Toronto.: John Wiley & Sons, Inc. pp. 137–8. ISBN 0-470-00379-0. OCLC 51720783.
40. Savir Y, Tlusty T (2007). Scalas E, ed. Conformational proofreading: the impact of conformational changes on the specificity of molecular recognition (PDF). PLoS ONE. 2 (5): e468. Bibcode:2007PLoSO...2..468S. doi:10.1371/journal.pone.0000468. PMC 1868595 . PMID 17520027.
41. Fersht A (1985). Enzyme Structure and Mechanism. San Francisco: W.H. Freeman. pp. 50–2. ISBN 0-7167-1615-1.
42. Warshel A, Sharma PK, Kato M, Xiang Y, Liu H, Olsson MH (August 2006). Electrostatic basis for enzyme catalysis. Chemical Reviews. 106 (8): 3210–35. doi:10.1021/cr0503106. PMID 16895325.
43. Cox MM, Nelson DL (2013). Chapter 6.2: How enzymes work. Lehninger Principles of Biochemistry (6th ed.). New York, N.Y.: W.H. Freeman. p. 195. ISBN 978-1464109621.
44. Benkovic SJ, Hammes-Schiffer S (August 2003). A perspective on enzyme catalysis. Science. 301 (5637): 1196–202. Bibcode:2003Sci...301.1196B.

doi:10.1126/science.1085515. PMID 12947189.

45. Jencks WP (1987). Catalysis in Chemistry and Enzymology. Mineola, N.Y: Dover. ISBN 0-486-65460-5.
46. Villa J, Strajbl M, Glennon TM, Sham YY, Chu ZT, Warshel A (October 2000). How important are entropic contributions to enzyme catalysis?. Proceedings of the National Academy of Sciences of the United States of America. 97 (22): 11899–904. Bibcode:2000PNAS...9711899V. doi:10.1073/pnas.97.22.11899. PMC 17266 . PMID 11050223.
47. Ramanathan A, Savol A, Burger V, Chennubhotla CS, Agarwal PK (2014). Protein conformational populations and functionally relevant substates. Acc. Chem. Res. 47 (1): 149–56. doi:10.1021/ar400084s. PMID 23988159.
48. Tsai CJ, Del Sol A, Nussinov R (2009). Protein allostery, signal transmission and dynamics: a classification scheme of allosteric mechanisms. Mol Biosyst. 5 (3): 207–16. doi:10.1039/b819720b. PMC 2898650 . PMID 19225609.
49. Changeux JP, Edelstein SJ (June 2005). Allosteric mechanisms of signal transduction. Science. 308 (5727): 1424–8. Bibcode:2005Sci...308.1424C. doi:10.1126/science.1108595. PMID 15933191.
50. de Bolster M (1997). Glossary of Terms Used in Bioinorganic Chemistry: Cofactor. International Union of Pure and Applied Chemistry. Retrieved 30 October 2007.
51. Chapman-Smith A, Cronan JE (1999). The enzymatic biotinylation of proteins: a post-translational modification of exceptional specificity. Trends Biochem. Sci. 24 (9): 359–63. doi:10.1016/s0968-0004(99)01438-3. PMID 10470036.
52. Fisher Z, Hernandez Prada JA, Tu C, Duda D, Yoshioka C, An H, Govindasamy L, Silverman DN, McKenna R (February 2005). Structural and kinetic characterization of active-site histidine as a proton shuttle in catalysis by human carbonic anhydrase II. Biochemistry. 44 (4): 1097–115. doi:10.1021/bi0480279. PMID 15667203.
53. Wagner AL (1975). Vitamins and Coenzymes. Krieger Pub Co. ISBN 0-88275-258-8.
54. BRENDA The Comprehensive Enzyme Information System. Technische Universität Braunschweig. Retrieved 23 February 2015.
55. Törnroth-Horsefield S, Neutze R (December 2008). Opening and closing the metabolite gate. Proceedings of the National Academy of Sciences of the United States of America. 105 (50): 19565–6. Bibcode:2008PNAS..10519565T. doi:10.1073/pnas.0810654106. PMC 2604989 . PMID 19073922.
56. McArdle WD, Katch F, Katch VL (2006). Chapter 9: The Pulmonary System and Exercise. Essentials of Exercise Physiology (3rd ed.). Baltimore, Maryland: Lippincott Williams & Wilkins. pp. 312–3. ISBN 978-0781749916.
57. Ferguson SJ, Nicholls D, Ferguson S (2002). Bioenergetics 3 (3rd ed.). San Diego: Academic. ISBN 0-12-518121-3.
58. Michaelis L, Menten M (1913). Die Kinetik der Invertinwirkung [The Kinetics of Invertase Action]. Biochem. Z. (in German). 49: 333–369.; Michaelis L, Menten ML, Johnson KA, Goody RS (2011). The original Michaelis constant:

translation of the 1913 Michaelis-Menten paper. Biochemistry. 50 (39): 8264–9. doi:10.1021/bi201284u. PMC 3381512 . PMID 21888353.

59. Briggs GE, Haldane JB (1925). A Note on the Kinetics of Enzyme Action. The Biochemical Journal. 19 (2): 339–339. doi:10.1042/bj0190338. PMC 1259181 . PMID 16743508.
60. Ellis RJ (October 2001). Macromolecular crowding: obvious but underappreciated. Trends in Biochemical Sciences. 26 (10): 597–604. doi:10.1016/S0968-0004(01)01938-7. PMID 11590012.
61. Kopelman R (September 1988). Fractal reaction kinetics. Science. 241 (4873): 1620–26. Bibcode:1988Sci...241.1620K. doi:10.1126/science.241.4873.1620. PMID 17820893.
62. Cornish-Bowden A (2004). Fundamentals of Enzyme Kinetics (3 ed.). London: Portland Press. ISBN 1-85578-158-1.
63. Price NC (1979). What is meant by `competitive inhibition'?. Trends in Biochemical Sciences. 4 (11): N272–N273. doi:10.1016/0968-0004(79)90205-6.
64. Cornish-Bowden A (July 1986). Why is uncompetitive inhibition so rare? A possible explanation, with implications for the design of drugs and pesticides. FEBS Letters. 203 (1): 3–6. doi:10.1016/0014-5793(86)81424-7. PMID 3720956.
65. Fisher JF, Meroueh SO, Mobashery S (February 2005). Bacterial resistance to beta-lactam antibiotics: compelling opportunism, compelling opportunity. Chemical Reviews. 105 (2): 395–424. doi:10.1021/cr030102i. PMID 15700950.
66. Johnson DS, Weerapana E, Cravatt BF (June 2010). Strategies for discovering and derisking covalent, irreversible enzyme inhibitors. Future Medicinal Chemistry. 2 (6): 949–64. doi:10.4155/fmc.10.21. PMC 2904065 . PMID 20640225.
67. Endo A (1 November 1992). The discovery and development of HMG-CoA reductase inhibitors (PDF). J. Lipid Res. 33 (11): 1569–82. PMID 1464741.
68. Wlodawer A, Vondrasek J (1998). Inhibitors of HIV-1 protease: a major success of structure-assisted drug design. Annual Review of Biophysics and Biomolecular Structure. 27: 249–84. doi:10.1146/annurev.biophys.27.1.249. PMID 9646869.
69. Yoshikawa S, Caughey WS (May 1990). Infrared evidence of cyanide binding to iron and copper sites in bovine heart cytochrome c oxidase. Implications regarding oxygen reduction. The Journal of Biological Chemistry. 265 (14): 7945–58. PMID 2159465.
70. Hunter T (January 1995). Protein kinases and phosphatases: the yin and yang of protein phosphorylation and signaling. Cell. 80 (2): 225–36. doi:10.1016/0092-8674(95)90405-0. PMID 7834742.
71. Berg JS, Powell BC, Cheney RE (April 2001). A millennial myosin census. Molecular Biology of the Cell. 12 (4): 780–94. doi:10.1091/mbc.12.4.780. PMC 32266 . PMID 11294886.
72. Meighen EA (March 1991). Molecular biology of bacterial bioluminescence. Microbiological Reviews. 55 (1): 123–42. PMC 372803 . PMID 2030669.
73. De Clercq E (2002). Highlights in the development of new antiviral agents.

Mini Rev Med Chem. 2 (2): 163–75. doi:10.2174/1389557024605474. PMID 12370077.

74. Mackie RI, White BA (October 1990). Recent advances in rumen microbial ecology and metabolism: potential impact on nutrient output. Journal of Dairy Science. 73 (10): 2971–95. doi:10.3168/jds.S0022-0302(90)78986-2. PMID 2178174.

75. Rouzer CA, Marnett LJ (2009). Cyclooxygenases: structural and functional insights. J. Lipid Res. 50 Suppl: S29–34. doi:10.1194/jlr.R800042-JLR200. PMC 2674713 . PMID 18952571.

76. Suzuki H (2015). Chapter 8: Control of Enzyme Activity. How Enzymes Work: From Structure to Function. Boca Raton, FL: CRC Press. pp. 141–69. ISBN 978-981-4463-92-8.

77. Doble BW, Woodgett JR (April 2003). GSK-3: tricks of the trade for a multi-tasking kinase. Journal of Cell Science. 116 (Pt 7): 1175–86. doi:10.1242/jcs.00384. PMC 3006448 . PMID 12615961.

78. Bennett PM, Chopra I (1993). Molecular basis of beta-lactamase induction in bacteria (PDF). Antimicrob. Agents Chemother. 37 (2): 153–8. doi:10.1128/aac.37.2.153. PMC 187630 . PMID 8452343.

79. Skett P, Gibson GG (2001). Chapter 3: Induction and Inhibition of Drug Metabolism. Introduction to Drug Metabolism (3 ed.). Cheltenham, UK: Nelson Thornes Publishers. pp. 87–118. ISBN 978-0748760114.

80. Faergeman NJ, Knudsen J (April 1997). Role of long-chain fatty acyl-CoA esters in the regulation of metabolism and in cell signalling. The Biochemical Journal. 323 (Pt 1): 1–12. PMC 1218279 . PMID 9173866.

81. Suzuki H (2015). Chapter 4: Effect of pH, Temperature, and High Pressure on Enzymatic Activity. How Enzymes Work: From Structure to Function. Boca Raton, FL: CRC Press. pp. 53–74. ISBN 978-981-4463-92-8.

82. Kamata K, Mitsuya M, Nishimura T, Eiki J, Nagata Y (March 2004). Structural basis for allosteric regulation of the monomeric allosteric enzyme human glucokinase. Structure. 12 (3): 429–38. doi:10.1016/j.str.2004.02.005. PMID 15016359.

83. Froguel P, Zouali H, Vionnet N, Velho G, Vaxillaire M, Sun F, Lesage S, Stoffel M, Takeda J, Passa P (March 1993). Familial hyperglycemia due to mutations in glucokinase. Definition of a subtype of diabetes mellitus. The New England Journal of Medicine. 328 (10): 697–702. doi:10.1056/NEJM199303113281005. PMID 8433729.

84. Okada S, O'Brien JS (August 1969). Tay-Sachs disease: generalized absence of a beta-D-N-acetylhexosaminidase component. Science. 165 (3894): 698–700. Bibcode:1969Sci...165..698O. doi:10.1126/science.165.3894.698. PMID 5793973.

85. Learning About Tay-Sachs Disease. U.S. National Human Genome Research Institute. Retrieved 1 March 2015.

86. Erlandsen H, Stevens RC (October 1999). The structural basis of phenylketonuria.

Molecular Genetics and Metabolism. 68 (2): 103–25. doi:10.1006/mgme.1999.2922. PMID 10527663.

87. Flatmark T, Stevens RC (August 1999). Structural Insight into the Aromatic Amino Acid Hydroxylases and Their Disease-Related Mutant Forms. Chemical Reviews. 99 (8): 2137–2160. doi:10.1021/cr980450y. PMID 11849022.
88. Phenylketonuria. Genes and Disease [Internet]. Bethesda (MD): National Center for Biotechnology Information (US). 1998–2015.
89. Pseudocholinesterase deficiency. U.S. National Library of Medicine. Retrieved 5 September 2013.
90. Fieker A, Philpott J, Armand M (2011). Enzyme replacement therapy for pancreatic insufficiency: present and future. Clinical and Experimental Gastroenterology. 4: 55–73. doi:10.2147/CEG.S17634. PMC 3132852 . PMID 21753892.
91. Misselwitz B, Pohl D, Frühauf H, Fried M, Vavricka SR, Fox M (June 2013). Lactose malabsorption and intolerance: pathogenesis, diagnosis and treatment. United European Gastroenterology Journal. 1 (3): 151–9. doi:10.1177/2050640613484463. PMC 4040760 . PMID 24917953.
92. Cleaver JE (May 1968). Defective repair replication of DNA in xeroderma pigmentosum. Nature. 218 (5142): 652–6. Bibcode:1968Natur.218..652C. doi:10.1038/218652a0. PMID 5655953.
93. James WD, Elston D, Berger TG (2011). Andrews' Diseases of the Skin: Clinical Dermatology (11th ed.). London: Saunders/ Elsevier. p. 567. ISBN 978-1437703146.
94. Renugopalakrishnan V, Garduño-Juárez R, Narasimhan G, Verma CS, Wei X, Li P (November 2005). Rational design of thermally stable proteins: relevance to bionanotechnology. Journal of Nanoscience and Nanotechnology. 5 (11): 1759–1767. doi:10.1166/jnn.2005.441. PMID 16433409.
95. Hult K, Berglund P (August 2003). Engineered enzymes for improved organic synthesis. Current Opinion in Biotechnology. 14 (4): 395–400. doi:10.1016/S0958-1669(03)00095-8. PMID 12943848.
96. Jiang L, Althoff EA, Clemente FR, Doyle L, Röthlisberger D, Zanghellini A, Gallaher JL, Betker JL, Tanaka F, Barbas CF, Hilvert D, Houk KN, Stoddard BL, Baker D (March 2008). De novo computational design of retro-aldol enzymes. Science. 319 (5868): 1387–91. Bibcode:2008Sci...319.1387J. doi:10.1126/science.1152692. PMC 3431203 . PMID 18323453.
97. Sun Y, Cheng J (May 2002). Hydrolysis of lignocellulosic materials for ethanol production: a review. Bioresource Technology. 83 (1): 1–11. doi:10.1016/S0960-8524(01)00212-7. PMID 12058826.
98. Kirk O, Borchert TV, Fuglsang CC (August 2002). Industrial enzyme applications. Current Opinion in Biotechnology. 13 (4): 345–351. doi:10.1016/S0958-1669(02)00328-2. PMID 12323357.
99. Briggs DE (1998). Malts and Malting (1st ed.). London: Blackie Academic. ISBN 978-0412298004.

100. Dulieu C, Moll M, Boudrant J, Poncelet D (2000). Improved performances and control of beer fermentation using encapsulated alpha-acetolactate decarboxylase and modeling. Biotechnology Progress. 16 (6): 958–65. doi:10.1021/bp000128k. PMID 11101321.

101. Tarté R (2008). Ingredients in Meat Products Properties, Functionality and Applications. New York: Springer. p. 177. ISBN 978-0-387-71327-4.

102. Chymosin – GMO Database. GMO Compass. European Union. 10 July 2010. Retrieved 1 March 2015.

103. Molimard P, Spinnler HE (February 1996). Review: Compounds Involved in the Flavor of Surface Mold-Ripened Cheeses: Origins and Properties. Journal of Dairy Science. 79 (2): 169–184. doi:10.3168/jds.S0022-0302(96)76348-8.

104. Guzmán-Maldonado H, Paredes-López O (September 1995). Amylolytic enzymes and products derived from starch: a review. Critical Reviews in Food Science and Nutrition. 35 (5): 373–403. doi:10.1080/10408399509527706. PMID 8573280.

105. Protease – GMO Database. GMO Compass. European Union. 10 July 2010. Retrieved 28 February 2015.

106. Alkorta I, Garbisu C, Llama MJ, Serra JL (January 1998). Industrial applications of pectic enzymes: a review. Process Biochemistry. 33 (1): 21–28. doi:10.1016/S0032-9592(97)00046-0.

107. Bajpai P (March 1999). Application of enzymes in the pulp and paper industry. Biotechnology Progress. 15 (2): 147–157. doi:10.1021/bp990013k. PMID 10194388.

108. Begley CG, Paragina S, Sporn A (March 1990). An analysis of contact lens enzyme cleaners. Journal of the American Optometric Association. 61 (3): 190–4. PMID 2186082.

109. Farris PL (2009). Economic Growth and Organization of the U.S. Starch Industry. In BeMiller JN, Whistler RL. Starch Chemistry and Technology (3rd ed.). London: Academic. ISBN 9780080926551.

Index